Powers of the Rational

Powers of the Rational

Science, Technology,
and the Future of Thought

Dominique Janicaud
Translated by Peg Birmingham
and Elizabeth Birmingham

Indiana University Press

Bloomington and Indianapolis

Preparation of the translation was assisted by a grant from the
French Ministry of Culture

Published in French as *La puissance du rationnel* © 1985 by
Editions Gallimard, Paris

© 1994 by Indiana University Press

The paper used in this publication meets the minimum
requirements of American National Standard for Information
Sciences—Permanence of Paper for Printed
Library Materials, ANSI Z39.48-1984.

Manufactured in the United States of America

Library of Congress Cataloging-in-Publication Data

Janicaud, Dominique, date
    [Puissance du rationnel. English]
    Powers of the rational : science, technology, and the future of
thought / Dominique Janicaud ; translated by Peg Birmingham and
Elizabeth Birmingham.
      p. cm.—(Studies in Continental thought)
    Includes bibliographical references and index.
    ISBN 0-253-33108-0 (alk. paper)
    1. Power (Philosophy) 2. Reason. 3. Science—Philosophy.
4. Technology—Philosophy. I. Title. II. Series.
B2430.J283P8513   1994
128'.4—dc20
94-6693

1 2 3 4 5 00 99 98 97 96 95 94

Nescis, mi fili, quantilla ratione mundus regatur.

—Axel Oxenstierna

"You do not know, my son, how little the world is ruled by reason." Cited by L. von Bertallanfy, *Théorie générale des systèmes*, trad. J.-B. Chabrol [Paris: Dunod, 1973], p. 120. There exists a different version of the same phrase: . . . *quantilla prudentia homines regantur* (. . . with how little wisdom men are ruled).

Contents

Preface to the English Translation

IN A RELATIVELY short time, compared to the history and prehistory of humanity, the sciences and technologies have transformed our planet, shaking immemorial ecological and ethnological equilibriums and, especially, causing human beings to doubt the meaning of their existence and their work, indeed their very identity. Two world wars and the acceleration of technological competition on a planetary scale have called into question the enlightened ideal of progress, and there is no satisfactory new model of "development" to take its place. Power has assumed unforeseen forms that have rapidly engulfed humanity. No ethical thought could have foreseen, predicted, or prevented the focus on weapons of massive destruction, the emergence of genetic engineering, or the universal computerization of a multitude of discursive and practical operations. Now that humanity has been exposed to the extreme temptations of cynicism and nihilism, has it lost control of its destiny?

This crushing knot of problems seems to exceed the scope and the means of classical rationality. No vision of history offered up to this point has comprehended the explosive multiplication of possibles that Nietzsche was, undoubtedly, the one person to predict. In *Powers of the Rational*, I wanted to confront this situation, without avoiding—under the influence of the present—the most fundamental questions about the meaning of rationality that have been posed since Kant.

At first glance the book's title might suggest that its goal is to offer an apology for rationality. In fact, "powers of the rational" must be understood in a phenomenological (and not a metaphysical or even simply a laudatory) sense. The point of view I have adopted is neither rationalist nor anti-rationalist; rather, it is that of an observer who claims that the rationality that binds the sciences and technologies ever more closely produces unprecedented *effects of power*. The most decisive lines of rupture are not necessarily as spectacular as the menace of thermonuclear destruction. Genetic engineering produces human mutations that threaten the very nature of human being. The computerization of archives and communication, the systematic development of "artificial intelligence," and the superspeed of worldwide electronic transmissions precipitate the instrumentalization of language. In the face of such phenomena, it becomes impossible to take shelter behind the old distinction between science (disinterested) and technology (applied); we are participants in a complete restructuring of the relationships between science and power, between rationality and domination. We must pursue new distinctions, from the perspective of power as well as from the perspective of the rational.

On the one hand (when one observes the effects of power), we must recognize the massive, incalculable, and irreversible deviations which cannot be explained by any of the traditional motives for power (a taste for money, an individual will to power, or even the struggle between social classes). On the other hand (proceeding from the rational as such), the production of power (or potentialization) is in no way related to a direct "tyranny" of logic. We can distinguish several *phases* of rational potentialization (which cannot be reduced to historical epochs). These four regimes of the growth of power are in principle compossibles, although in our time an ever more marked disequilibrium operates to the almost exclusive benefit of the phase of automatization of the techno-scientific complex (which I call Phase IV).

Use of a "phenomenological" approach in the first stage of my research may seem surprising, since neither the core of my thesis (rational potentialization) nor its development (the development of the effects of the powers of reason) seems to be the object of a Husserlian phenomenological critique, strictly speaking. Power is not an essence that the noetic gaze can stabilize as a correlate; rather, it is a differential relation constantly in motion, whose exteriority vigorously reminds us of the ultimately *unthinkable* character of the self-presupposition of rationality. While conceding that the present study does not reflect the large or totalizing sense of phenomenology, I claim here a neutral gaze, welcoming the effects of rational power such as they occur, without imposing a value judgment. A *minimal* phenomenology permits us to establish and describe the most massive effect, that of contemporary power, and to provide a more nuanced analysis. Such an analysis can take place when phenomenology adheres to a diacritical genealogy of the phases of potentialization). Nietzsche spoke of the "terror of the incalculable, as the proto-instinct of science."[1] Such limits are indicated when phenomenology is involved in a gigantic historical movement, seemingly predestined, whose "results" appear to exceed all "intention." Nevertheless, diagnosis and reflection must be based upon observation of phenomena in as "neutral" a manner as possible. Phenomenological attention is more important than ever in this time of accelerated technological innovation. Ideologies prohibit an openness of spirit (Marxism is an overwhelming example); less obviously, philosophical arrogance can lead one quickly to become a partisan. Perhaps that was the case with Nietzsche's apology for the *will to power*.

Have we practiced an *epoche* that is radical enough for every reactive or metaphysical reading of rational potentializations? And, moreover, has there been enough critical distance in relation to the Heideggerian interpretation? Certain readers have suspected *Powers of the Rational* of being a so-called soft version of the Heideggerian destinal thesis on the essence of modern technology. Such a view fails to recognize the numerous passages in which distance is maintained in regard to reducing to *Gestell* the complexity of the phases of rational potentialization, and in which we also warn readers of the danger of a Manicheanism separating the technological world from its profound and reserved truth. Thus, this book engages in a tense dialogue with Heideggerian thought. The title itself testifies to this. To start from the "powers of the rational," always in an ambiguous sense, is to take the opposite course from a pure and simple withdrawal into meditative thought. This dialogue is extended, in "Rationality as Partage" to the Hegelian heritage and to hypercritical lucidity. The book's primary divergences from Heideggerian thought can be summarized as follows: we no longer presuppose the sending of being (as appropriating unity) or an awaited historical Event ("eschatological" recourse); the mastery of being is no longer understood from the exclusive viewpoint of metaphysics, but differentiated into phases; as a result, the Greek world must no longer be interpreted solely according to the withdrawal of authentic *A-lethia*, and the contemporary world is no longer strictly united around an "essence" such as *Gestell*. The "recourse," if it is unthinkable, does not simply involve waiting (even if this is respectable): it also implies substantial effort on our part to understand.

Another great reference whose shadow looms over this work is Nietzsche. It has been called to my attention that the book lacks a thematic and searching explication of this great body of thought. This observation, justified in principle, requires a twofold response. First and most evident, such an explication would have demanded taking into account some very important developments, in particular Heidegger's *Nietzsche*, which would have diverted the purpose of the book toward the hermeneutics of modern metaphysics. Second, Nietzsche is not really absent from this book, but is intentionally held at a distance. One of my major tasks in the future will be to clarify all the underlying reasons for this distance. What is already evident is that *Powers of the Rational* undertakes a movement of thought freed from all prophecy and a process of *catharsis* regarding the will to power.

But, there is a point where the present research converges with an orientation found at once in Hegel, Nietzsche, and Heidegger. Taking account of the dominant historical movement of the West obligates us to play down the role of individuals. Does the present dynamic of potentialization totally escape human control? Here we must distinguish between the irreversibility of a process (there is a consensus about the fact that one cannot purely and simply return to the past, as if technological progress were annullable) and, on the other hand, the automatization of the techno-scientific sphere (I have established a tendency; I have not claimed that the global system, closed upon itself, would escape every form of human control). Although irreversible, the process remains open.

What is certain is that a schematic rejection and an overly rapid unifying of technology run the risk of producing an inverse effect from that which the best-intentioned would see: rather than preparing a "salvation," one risks accentuating further the separation between intellectuals and the effective course of the world; one will move even further from the possibility of articulating the possible upon the real (above all from the political point of view). We have, therefore, wagered on escaping both the soteriological vision of the later thought of Heidegger ("only a god can save us") and the purely and simply technicist reduction. It is not a question merely of a subjective choice; it is necessary here to see above all the logical consequence of the theory of the phases of potentialization, which replaces a unified philosophy of history. Phase IV is ultimate in the sense that it is the last to appear and that it is the phase which releases the most power, but it does not exhaust completely (for the moment at least) the other phases or the possible which remains open to the unforeseeable. The historical possible exceeds the logical possible, which itself does not remain as soberly enclosed in noncontradiction as Kant had believed. There is, hence, no "end of history" in any sense; the global philosophies that spread this myth have lost their validity and even their plausibility, in spite of all the attempts at artificial resuscitation.

According to the same internal logic, it is not necessary to understand the term "techno-science" in a substantialist or excessively unifying sense. The coexistence of many phases implies that the scientific project can preserve a relative autonomy in relation to purely technical problems. As Jean Ladrière has noted,[2] science represents the inverse of technology insofar as it "informs organizations," whereas technology organizes information; but this tension and difference maintained between science and technology is inscribed at the interior of a global and effective convergence between sciences and technologies. In this regard, it would have been more correct to speak of techno-sciences (in the plural) rather than techno-science in an absolutely unified sense.

Thus, the occasion presents itself to open the philosophical discourse to the scientific realities that exist. I very much hope to be read (and critiqued) by the English-speaking scientific community, even more than I have been in France. The obstacle to the dialogue resides, to a great degree, in the misunderstanding that persists around the meaning of "power." When the physicist Georges Waysand recently remarked that "the power of science is constituted by eliminating the power of the scientific,"[3] he put his finger on a reality that is entirely other and much more decisive than the direct power of an individual: all scientific progress results from a cumulative process which, in its turn, *makes possible* other scientific discoveries, and new technological applications.

The vector of this global inflation of power is no longer an individual, however inspired, nor even a group. It is in principle the "scientific community"; it is, at the level of effective enactment and direct benefits, a state, a multinational society, an *ad hoc* institution such as NASA or CERN.

The subordination of the individual, therefore, takes place according to two factors: objectification and giantism. Since the time of Greek mathematics, the first has governed the truly epistemic development: as soon as a theorem and a series of demonstrations are understood and formulated, they no longer belong to their author. They have validity through the internal coherence which an individual merely lets appear. Euclid, in this regard, is no more "personalizable" than Bourbaki. In gaining, since the seventeenth century, knowledge of the physical world, then progressively of new sectors of reality (including "human"), the exigency of objectivity has reinforced the constraints that constitute what Popper calls the "scientific tradition,"[4] and which act as so many guarantees against a regression to subjectivity or to arbitrariness.

Giantism is not in itself an invention of the twentieth century, although the twentieth century has carried it into new domains, at least according to two points of view: the considerable growth, especially in the last half century, of dimensions of the scientific community and, in particular, of a number of publications; the launching, since the Manhattan Project (the secret plan for the first American nuclear bomb) of enormous techno-scientific programs (military, space, computers, medical), where scientific knowledge is not simply utilized "among others," but where this knowledge is put to work according to very specific interests.

We must draw out the consequences of these familiar facts in order to increase our understanding of the problem of rational potentializations. The question of the scientist's small power as an individual is no longer at all at the level of what truly occurs. To the first subordination (noble and consciously assumed) issuing from objectivity are added the specific constraints of *Large-scale Science*. The result is this paradoxical and frustrating situation for the scientist: he or she appears to have even less importance (at least as a general rule) than science as a global process, profiting in possibilities and in power!

The temptation, therefore, is great for the scientist to become disinterested in the problems of the insertion of science in society, and in the questions of the future (moral, cultural, etc.) that depend on it. There is the temptation to have a purely functional conception of scientific work. This temptation is a variation of technicism. It ought to be critiqued for the same reason as this latter.

Since we are speaking of critiques, I should emphasize that I recognize the necessity of constantly revising and refining the inquiry that I am pursuing. Although I confirm the general orientation and the fundamental theses of this book, I must concede that I would not write it entirely in the same manner today. The most important corrections I would introduce concern the following points. The reflection on "Technology and Language" displays an excessive pessimism about the technicization of natural languages; it is surely possible to strike a balance between this pessimism and the extreme optimism of an Umberto Eco.[5] "Gabor's rule," formulated at the end of Part I, has a heuristic value, but it does not function automatically. The contradiction between technicalization and ethics is not always direct. In particular, it is necessary to bear in mind that economic factors do not automatically work to the advantage of the most advanced technological innovations (the profitability factor is not necessarily more "moral," nor is it strictly technological). The tone of what I wrote about the New Science was certainly too polemical. I should at least warn readers of the mythification of contemporary science, the emergence of a possible new form of scientism.[6]

Moreover, it was necessary to correct allusions to contemporary events. In less than ten years, the worldwide political situation has changed astoundingly. Do the upheavals that intervened invalidate a critical examination of rational potentializations? Not at all. But we must note an important modification. The irrationalities of the worldwide shift of power have displaced the East/West axis toward a North/South axis and a balkanization of conflicts. The precariousness of attempts at regulation only makes this fragility more apparent (the Rio conference on the environment was a case in point—a sad acknowledgment, which must not discourage new efforts). The alarming situation of the former Soviet Union and its former satellite nations with regard to nuclear danger is equally urgent: the truly political aberrations of a specific form of totalitarianism have intensified the technological dangers which also exist potentially elsewhere. If it is true that the major technological risk is henceforth a planetary reality, the problem of its control is not always posed in strictly technological terms. The analysis of the powers of the rational (and of their irrational reversals) must be constantly rethought insofar as hypermodern power re-potentializes unceasingly. The great error of the Marxists was the belief that Marx's work could be managed like capital!

These critical remarks certainly provoke us to redouble our prudence in formulating concrete analyses and especially in any projection of the future. They also show that the question of the powers of the rational must not be judged according to the restricted dimensions and the fluctuating directions of contemporary reality. Rather, the reverse is true: the process of questioning illuminates the present circumstances and obligates us to measure the distressing depth of our destiny of power.

This book, with its insights and shortcomings, is intended as a contribution to fostering continuing research, in particular, on the link between rational potentializations and the most advanced forms of capitalism. The collapse of state socialism makes this reflection even more urgent. My hope is that the workplace opened by *Powers of the Rational* is not closed soon and that the elegant translation of Elizabeth and Peg Birmingham inspires the desire of new inquirers to come and work here.

Powers of the Rational

Introduction

The power or the impotence of the rational?

THE POWER OF the rational? It would probably be better to speak of its impotence—its inability to master the passions and follies of twentieth-century human beings, to govern states and to guide the planet toward a stable and definitive peace, to lead scientific-technological progress, to establish beyond question its own essence as well as its relation to Being and to values.

This first observation already has two consequences. The first is that it has become impossible to organize knowledge and action on the basis of an alleged omnipotence of rationality. In regard to the four strategic areas just mentioned (the individual, the State, techno-science, philosophical sovereignty), complete rationality is failing; even a recourse to the most ambitious metaphysical system of rationality, that of Hegel, cannot reverse this situation—quite the contrary. The purely rational elaboration of a mastery of self-consciousness is more than ever an exceptional and fragile undertaking, the privilege of true philosophers—today as in Hegel's time. The rationalizations of psychology do indeed produce a pervasive conditioning, but this either stifles the personality, or merely allows it to assume its most subtle peculiarities. At the political level, Hegel himself claimed to delineate a perfectly organic rationality only through the production of the *Idea* of the State, of civil society and of abstract right; in our time, the planning involved in the methods of government, of administration, and of the wheels of state leaves the responsibility of a leader or of a group of leaders intact when vital political choices are necessary. This is done without any scientific agency being able to recommend—and especially to apply—solutions that might appear to be indisputably rational and solely adequate. Hegelian ontotheology could still claim to possess the finality of a scientifico-technological progress that was neither autonomized nor accelerated; it was not yet evident that all sectors of scientific knowledge could enter into speculative logic, or that the alienations induced by the birth of mechanization were integral to a system of needs compatible with the flourishing of a civil society worthy of its name. Finally, as concerns the very essence of rationality and its relation to Being and values, a systematic and sovereign unity is still lacking. In becoming specialized and formalized, scientificity sets aside the ultimate questions and concerns itself less than ever—except in marginal attempts—with the eternal. Interdisciplinary study cannot pass for even a pale shadow of the triumphant certainty of Hegelian *Wissenschaft*. The rationalizations of the twentieth century hide behind scientific and technological successes in order better to avoid every question that is too originary or metaphysical, or better to protect themselves with some reassuring or mobilizing ideological formulas. But there is no definitive mastery of the rational as such.

Ought the second consequence to be an observation concerning the impotence of the rational? Not at all. Once its omnipotence is excluded, it remains to determine—and this is no small task—the limits within which the power of the rational is exercised. But are there definable limits? Is there a region of reality today that is safe from rationalization? Within the four strategic areas mentioned above, rationality continually "intervenes."

Even in this claim that the *reign* of reason is absent, regions of rationalization have also been recognized. Threatened by conditioning or techniques that seek directly to control him or her, the individual is equally endowed—at least if he or she is not socially disadvantaged—with analytic procedures, with therapeutic and cultural recourses, with an unprecedented variety and quantity of information. Whatever the degree of socialization or integration, the individual is incontestably subjected to various rational "influences." This is even more obvious at the political level: the modern State—whether one celebrates it or not—has at its disposal an impressive collection of procedures for inventory, control, and "incentive," combining the contributions of the social sciences, communication techniques, and technologically sophisticated equipment; here too, rationality plays the role of a current electrifying the body politic, above all its intellectual elite. Although techno-science is not mastered by a sovereign reason, or even carried off in a perhaps catastrophic direction, it is the very place of the principal expansion of rationality, a formidable force that benefits from the cumulative effect of various fields of knowledge and procedures as well as from a colossal concentration of resources in the conduct of the "politics of science" and industrial strategy. Finally, how can we deny that philosophical intelligibility, as neglected as it is by political leadership, as diverse and as fragile as it appears to philosophers themselves, had to and must still absorb all of its mutations, draw up a new map of rationality, and testify, perhaps through its own ruptures and uncertainties, to the power of the rational?

One point seems already established: reason reigns neither directly nor completely in the world. It reigns neither as a monarch nor as a Great Watchmaker, nor even as its image deified by the French Revolution; but everything remains to be done to resolve the enigma that the present world proposes to a mind that is still capable of wonder. The immense techno-scientific transformation that unceasingly exerts and expands its power and executes its effects does so, in one sense, in the name of rationality; in any case, it is certainly through the mediation of innumerable rational, theoretical, technological and practical procedures that it assures its hold on the world. "Modern Science," writes Jean Ladrière, "is closely associated with a power over things, and over man himself, and this is why it appears linked to technology to the point of being indistinguishable from it."[1] Thus, we come to endow the rational—at least in its techno-scientific configurations—with the effective attributes of Power, because, substituting it for myths, for traditional rites and cultures, it becomes the refuge, the sign or the model of universal order and even of the meaning of life. Consequently, man (individual, class or genus) is honored to be its agent and it is from this source that man comes to be intoxicated by the delights of power. This global operation henceforth expands without cult instruments: infinitely more efficient than positivist or Masonic temples, big laboratories and industrial complexes, research institutes and administrative councils coordinate and plan the "rationalization" of this world, which the Frankfurt school describes as dedicated to total Administration.

Antinomy: the world has never been so rationalized; reason has never been so impotent. The power or impotence of the rational? Rather than fitting this question into a definitively formalized configuration—whether purely antinomic, factually contradictory, or simply undecidable—it is undoubtedly advisable to return to the conditions of its position, asking what exactly we understand by reason, the rational, and rationality in these different propositions and observations.

The rational in question

If this discourse concerning power and impotence of the rational has a minimal plausibility (to restate Pascal, we have an idea of rationality "invincible to all phyrrhonism," which—it seems—is continually informed by experience), it nevertheless remains susceptible to an objection of principle: do we not presuppose a definition of the rational? How do we recognize or ascribe power or impotence to a "notion" or a "reality" that is still vague and undefined?

To be sure, the rational is not an ordinary *content*, one thing among things, a being among other beings. I do not find a ready-made rationality presented to my wondering eyes, in everyday experience or even in the "journey of life." More precisely: the rational that *can* be presented to me in everyday experience is eventually only a kind of intellectual "ready-to-wear" that follows from a considerable sociolinguistico-historical *elaboration*. A basic conclusion must be drawn after Kant: the rational is not an object of experience; more generally, it is not of an empirical origin, at least not from a uniquely empirical origin. Certainly, empiricism does not acknowledge total defeat; does not that which appears to me as nonempirical result from "habitual connections," as Hume said? We immediately see that the real definition of the rational (as distinguished from the too simplistic nominal definitions that teach us nothing: "rational: faculty of reason," "coherent exercise of rationality," etc.), eludes us at the very moment when we believe we have grasped an obvious distinction. If the rational is not an empirical content, is it not the *form* of all "objects," "entities," "realities" that I identify as rational? But whoever speaks of identifying form speaks of judgment; more precisely, at the philosophical level: a definition of the rationality of the rational will imply a theory of judgment, as in Kant once again. The first conclusion to draw from this brief analysis is the following: the rational cannot be defined without itself taking part in form and not simply in content. But this intervention of the *form* of the rational in its own definition paradoxically indicates from the outset the insufficiency of any formal definition that would fail to take into account this circularity of the rational, this self-anticipation of a concept that overruns all concepts since it proposes the norm for them.

On the other hand, but by the same token, if the definition of the rational cannot be immediate, this is true for any "delimitation." How can we decide what is rational and what is not, when this undertaking depends on a definition that we do not possess? If we decide on this a priori, we again presuppose as a criterion a condition of the possibility of experience; we decide that a priori judgments are possible, a possibility by virtue of which the fundamental philosophical judgment that articulates this discussion is effected. We return to a transcendental theory of judgment and of rationality.

The nonempirical, nonimmediate character of rationality indicates, therefore, an opposite direction from the pure and simple definition: "autonomization." What do we understand by this? Given that the definition of the rational proves to be circular, implying through a determining form that which one wants to define, it is tempting to begin now with the act of defining and make it the decisive moment in two forms: transcendent or transcendental. The "autonomization" is none other than this immanent tendency toward reason that Kant names "natural metaphysics": the actualizing projection of a circle through which rationality presupposes itself. In its transcendent form, this autonomization gives rise to a substantial moment, a divine *Nous* (or *Logos*) whose efficacy man more or less completely appropriates, and in relation to which he will define, as did Plato, a hierarchy of levels of knowledge. The transcendental form is the ultimate version of the anthropolization of this supreme moment; a more subtle solution imposed by the contradictions of a substantialist rationalism, it critiques the objectifying autonomization in order to return the autonomization to the side of the subject, though on the side of a transcendental subject. We again find the transcendental, the condition of the possibility of experience in general and of its objects. But what is meant by "condition of possibility"? If we understand it rather easily in relation to categories, do we not come up against a *lucus a non lucendo* when we turn toward rationality itself, as such? Transcendental rationality then appears as its own condition of possibility. Even if this circularity is "mortgaged" through the relation with experience—insofar as one has left the sphere of pure reason and is preoccupied with a specific knowledge—it does not reveal any less this self-anticipation that rationality immediately displays.

To be sure, one always has the right and the possibility of refusing to take this autonomization of the rational into consideration. One gains this right, if one assumes this self-presupposition of rationality as the very condition of every scientific endeavor, and if one "proves" the rationality of the rational, not in general, but through the determined exercise of the aforementioned rationality within a formal or experimental domain. Thus Quine, at the beginning of his *Methods of Logic*,[2] does not waste time defining the rational in itself or even Logic. After recalling the object of logic as quickly as possible,[3] he passes immediately to its effective exercise in the form of an exposition of the formalization of "truth functions." To what extent is this way of proceeding justifiable? To the very extent that all science, in the modern sense, is first of all *operative*, that is to say—as Ladrière[4] reminds us—is defined neither by an inquiry into its own foundations nor by the explication of the totality of its implications, but by the enactment of precise procedures of transformation, formalization, thematization, generalization, and the constitution of a network of mutual connections. Modern science proves movement by moving; by working, it demonstrates its rationality. Proceeding in this way, it acts *as if* it knew (which is the case with Quine, but certainly not with the great majority of scientists) that the thematic study of its own epistemological conditions of possibility must lead to an endless inquiry and to the neglect of its own objectives. In more familiar terms: its refusal to philosophize is constitutive, whether it be conscious of this or not, whether it has the opportunity to express itself or not.

The refusal to take the autonomization of the rational into account is less justifiable, though just as possible and widely practiced, in those disciplines broadly called the "human sciences," because the very results of the investigation depend on the conception of "rationality" injected into the complex circuit of hypotheses, inquiries, proofs and counterproofs. In mathematics, physics, chemistry, and even biology, it is of little importance that the rationality of the rational remains in the background, since this notion—even if it is normative—does not directly intervene as such, in neither equations, nor formulas, nor experiments, etc. In contrast, despite all the efforts of economics, sociometry, and experimental psychology to quantify their data and to objectify their operations, the reading of the results—no matter how rigorous they may be—is subject to a margin of indetermination and interpretation infinitely greater than that which inevitably exists in the exact sciences. An opinion poll, for example, may well be conducted with the greatest mathematical and practical rigor (analysis of factors, systematic calibrations, seriousness of the interviewers, etc.), yet the context of its appearance and its "destination" are encumbered by a not insignificant burden of uncertainties that are going to overdetermine its "rationality." Is it even only a question of uncertainties? If a poll is not *interpreted*, it simply lines up columns of figures alongside other figures. It ought thus to be *read* in view of its larger intentions: if not those of the subjects being investigated, at least the objectives of its authors. A notion such as profitability comes into play in relation to the dictates of the poll. Will it be possible to avoid questions of this type: Is this poll the most rational under the pretext that it is the most profitable? How do we define profitability itself in this domain? Whatever approach we take to the problem, we are obliged, as Jon Elster recently emphasized, to integrate not only *functional* explanations, as is still the case in biology, but also *intentional* explanations into the social sciences.[5]

Certainly, a reductionism exists by means of which behaviorism, sociobiology, and their various offshoots attempt to deny this distinction. This reductionism is the extreme form, the most scientistic, of the refusal to consider the "autonomization" of the rational. It is not for me to critique this reductionism from the point of view of the methodology and practice of the social sciences themselves; this task has been more than sufficiently undertaken by specialists.[6] I would like to restrict myself to recalling that this presupposition that reductionism refuses to make globally concerning human beings is made *nolens volens* concerning rats or other tested populations. It itself defines the terms of measure to be applied to behaviors whose *basic rationality*—be it quantified or not—must be well defined. Even at the most elementary level of running through mazes, this basic rationality is necessarily presupposed. The behaviorist attitude that projects an absolute exteriority or neutrality on every object is itself the absolutizing of a project of objectivity that refuses to critique itself (and that ought to be, in our view, all the more suspect in that it retains only the certitudes of science and not its critical sense). In his time, Merleau-Ponty already demonstrated this mechanism.[7]

But if this reductionism deserves to be refuted, its importance ought not to be overestimated: it represents only a peripheral phenomenon in the contemporary scientific-technical movement, the symptom of a refusal to think the rational—whose "noble" form is found in rationalism. It is also a symptom since the exercise of rationality becomes more complex when it takes for its object the "rational animal" that is the human being. The most decisive contributions to the rationalization of actions and of behaviors have not been made by neglecting this complexity. Thus, von Neumann and Morgenstern, whose game theory, as we know, has been extremely important in political economy and strategic studies, analyze the "problem of rational behavior" at the outset of their work, *Theory of Games and Economic Behavior*, recognizing that there exists no "satisfactory treatment."[8] Why approach this question if it is insoluble? Because every game theory must be founded on a minimal presupposition: the accepting of the rules by the players and their desire to win. In economic terms, "the consumer wishes to obtain maximum utility or satisfaction, and the entrepreneur wishes to obtain maximum profits."[9] Although profit is quantifiable, the notion of "utility" can appear merely verbal. For this reason, von Neumann and Morgenstern make the methodological decision to consider money to be the goal of *all* the participants in the economic "game." A common term functioning as universal quantifier is thus found. Now at least the problem of the rationality of behaviors, without being fundamentally settled, has a methodological solution sufficient for the development of game theory: "The individual who seeks to obtain these respective *maxima* (profits or "utility") is also considered to be acting "rationally."[10]

We see that, from this perspective, rationality becomes an operative notion. We discover an analogous conception in the logic of deterrence. Each person is assumed to cling to life and act accordingly, but it takes only one player to be carried away by a suicidal will or ceding to a monstrous whimsy in order for the ingenious equilibrium of threats and "bluffs" to collapse. More generally, the employment of such a standard appears at once convenient and logical in the network of the social sciences in which the strategic approach tends to become the rule. This is the famous "charity principle" dear to Donald Davidson. This principle consists of assuming, as a sort of working hypothesis, the "rationality" of every action proposed in the field of research. In von Neumann's economic terms, this would mean that every "event" provoked by one of the participants can be explained by his or her desire for utility or for profit, that is to say, money. In Davidson's problematic, this means, in a more comprehensive manner (insofar as this implies nonquantifiable intentions) that every apparently "irrational" behavior can, in the end, be explained with a more complex strategy (the goals of the partners could be multiple, long-term, etc.).[11]

If we are pleased to see Davidson acknowledging the complexities of human behavior, drawing the theoretical consequences and severely criticizing the reductionism of experimental psychology, we should also see that his "charity" proves to be a double-edged sword. On the one hand, he refuses to classify man as a collection of "responses" to *stimuli,* and he is not content merely to outline their parameters. On the other hand, he presents the strategic hypothesis as the ultimate solution because it is the most comprehensive. Therefore, under the pretext of refuting behaviorism, he is led, with Elster, to think of human beings as "globally maximizing machines" as opposed to animals (locally maximizing machines).

Our problematic is not to determine the intrinsic truth of this thesis, but to discern "what happens" in this methodological operation with regard to the rationality of the rational. First of all, the "principle of charity" *generalizes* this strategic approach. What a restriction of the human field (the reduction of human beings to "economic agents") so that von Neumann's theory can function! Is the "principle of charity" any less reductionist? Hardly; it too brings about a more universal reduction: the reduction of humans to "strategic agents" or (in cybernetic terms) optimizing operators, setting aside or not even concerned with knowing whether their links to other dimensions (at the heart of language and silence) are also important. Is the rationality of these reductions the same as that which operates from the outset as a "working hypothesis"? Although implicitly, it goes further, carrying what was initially only a simple hypothesis to a higher, properly philosophical level. This is an operation emblematic of contemporary scientific rationality. Nothing is more respectable than its operative "modesty," considering the difficulty of determining rationality in and through itself; but nothing is more intrusive than this virtue (another version of the "principle of charity") whose dynamic is going to replace everything.

This analysis, therefore, reveals at once the philosophical *relativity* of rationality as soon as it is removed from the initial rigor of its logical and epistemological con-

ditions and is globalized to apply to human behaviors, as well as the effective omnipresence and ultimate power of this very particular type of intelligence: essentially *operative*. No one knows, in the end, if this is truly the most rational. It is of
little importance: the less this rationality is questioned at its foundations (and, consequently, at its limits), the more it is extolled and practiced without limits.

If rationality is considered as only a web of self-referential relations, we completely disregard that which is classically called its "applications," wherein so many
of the effects of power are revealed. Or rather, we diligently endeavor to retain only
those effects that denote or confirm the internal efficiency of rationality. How can
we deny that we are engaged with a sovereign operativity in its instances, essentially
in the logico-mathematical domain? This internal power of self-regulation was the
only one that was seriously acknowledged by classical rationalism when it ceased
casting its eyes toward the ideal of pure truth. When a more exterior rational domination was considered—for example, by the "practical philosophy" of which Descartes speaks in the sixth section of his *Discourse on Method*, it was not yet truly
detached from the self-regulative model; it was a rationality focused on the truth, a
truth *index sui*.

Consequently, we can better understand how the aforementioned antinomy between the rationalization of the world and the impotence of reason has been able
to solidify. The latter, under the guise of a faculty, substantialized the self-anticipation that we have already recognized as constitutive of rationality: it was a privileged metaphysical expression of it and nothing more. The former is de-centered
toward an effectuation that it pursued independent of its metaphysical origin, following a logic all its own. We can now also better assess this task: to understand the
situation in thinking this scission of the rational.

This questioning of the power of the rational will not proceed from a definition
any more than does today's scientific and technological practice. It will gather the
rational in its results, such as they offer themselves in the world today. And it is only
from such a phenomenology of this power of the rational (in its effective operativity) that the distinction between its phases, its genealogy, can take effect.

We must establish that the genuine Relational does not account, neither as a
direct cause nor as a principle of explanation, for the expansion of scientifico-technological power.[12] Instead of casting the latter into the outer darkness—a very un-
philosophic Manicheanism—we will be led to ask whether Power (in the modern
sense) does not put into play a specific kind of rationality, and whether this specific
rationality breaks with rationality in general or with the classical discourse on truth
(Aristotelian or Thomistic). Thus, the beautiful homogeneity of rationality is going
to be traversed by a very radical rupture between the true and the operative, a still
insufficient characterization of a caesura—or revolution—whose ontological, methodological, and scientific tenets and outcomes must be thoroughly examined.[13]

Unfortunately, there is no assurance that rationality is capable of elevating itself to this critical self-differentiation by the sole means of argumentation or by for-

malization according to its rules. On the contrary, experience teaches that the spontaneous philosophy of rationality is a belief that redoubles the virtue of linking objective adequations by founding it on the side of the subject—the belief in the "value" of reason, whether through methodological doubts or, as with Monod, through the disillusioned stoicism of an "ethic of knowledge." Far from detaching itself from its direct functional norms, rationality seems rather inclined to turn back to the coherence of its own systems or of its analyses by valorizing its global project once and for all. It is thus, facing the most brutal dominating effects of techno-science, that the montage of what we call the rationalist double play will occur.

The double play of the rationalist sublimation

A rationalist thought is one that, while not being content to carry out operations conforming to logical rules (in a formal way in the positive sciences or in the practical field), considers reason as a universal model and thus extends the use of rationality to the totality of the real, unconditionally making it the supreme norm. Classical rationalism led to an a priori valorization of reason, without considering its effects. Facing the ambiguous deployment of the power of the rational, this attitude became difficult to maintain in its entirety.

Today's current rationalism, therefore, combines several systems of defense and, in any case, takes into account the power of the rational. Its resiliency yet remains the idealization of rationality. Contemporary rationalism is certainly no longer what it was between the two wars, exemplified by Lalande and Brunschvicg. It is, nevertheless, not at all certain that the presuppositions and prejudices of that rationalism have completely disappeared from the minds of a certain number of our contemporaries. To verify this, I am prompted to go so far as to unearth *La Raison et les normes*.[14] No student now disturbs the dust on this tome. From the first pages, Lalande defines reason very narrowly—o merciless restriction—as a "mental function" of man, opposed to other faculties, passions, impulses, "whims": "Rationalism consists in taking the side of existence and the value of reason thus understood."[15]

Everything is therefore set out in the first lines: the psychologistic deduction, the moralistic reaction. The first artificially stabilizes the "reason-faculty" in an essence whose origin and legitimacy are never established. Thus boxed in by a totally implicit metaphysical humanism, reason will be *celebrated* without regard to its ontological field, its operativity, its effects. What we call, on the other hand, the moralistic reaction consists of immediately taking the side *of* reason without even having dissociated the question of value from that of existence. It "is necessary to trust in reason", the good apostle keeps repeating,[16] like a lawyer so hurried to prove his client innocent that he forgets to examine the evidence.

Though obsolete, this academic rationalism has the advantage of offering a "symptom" that is prevalent among rationalists, even the most vigilant: wanting to immediately display certitude, and undoubtedly thinking that the best defense is a good offense (but is reason attacked? And why?), they too make the leap from *existence* to *value*. For the rationalists, everything occurs as if rationality could be developed and propagated only in a communicative optimism, as if arguments, proofs, and experiments demanded—beyond their own validity—the *electricity* of enthusiasm, switched on by intersubjectivity, interconnected in the space of communication. And, therefore, from *functioning* to *valorization*, the outcome is favorable. The conclusion being reinforced by the "benefits" of technology and all the reflexes that correspond to it is "it works, therefore it is good for us."

But why is it necessary to defend or to celebrate reason? Why exempt it so quickly from the examination of its effects, be they ambivalent and agonizing? Why cast embarrassed glances toward the vocabulary—and the realities—of power, force, and domination? Why this rush to use the water of reason to wash the hands of so many modern-day Pontius Pilates whose beloved research "results" are frequently questioned?

To raise these questions is not to have a ready answer; certainly, the concern for security, the desire to maintain a good conscience plays a role in these attitudes. But there is a deeper reason that prevents most rationalists from considering the *power* of the rational: they develop their argument from the fact that this power is comparable to a halo (or a cloud of fallout) that is not part of the phenomenon defined *stricto sensu*; or more precisely, according to them, this power, in its most massive, brutal and revolting form, cannot be analyzed *in terms of the very operativity of the rational*. Now we will see that herein lies a very formidable "ruse" of a specifically modern rationality: "effects of power" are discernible at the very interior of this operativity, provided that the latter is not understood in purely formal terms—but in relation to its project or strategy. We will, therefore, be led to confront rationality with itself and to redefine it on the basis of its goals and its articulation of the real, thus heading off the rationalist tendency to revert to the pure and certain terrain of logico-mathematical coherence.

Because the valorization of the rational is more subtly idealized in Husserl, it is going to resurface in his work in a more interesting form. Confronting the power of the rational—the successes of its inner operativity, its experimental schemas, the expansion of technological networks—the *valorization* which takes place, even though eminently positive (that is to say, not casting any doubts upon the reality of the progress induced by reason, and continuing to place great hope on its progress), will take into account the (apparent, provisional, or real) negative effects of rationalization. This correction will be realized thanks to an *idealization* that becomes all the more heightened as the analysis of the present effects of rationalization appears ever more bitter.

When Husserl meditates on the crisis of the West, he never ceases to turn toward rationality to rediscover that which founds the West as such, the cause of its maladies, and the ideal that will save it from them. It is universal and objective knowledge, originating in Greece, that henceforth opens humanity to an indefinite progress in which we are still being swept along. If, however, the contemporary crisis is so acute, it is because, having broken with its ancient gods, humanity does not yet fully accept the exigencies of rationality: "The reason for the failure of a rational culture, however, as we said, lies not in the essence of rationalism itself but solely in its being *rendered superficial*, in its entanglement in 'naturalism' and 'objectivism'."[17] The modern human being is rational only negatively and abstractly, in order to destroy more than to build. And yet, it is through an increase of rationality that a new human being will be constructed. The crisis has a status akin to error: reflection on its causes ought to eliminate it; and the love of knowledge exonerates it of all the errors to which we have surrendered in its absence. Paradoxically, the crisis of reason leads Husserl to completely exonerate reason, and even to idealize it all the more purely because it has been so radically attacked.

Thus, even if Husserl does not make the power of the rational absolute in the way that Hegel did, even if he no longer builds an explanatory and totalizing metaphysics, rationality remains for him principle and end: the principle of all valid knowledge, the end of this very knowledge. Indeed, he speaks of a "teleology of European history." We must pay attention to this finality. It is no longer in any way exterior or predetermined; it is the truth by which scientific knowledge is regulated and verified; it carries within itself the criteria for all validity, provided that logical rules and experimental procedures are continuously applied. Rationality indeed remains an absolute, in the sense that there is nothing beyond it. It is the ultimate horizon, no longer theologically, but programmatically.

Husserl expresses with coherence and clarity the spirit of rationalism of our century. *The Crisis of European Sciences* has a canonical value beyond the conflict of multiple variants, of numerous distortions or misunderstandings. Why? Because, while recognizing the power of the rational (a sound base for knowledge of the present), it *valorizes* rationality absolutely, elevating it—idealized—to the pinnacle of knowledge and action.

And here the major orientation of the philosophy of the technological age is already outlined: *the power of the rational is recognized only in order to better intensify its idealization*. The contemporary mind says, "See what rational work has produced. The scope of these transformations allows us to predict what will come. We must have yet more rationality! Errors rectified, new discoveries, a generalized rationalization of the practical such is the shining future of humanity."

In fact, Husserl offers a version of a prevalent rationalism that is not only canonical, but even angelic.[18] The power of the rational is hardly envisioned in the totality of its effects and in their complexity; it serves rather to vindicate rationality itself. As an evident and obligatory consequence of the qualities of the rational, it reinforces its *idealization*. For the community of mortals of the twentieth century, the efficacy of reason is the gauge of its intrinsic value and of the benefits that it brings; this widely accepted attitude is shared by the scientific community, with, of course, many variants and in more or less subtle forms. One will object, as has Bouveresse, that Wittgenstein is an "anti-Husserl," and that this holds true for his heirs as well as for those of Frege. Is there not therefore an epistemological rationalism that escapes our objections in the authors just mentioned, as well as in Popper or, more recently, Newton-Smith?[19] We do not intend to respond dogmatically to this objection; it is a matter of examining in each case whether the "double play" functions to the benefit of metaphysical rationalism. If Wittgenstein's thinking can withstand this test, it is perhaps a question of an exception that proves our rule: does "epistemological rationalism" escape metaphysics by the pretext that it does not mention it?

Originating from an ideal model, rationalism is continually captivated by it. The golden age of triumphant scientism has ended. In that epoch, an inventory of the practical triumphs of scientific and technological reason was sufficient to reinforce idealism—to the highest degree. Since the power of the rational has revealed its cruel, dark face, the untroubled recourse to a double register (ideal and real) is no longer possible. The double register becomes a double play [*double jeu*]. When the consideration of the effects becomes awkward, worrisome, unbearable, rationalism turns towards the ideal, reactivates it, intensifies its light, even if it implies exhibiting anew the benefits of effective rationalization as soon as the latter proves itself justifiable and demonstrable.

Undoubtedly, one will say, you evoke the garden-variety rationalist; but Husserl is more consistent. Having never made the rational pass the acid test of technico-practice, he, nonetheless, ascertains a crisis of the West, which he explains by a devaluation of rationality itself; he therefore does not place the responsibility for the crisis on exterior factors; it is, indeed, rationality that must act to correct itself. I, however, respond that this examination of conscience, far from placing in question—even provisionally—the foundations of logic and science, leads to a programmatic reinforcement of rationalization and to an idealization of it that is all the more intense, insofar as the realities are suddenly seen to be grim. Even if Husserl himself does not succumb either consciously or explicitly to the double play that is too frequent in rationalism (the ideal on the one hand, the "good results" on the other), he nevertheless practices a unilateral revalorization of the aforementioned rationality. *The ought-to-be of the rational serves thus to obscure its being* and the fact that rationalization and power are henceforth largely linked. Facing the power of the rational, we would like to avoid this angelicism.

The rationality or irrationality of Power?

After acknowledging the constant interaction between science and technology, Jean Ladrière remarks, in *Les Enjeux de la rationalité*, that they tend "together to constitute a kind of unique super-domain or super-structure, at once conceptual and practical, having a dynamic character, evolving in the direction of a growing complexity."[20] This super-domain, whose development increasingly conditions our world and our existence, is what I will here call Power—whose dominating and ever more alienating character has already been established by Ladrière: the *logos* of the scientifico-technological super-domain "becomes like an external power" that attempts to impose its own law on human beings.[21]

Is this power, which increasingly transcends the individual will, rational? All the while objecting that rationalism secures the essence of rationality and continues to idealize it, we are prepared to concede to rationalism the following (which apparently concerns rationality itself less than it does its inscription in the current state of the world): nothing proves, in principle, that this Power—so violent (and "violating")—is *purely* that of the *rational*, that there is a relation of cause and effect or of a principle and a result between the second and the first.[22] Unless, of course, the rational is considered from only one level, from one single point of view. But this is impossible or absurd in our century, considering the mutations of logical rationality as well as the extraordinary diversification of the scientific field.

Therefore, with our backs to the wall, it will be necessary to ask: What is the principle of this power? What rationality is played out here? To what degree and under what auspices? Also, which irrationality? The path we have embarked upon distances us from the representation of a unique rationality that would produce, through a progressive and cumulative development, a certain number of effects quasi-deducible from its essence. We have already seen that there is no essence of rationality which would extend through history and lead to our technological world. I argue, and this must certainly be radicalized and subjected to proofs, arguments and meditation, that Power results from an upheaval within rationality, and that the zigzag course of the rupture, historically identifiable, is still marked in our goals and our modes of thought. It is too early to draw all of the consequences of this, but already we see everything revolving around this question: Does a specific rationality propel Power? Or does this *specific* motor have an origin that is not intrinsically rational? Or again: Is this specificity too complex to be decided merely on the basis of these two alternatives?

If, instead of placing the emphasis on the rational, we were to place it on power, and if we were to question the degree to which it is rational, the same swing of the pendulum, the same uncertainty, and finally, the same antithesis arises in relation to the reign of reason. Such Power (capitalized when it designates the current scientifico-technological complex as opposed to power in general) is certainly made possible and is continually nourished by scientific rationality; its technical efficiency and economic feasibility require the utilization of models and procedures that, in the global sense, are equivalent to the *rationalizations* of research, of production and of consumption. But neither finality, nor unity, nor even results account for rationality. How does this Power fit in? Does it even converge toward a coherence? Does it not liberate the worst as well as the best?

These questions are not academic. Even if they must be rooted in history, we will quickly see that they erupt into our present. Indeed, if we come to demonstrate

in a somewhat convincing manner that the power of the rational is, in its living resources, not equivalent to that of Power and that, conversely, Power posits itself as
the best representation of rationality by a kind of *captatio benevolentiae*, then certain consequences follow. The *spirit* of rationality in general (if it can be uncovered,
which we believe it can) in no way explains (in any case, not by itself) techno-
scientific domination; this domination is made possible and produced by a *will to
objectivity* (or to certitude) of which we must ask whether *it is itself purely rational.*
As Kant observed, pure rationality—left to itself—leads neither to certitude nor to
apodictic objectivity. Even if it is not my task here to reconsider the transcendental
dialectic (because my purpose is at once larger and more originary than the question of the legitimacy of special metaphysics), nor to hold to a critical, even recon-
structed, topology (because my approach does not wish to be purely transcendental,
but intends to be articulated with the effective reality of the present world), the
Kantian legacy will be appealed to insofar as the critical topology of pure reason
not only leads to a theory of the limits of knowledge, but also contributes to unsettling—at least partially—the Cartesian and modern convergence between truth as
certitude and the all-encompassing project of understanding (and of seizing) the
world. This unsettling—although Kant's thesis on being is reflexive—is produced
not only in the schematism, as the early Heidegger believed, but also in the approach
to the beautiful, to the sublime, and to life, on the one hand, and above all in the
uncovering of the moral law as the imperative of *pure* practical reason on the other.
In this regard, how else could Hölderlin have saluted Kant as the "Moses of our
nation," if, in his eyes, this modern prophet had been only one more messenger of
the Enlightenment? The conception of the pure moral law as rational comes close
to the conception of a rational *dynamis*, in a sense that does not at all converge with
its technical-operative power. Does Kant think "contiguity" such as we understand
it, which is to say, the co-possibility between the dominant and the reserved, between the language of power and that of the ontological enigma? Yes, if the critique
liberates its hermeneutic potentialities rather than being reclaimed by scientism
and positivism.

Thus, rather than consider the objectifying goal as self-evident and rationality
as a great force that travels a straight path toward its completion (indeed, we have
seen that Husserl speaks of a teleology immanent to rationality), I will invert the
perspective by showing that the great rupture *that caused the rational to tilt toward
the objective* in the seventeenth century is not yet completed and that, consequently,
Power illegitimately claims everything that is possible for rationality. This does not
mean only that there is theory that can be non-operative, speculation that—thank
God—can and must escape the world of organization and the laws of the marketplace. More fundamentally, the project of Power must *constantly reactivate its original **coup de force*** in order to support its legitimacy (and it often finds among its
"adversaries" a relative, involuntary complicity, to the degree that these adversaries
do not mark the break between rationality and the will to power). The rationality

that is effected in techno-science is a *specifically excessive rationality* [*surration-alité*],[23] which, through a play of retrospections and retroactions, succeeded in imposing itself, even in its philosophical genealogies, as the goal of progress, thus taking the place of the Platonic Good that it too often improperly placed at the completion of History.

But, on the edge of the great road of Power or at its margins, the rational still has something to say, or, more precisely, it still reserves a possibility that is—and for a very good reason—too little cultivated by techno-science itself, but also by meditative thinking that has perhaps believed too soon and too quickly to be able to bring about a global retreat vis-à-vis *the rational and the metaphysical*. When Heidegger examined the genesis of the principle of reason, he rightly insisted on an incubation period (more than twenty centuries) that preceded its formulation by Leibniz; but he lets it be forgotten perhaps a bit too much that this principle could just as well not have imposed itself in its modern and efficient form, without the *method* "taking control" at the very heart of rationality, initiating and announcing Power. To be sure, the affirmation of the decisive importance of this rupture does not contradict Heidegger directly: I even claim to follow him in reinterpreting all scientifico-technological modernity on the basis of the will to certitude that is immanent to the *cogito*. It is not a question of denying our debt in this regard; but the lighting changes somewhat if the beam of the projector is focused on the immense resources of a *rational possible*, which is always near to us in spite of the jealous exclusivity of Power, without—moreover—this power of the rational being severed from the truth of Being. Historically, to recall a not insignificant example, the empire of Hegelian rationality is henceforth comprehensible as fallow territory and toothing-stone (in relation to the dominant course of the West), rather than as a relay of metaphysics-completed-within-planetary-technology. Currently, there is no longer an expectation of a "turn in time," although respectable and respected, the sole conduct worthy of thought: there are also the rational critique of unwarranted or totalitarian rationalizations becoming necessary, the meditation that carries rationality to its limits, acknowledges the constantly offered resources of the ontological enigma, and interprets it.

"In noting at this point the caesura between purely relational rationality and objective rationality, you sever—an untenable paradox—the course of the rational, without being able to prove that the latter is not the accomplishment of the former. Will you deny that the boundary that you are uncovering henceforth definitively separates the prescientific from the scientific in the modern sense? As you will be incapable of denying this, you risk supporting an extreme and very fragile paradox, refusing to see in rigorous scientificity the outcome of centuries of a more or less pure, but uncertain, reason, finding its way. Emphasize as much as possible the importance of the philosophical and epistemological revolution that founded modern science, but do not use this caesura to advance outmoded forms of rationality!"

This objection is legitimate, and it would be necessary to address it if my intention were indeed to deny the rationality of the scientific *method* that has been in place and has revealed its efficacy since the beginning of the modern era. This is absolutely not the case: I have spoken of a specifically *excessive rationality*, and I will not use the irrational reversals or absurd consequences of many technico-scientific accomplishments in order to contest their rationality (the irrational has meaning only in reference to rationality). I am concerned even less with substituting a secret or aesthetic *valorization* for a rationalist, scientistic, and technicist *idealization*, or of advocating a nonconformist or "wild rationality." On this point, the critique of the myth of the "new science" will allow me to show what separates me from Serres, Prigogine, Morin, and several others, to disabuse them of their true or feigned optimism, all the while acknowledging certain of their insights.

Is it possible to uncover the rational possible beyond all rationalism, to take a critical distance from the dominant rationality, beyond all irrationalism? The wager is strengthened by taking into account reserved language-worlds, in particular that of the ontological enigma. How will this all hold together? My response will only be revealed in the light of an exchange with Heidegger's thought, which the following pages will outline.

The ambiguity of Heidegger's position

Taking up the ambiguity of Heideggerian thought, I refer neither to *Zweideutigkeit* in its particular sense in *Sein und Zeit*[24] as a qualification of "thrown-ness," nor in its very general sense as the hermeneutic overdetermination, such as it constantly runs through the work of the Freiburg master (from the ambiguity of the "hermeneutic circle" that opens *Being and Time* to the *Fragwürdigkeit* of his last writings). Not only has Heidegger taken up "ambiguity" in this latter sense, he has made it the theme of his meditation, whether pertaining to Being, truth, time, etc., from the viewpoint of an always deepening meditation on the "unthought" in Western metaphysics.

The ambiguity that concerns me is much more definable and yet, it seems, was not truly explained by Heidegger himself. Before defining its contours, I note that its importance goes well beyond a simple taking up of a rationalist, antirationalist or irrationalist position. Heidegger has always, rightly so, rejected such taking sides. Those who accuse him of irrationalism have, usually, hardly read him. His most typical manner of confronting rationality is found in these few lines from the *Logos* lecture: "Yet what can reason do when, along with the irrational [*Unvernunft*] and the antirational [*Widervernunft*] all on the same level, it perseveres in the same neglect, forgetting to meditate on the essential origin of reason and to let itself into its advent?"[25] In this passage, as in many others, Heidegger rebels against a recourse to a rationality that would claim, in the name of logic, to dispense with a supposedly more essential meditation. There is, on Heidegger's part, a clear and constant refusal to allow the respective domains of logical thought and thinking as such to overlap.

Rationality, like science, cannot go back to its own foundation. Certainly, metaphysics attempts, in absolute idealism, to found itself; but this totalization does not prevent it from persisting in the properly metaphysical forgetfulness of the ontological difference. I note, moreover, that in the passage cited, the negligence of reason is not presented as a necessary consequence of its essence, but merely as a dangerous eventuality that one need not accept as a foregone conclusion. Finally, in a move very typical of Heidegger, a reason that constricts itself with its own rules degrades itself to the level of that which it attempts to denigrate: the absence of rationality, and even madness. This is why rationalism and irrationalism are placed together, the second being only the reverse side of the first: " . . . irrationalism, as a denial of *ratio*, rules unnoticed and uncontested in the defense of "logic.""[26]

Reason, content with applying its rules, *does not yet think*. If Heidegger appears to link it with irrationalism, it is because he claims to situate himself as far from irrationalism as he does from rationalism. He is far from irrationalism because it is at no time for him a question of supporting the passions or subjectivity against objectivity; and if one believes that the return of *Sein und Zeit* to the prepredicative, as well as the quest for a "resolute decision," is removed from rational mastery, it is necessary to recognize how much this return and this quest are carefully connected to the royal road of philosophy (for example, in paragraph 44, the founding of adequation upon unconcealment); it is, moreover, necessary to underscore that the later *Kehre*, so sober and meditative, ought to erase any suspicion that remains concerning Heidegger's "irrationalism."

I claim this, despite the famous phrase in *Holzwege* wherein reason is characterized as "the most stiff-necked adversary of thought."[27] This rather provocative phrase, which I will not take up (one should begin to understand why by the end of this introduction), can appear antirationalist, but, certainly not irrationalist. One finds repeated in a more poignant form than usual in this phrase, Heidegger's—very coherent—thesis that pure and simple *rationality* is not a sufficient guarantee of *thought*, with, moreover, the emphasis placed on the danger of "self-satisfaction" to which the exercise of rationality can lead. Heidegger implies that, it is, therefore, necessary to break with the arrogance of reason; but there is nothing scandalous or even novel in this gesture: Galileo, Descartes, Pasteur all confronted impeccable reasonings. There is, as Bachelard insists, an antirationalism that is beneficial to rationalism itself; furthermore, Heidegger maintains that thinking ought to be able to think "against itself." Thus, Heidegger does not oppose the rationality of reason (whose scope he perhaps, despite everything, misunderstands); he opposes the *exclusive* character of this rationality and the claim that it is the *ne plus ultra*.

I must make one more concession to Heidegger before coming to the very heart of the question. Are not his misgivings honorable? Is it sufficient and precise enough to refer to rationality? Are the outlines of *partage* thereby grasped any more precisely? This objection could put my own enterprise in peril. We must examine it with all the seriousness it deserves.

If Heidegger did not develop the theme of rationality, and even resisted using this term, it is clearly by design. The beginning of an explanation can perhaps be found in the following passage in which he indicates that the difference between the rational and irrational is not clear: "a difference which is as far as it is concerned based on an instance of rational thought, itself not clarified."[28] Let us agree on this: Heidegger does not avoid going deeper into the question of reason as foundation, especially in the remarkable *Principle of Reason*. But the preceding quote reminds us on the one hand that he never considered the rational-irrational distinction to be crucial or central, and on the other hand, that he continually refused to invoke rationality as an all-embracing supreme power or as ruling all of thought. Simply stated, it is—of course—a refusal of doctrinal rationalism, but also a refusal of that complaisance which, abandoning the meditation on the essence of thought, entrusts it to a "rationality" reassuring by its implicit reference to logical principles and arguments, to scientificity in general, but a rationality that is vague enough to welcome all "non-scandalous" forms of thought. At this point, certainly repugnant to rationalists too sure of themselves to realize it, Heidegger exerts a critical function, calling upon us to shed more light on what is called rational thought.

Heidegger, thinker of difference, is reticent to allow thought to settle into a *space of generality* or of conformity where the question of origin (of reason as foundation) is obscured by the setting-in-place and the deployment of the *relational*. The "destruction" of the ossified tradition is precisely the questioning of that which seems (unduly) evident to the Western episteme: the rational of its reason. In turning against rationality the will to clarity that in principle it ought to assume to the end, Heidegger wields the weapon of irony. In so doing, he has recourse to the following argument: the sciences "objectify" all beings; only the question of their essence escapes their grasp and is sheltered from all questioning. Rationalism only generalizes this attitude; it installs an unwarranted assurance in this fundamental place where interrogation ought to occur.

It is now time, having recalled both the "dignity" and coherence of Heidegger's position, to speak of the ambiguity that I intend to isolate and whose first analysis—brought up as an introduction—will allow me to show to what degree my own project is at once inspired by and demarcated from the Heideggerian meditation on rationality.

Ambiguity, which often tends to harden into dichotomy, can be understood either in the approach of rationality, or in that of a thinking yet to come. But it would be perhaps better to detect it at its commencement—in the very differentiation between reason and Being. On the one hand, "Being and reason (*Grund*): the Same"[29]; on the other: "Being . . . remains without reason because it is itself reason, the foundation."[30] These two positions are evident to whoever is capable of emphasizing the principle of reason, either as grand principle (the "reason-giving" [*rendre raison*] of representative thought) or as "petit principle" (the word of Being which gives itself as reason). One will object to us, arguing that there is no ambiguity here, but an explicit meditation on *difference*: not to reduce the "petit principle" to the grand principle is to refuse to "represent Being as being."[31] There is the same movement of thought in *Identity and Difference*, where the Heideggerian distancing permits the Being of metaphysics to appear in its double foundation and, at the same time, allows the onto-theological constitution to appear as the forgetfulness of the difference. It is always a matter of freeing thought from the circle of a rationality enclosed upon itself. "Expelled from the truth of Being, man everywhere circles round himself as the *animal rationale*."[32] Heideggerian thought is *differentiating* in respect to an exclusive reign of the rational. Where is the ambiguity?

It appears only when we attempt to summon [*sommer*] (in the double sense of adding up and of compelling something to appear) *differentiation*, in order to draw up conclusions, and to constitute a "new thought." Heidegger both allows and forbids himself this. He allows himself to attempt to reestablish a more originary philosophy in *Being and Time*; but the *Kehre* is going to meditate on the impossibility of this refounding—not without "putting to the side" a "thought of Being" whose autonomy is at once affirmed and denied. *To think an ontological difference that would be continually differentiated*, without being hardened into identifiable difference, is the very difficulty of the *Verwindung*, the feat of this "overcoming" of metaphysics, which we should not even call an overcoming.

In doing this, are we forgetting the relationship to rationality? No, because the ambiguity that we are attempting to isolate is always the same aporia, the same *thought by paralipsis*, but which is now turned toward rationality.

Is rationality simply the "reason-giving" [*rendre raison*] that leads to the technical exploitation of the earth and that reduces everything to a calculable object for representative thought?[33] Or is rationality more: the sending (*Geschick*) that constitutes and delimits us as Westerners, from which we cannot escape without falling again into the most forgetful errancy? Heidegger responds positively to these two questions: herein lies the ambiguity. Heidegger does not deny this ambiguity; he takes it upon himself. Is it not the ultimate treasure of our destiny offered to thought? Without any doubt. Do not misunderstand my approach: I do not intend to "refute" Heidegger or suppress the ambiguity that I just exhibited. I intend to start from *the same ambiguity reformulated from the point of view of rationality*, that is to say from the Relational.

The Relational has already presented itself to us as the field in which intelligibility is constituted—producing identifications, constructing and combining logical and scientific articulations. In a second-order language, that of linguistics, we would say that the Relational is presented with the diacritical sign. The Relational designates that which the Heideggerian "there is" dismisses in advance, without being able to suppress it: being, the "this," the limit, and the other beings with which they in one way or another "correspond." Heidegger goes back to what is Primordial, arguing that there is no being without Being, no presence without originary temporalization, nor more pure unconcealment, etc.

One will perhaps object that my response is that of metaphysics, a return to the anteriority that Heidegger had claimed to "destroy" and, consequently, it is an annulment of what the Heideggerian hermeneutic has established. The wager of my endeavor is to deny this and to claim the following: Heidegger's thought could appropriate metaphysics and weave its way within it only by giving evidence of listening to it and grasping its *intelligibility*. If Heidegger never *completely* severs his thinking from the tradition, it is because what *he shares with it is rationality itself*; it is because it is impossible for him to dispense with the Relational. When, for example, he writes, "Being itself is the Relation"[34], he intends to show that it is not man who constitutes being, but rather the reverse.

Nevertheless, these few words warrant attention: it shows the ambiguity that we seek. On the side of Being: at once *Lichtung*,[35] a "lighting" of all unconcealment and relation. On the side of relation: at once Relation par excellence (the condition of possibility of all the others) and identifiable relation—which it is necessary to refer analogically to the Relational. Now, just as Heidegger carries out an interrogative reading at the heart of metaphysics, calling our attention to an "unthought" at its interior, so we will install ourselves at the very heart of Heideggerian language in order to reveal an *intelligibility* that its author leaves undeveloped. Indeed, whenever it is a matter of relation, of primordiality, or, above all, of difference, there is always postulated the complicity of a *sustained relation* with meaning, although the gaze of the thinker claims to go above or beyond this. Likewise, inviting us "to think Being without being."[36] Heidegger could not have overlooked the fact that such a backward glance had always ruled over his own undertaking and continued to reappear in his very refusal (it is this kind of approach that I earlier called "thought by paralipsis").

To think is to enter into relation (in a sense, of course, larger than the categories of relation). In proposing this, I do not exclude Heideggerian thought as *Danken*, thankful knowledge that welcomes and gathers, whereas the concept tends to capture; but I refuse to allow a strict separation between the concept and the gathering. Certainly, we must recognize a limit within the Relational: I call it *partage* (in consonance with the Heideggerian *Geschick*). The thought of "contiguity" consists at once of testing this *partage* as an unavoidable "there is" (the pre-relational) and of identifying it as the condition of every relation.[37] But rationality itself is *Geschick* for Heidegger, as it will remain for us. What, then, is the difference? In turning away from the Relational in order to think only the unthought Primordial, Heidegger retains from the current rational *partage* only the dominant evaluation (manipulated by a uniquely representative thought), and reserves all future potentiality to the thought that is *other*. The present of rationality is stripped of all potentialization. On the other hand, the thought of "contiguity" sees difference inscribed in *partage*, traversing the rational; far from confining the Relational to the side of techno-science, it evokes it as it is looking for an ontological dwelling and seeking, close at hand, to enter into an understanding with the possible.

Addressing these successive points has allowed, I hope, to determine in what ways I remain Heideggerian and in what ways I separate myself from this inspiring thought (not by abandoning Heidegger's questioning, but through a return to the *side* of it from which he turned away). To travel upstream against the tide, from the pseudoevident exercise of rationality to the question of its origin, to consider reason as *a partage*, its foundation as without ground, is to agree with Heidegger's refusal to forget the ontological enigma. It is to refuse a relentless identification between rationality and Being. It is to leave exposed and irreconciled the *sending* itself by which the *logos* imposed itself and, as Husserl said, opened the indefinite field of its progression. But to take this destination seriously to the point of assuming it in its entirety, to recognize *what is our lot* is the daily and massive exercise of rationality (that is to say the covering over of its origin by the homogenization of the region where rationality developed), is to look in the face and loudly articulate that which results from Heidegger's explicit thought as well as what it tacitly admits: living in an epoch of planetary technology, we escape neither its logic nor its practice, even if we prepare other paths and listen to its unthought. The dominant language (that of techno-science) does not stop simply because some have scrutinized its limits; the forgetting of Being will undoubtedly become still more extreme and the task of thinkers even more difficult.

This turn toward rationality as *partage*, and the espousal of its acuteness in order to better mark its limits (on the side of the rational, but also in the terms of the languages that it represses), does not mean a return to the forgetting of Being, to the pure and simple will to truth, nor does it nullify the hermeneutic gains (and the imponderable) of the Heideggerian step back (*Schritt zurück*). To be sure, the style, the tone, the "attitude" will be different, because the undertaking will be pushed to the point of asking whether the time has not come when the *self-position* of the rational, leading to a forgetful and destructive exclusivity, does not yield to a *self-limitation* of the "space of conformity" (without purely and simply leaving, as Heidegger increasingly did, the dominant language alone). In spite of this divergence, which the course of this inquiry will intensify even further, the thought of contiguity is located in the field of *critical differentiation*, in the sense that Heidegger situates it between "that which demands a proof for its justification" and "that which claims for its test (*Bewährung*) a simple look and apprehension."[38] But this thought will not be content to affirm and refine its methodological originality; that would be to forget that the world of Power—itself incapable of resolving the enormous questions that it imposes upon us—is a world of urgency.

An outline of the new approach

Before beginning my inquiry into Power, it is necessary to give a more positive and precise idea of the direction followed. As always, anticipation is dangerous, allowing one to believe that everything has been decided in advance, and therefore breaking with the spirit of inquiry that, in fact, never ceased to animate me. For the moment, what is possible and desirable is to indicate the initial hypotheses and their probable consequences, if the course of the inquiry confirms them.

It would have been tempting to construct a general theory of domination. All-encompassing and subtle, it would have allowed for understanding once and for all, and undoubtedly justifying, the course of the world, its twists and turns and even its retrogressions. But it is apparent that this undertaking—already attempted by Hegel[39]—would have presupposed a metaphysics of the rationality of the real, which itself is in question, and moreover, such a theory would have run the risk of being much too general. To begin from the phenomenon of Power such as it is given here and now to "natural consciousness" is already to cease confusing it with what it is not, and to begin to view it in its specificity. This is why I propose to articulate, in the first part of the book, an analysis of modern technology within a phenomenology of Power.

This is an apparent, but necessary, detour before entering the kingdom of science. This kingdom, which began in Greece, is not completely "of this world," but sounds the possible, knocks at the door of Power without opening it to the scene of History. We confront a formidable enigma with the fear that tests every person before the Sphinx: how can the rationality of episteme, already having come so far, not trigger the irreparable course of Power? The central hypothesis is thus announced: the power of the rational and Power are two. If we establish that Power as such is not born from rationality in general or from purely epistemic rationality, but from a particular rationality or, more precisely, from a specified excessive rationality [*sur-rationalité*], we must then isolate the *operator of power* that allowed Western rationality to achieve the prodigious "forward leap" of modern knowledge-power. This step is not sufficient. We will then follow this excessive rationality in its most violent and dangerous "effects of power" and ask whether these "results" (more or less disavowed by rationalism) always spring from the same operator of power.

The *unfolding of the rational* increases both the possible and power, while the strategic development of a certain type of rationality increases only Power. The totalization of this excessive rationality carries this power to the ultimate, intense, and perhaps suicidal stage that we know today (planetary imperialism, the ultimate stage of technology). Are these so many stages of the same linear process? I doubt it. What was this *supplement* which, metamorphosing pure and simple rationality, allowed it to arrive at a truly operative science: the will to power or a more subtle conjunction? No one can avoid the question of knowing whether the will to certitude is, or is not, rational, whether it does not hide an unacknowledged irrationality in the desire for excessive rationality. We must examine this fully, asking whether the ultimate phase does not take a new step in the Reversal of the rational into its opposite. Undoubtedly, a new step; but of the same kind? Or is it new, above all, because of its irreversible character?

The affective responsibilities and the quasi-religious investments in rationalism strangely allow it to forget a truth of experience: in reasoning too much (or reasoning tangentially to that which is in question), one falls into the absurd (or even worse); excessive rationality—one could say—is paranoid. Rationality itself *can* be paranoid; does its hubris lie in this originary potentialization—or elsewhere? Although the positive procedures of science and technology believe that they are protected from these deviations (and they are, indeed, when isolated in their internal logic and their areas of well-delimited origin), they are not at all *insofar as they are pieces of a total system* whose excessive rationality is imposed by its own dynamism more than by its internal coherence or its link to Being.

One witnesses in this reversal of situation, incomprehensible even to its perpetuators, an unforeseen revenge of the dialectic (and perhaps of a more originary Law of Reversal).[40] Once it has crossed a certain limit (to be defined), rationality reverses to irrationality—its opposite, or rather, its caricature. Let us not accuse only the metaphysical spirit! Abusive totalizations—for example—are, for the most part, the result of a technological society impatient to export and to universalize its local successes. Indeed, we must recognize in the most rigorous and most efficient rational processes a persistence of the tendency of reason, denounced by Kant, to proceed to the absolute. Even if this tendency is no longer at work in the sense of transcendence, but rather in the practico-technological domain, it is no less vigorous and does not corrupt any less from within the exercise of rationality. As much as one might claim that these "flaws" are to be ascribed to psychological failures of scientists and engineers, one cannot deny the consistency of their reappearance. And if one maintains some intellectual honesty, one will have to recognize that the contradiction is inscribed in the structure of a certain rationality, and that it is not necessary to be an expert in dialectics in order to isolate its core.

Must we be content with the apparent modesty of a rationalist relativism simply advocating a correct exercise of rationality? Thus, for Carl Hempel,[41] the rationality of a reasoning, of a process or a behavior, is always relative to the *goal* established at the beginning (if such a goal exists and is perceptible). The fact of going to buy bread at the corner is in itself neither rational nor irrational; it can become absurd if I clearly know that this particular corner is occupied by a tobacco store and, nevertheless, persist in wanting to buy bread there. All reasoning depends on information and on the anticipated conclusion: this rationalist empiricism plays the game of *Zweckrationalität* at the elementary level. It is tempting to conclude from this that the adequate exercise of rationality is always *relative* and that consequently it is sufficient to focus on that which is imposed by the specificity of each type of problem, each particular case. Is this not exactly what every "scientist" does in his or her domain? Auguste Comte had already posed this essential principle of the positivist spirit: "Everywhere substitute the *relative* for the absolute."[42] But, besides the fact that there is a contradiction in terms here (a relative that is universalized is no longer relative, and is *absolutized*, even if it is in a different sense than ancient metaphysics), it is not enough to proclaim oneself a *relativist* in order to be immune from every deviation. On the one hand, error always remains possible in the punctual exercise of rationality; on the other hand, and more fundamentally, the definition of the very terms of a problem is the most formidable task there is, when one leaves familiar territory. If the relativist adage had been sufficient, the history of science would have ended with Comte, or before: there is no ignoring the fact that it is full of impeccable but empty reasonings, perfect—but useless—experiments. As for History itself, particularly the tumultuous and distressing history of our century, how can we not discern multiple examples of the insufficiency of rationalist relativism? It is a century of sensible compromises and impeccable plans resulting in catastrophes, from the signing of the Munich accords to the "pacification" of Vietnam!

The weakness of rationalism is, therefore, inscribed in the structure of rationality: it is the ransom of the power of the rational at multiple levels where it is determined and applied, and proves its operational ability. The formalization of this internal dialectic (continually verified in experience) will not be absent from my inquiry, but it will not constitute the ultimate horizon. It will not dispense with a meditation on this fundamental truth: rationality *as such* offers no other limit, no other guarantee than itself. After obtaining the certitude of its necessity, it remains for it to make an inventory of its fields of application, but also of its stages, its degrees, and its limits. A rationality that immediately claims equivalence with the totality of the real, as in Hegel, also gains a certain idea of the absolute, exclusive in its unconditioned presupposition. Yet, one will say that science now knows how to moderate itself and sends absolute knowledge back to the museum of antiquity. Granting this, can it prevent the technological and practical totalization of the rational? Is not its methodological modesty the strategic reverse of a constant will to conquer? We find here the incessant reactualization of the Baconian ruse: to obey in order to command.

When the excessively rational [*surrationnel*] reverses into the irrational, a limit imposes itself on rationality, which is suddenly dispossessed of its sovereignty. Reversal imposes its law. But *what authority* will assume this capacity to recognize its limits and control it? A few lucid individuals will never be enough to correct such an enormous process, since minor corrections will not be sufficient. On the side of Power, the effects have multiplied, the initial "choice" unfolds its consequences. Rationality, in the grips of the complexity and irreversibility of real processes, in individual or collective history, is thus often condemned simply to note its impotence—which in some way adds a touch of derision to the absurdity *in principle* of a totalizing excessive rationality [*surrationalité*].

No doubt, the excessively rational is an indispensable concentration for the extreme potentialization of spirit. It is impossible to attain the unprecedented without raising the stakes and entering into high-risk zones. Although reason can loudly object, it still does not escape the reversal. It is hardly surprising that here Pascal lends his terms and offers the help of his thought on *order*. Just as the supreme intelligence no longer resides in the order of grace, we must recognize that an intense and organized rationality fails in the face of the most simple and disarming demands placed upon it by life: childhood, love, death come to stare at our developed societies and do not find anything better than uneasiness and disarray in every response. We celebrate, and perhaps rightly so, the prodigies of techno-science, but nothing is more successful than the operation carried out in isolation—within the strictures of the Plan, of the State, of a given society or party, in the intellectual comfort that the reading of a scientific review provides. Placed in relationship to its conditions of appearance, inserted into the exterior world, every techno-scientific marvel is relativized (is it not necessary to let it take up the game of rationalist relativism that it implicitly implies?): and it is not a bad thing for this to happen. Otherwise, *the vertigo of the complete rationalization of life is given free reign*; civil engineers do not want to "be reasonable" (they *are* always right, having a monopoly on a certain rationality, which our societies have a tendency to recognize as having all the prerogatives). And one notes fearfully, if there is still time, the arrogance of the former *homo sovieticus* before his underdeveloped pseudo-comrades, just as one notes the impossibility of *homo americanus* entering a dialogue with other cultures, at least a dialogue that is not couched in terms of statistics or weapons. More generally: one observes the impotence of Western man ("developed" and therefore thought to be reasonable) to master the problems and, above all, to pose the questions resulting from his own expansion, henceforth planetary.

Whatever the power of the rational, and precisely *by direct reason* of its power, a lucid (still rational) thought can and must proclaim that the *complete* rationalization of life is the most insane project of History. Reversing the sign of all rational benefits, it precipitates the ruin of that which it sought to save. From being positive, it becomes suicidal. Anyone can find, unfortunately without difficulty, in contemporary history, fresh examples of these pacifications that degenerate into genocides, these development missions that produce ethnic genocide, etc.

But is it enough to show restraint and common sense? Without being among those who derisively slash these good old qualities, I believe the present development to be too profoundly, massively and irreversibly established to be curable by good intentions or amenable to partial corrections. The actualization of operative rationality has passed the point of no return. Can it still avoid the ultimate perils, among which is the apocalypse? Has it outside itself or within *itself* the resources that would permit it to master the mastery, or better still: to open itself to the reserved and still closed dimensions of language and of the Enigma? By following these questions, I would like to show that it is not enough to mourn the disappearance of the restraint of the ancient Chinese and the scruples of Archimedes, or even to repeat the advice of Francis Bacon in order to confront the thought of the power of the rational today.[43] If "right reason and sane religion"—these Baconian recourses—no longer suffice, to what authority do we entrust ourselves, to what task do we devote ourselves? Our efforts will be in vain if they do not contribute, to some degree, to giving to this question the beginning of a response.

PART I

Impotence in the Face of Power

1

Facing the Incalculable

... increased "power" is also, *ipso facto*, increased impotence or even "anti-power,"
a power giving rise to the contrary of that which was the original aim; and who
is to calculate the final balance sheet, in what terms, on what hypotheses,
and for what time horizon?
—Cornelius Castoriadis[1]

Approaching the incalculable

FIRST WORLD WAR: 8.7 million dead; Second World War: 40 million. In Hitler's camps: approximately 7 million victims; in Stalin's camps: 30 million, according to Solzhenitsyn. The incalculable is there, in numbers at once terrible and meaningless, which in their feigned neutrality mask the unbearable scourge of these events. The incalculable is also to be found, more radically, in our impotence to say *these things* (*cela*) other than through statistics. We calculate, for want of something better.

In the future, perhaps a worldwide "conflict," a euphemism for an unheard-of, unthinkable catastrophe. Hermann Kahn alone consoled himself before the event by arguing for a few survivors. In the future, if the worst scenario is avoided, unprecedented genetic manipulations, the symbiotechnology of human beings and of cybernetics, the almost total control of geographic space and the informational field (perhaps also the mental field) by the superpowers, armed with satellites, lasers, super-computers, data banks, etc. Once again, it is a question of computing machines, the margin of *overkill*. We will assess the "advantage" of one technological potential over another—for want of something better.

What has happened? What is threatening to take place? Past catastrophes and potential risks confront rationality with the effects of its own power: rationality learns that it is easier to generate effects than it is to control them totally; it discovers, more radically still, the dark side of total mastery and the terrifying "logic" of a surplus of power. What is to be done? Take this major technological risk into consideration, suggests—for example—Patrick Lagadec, who has both the lucidity to recognize that "rationalization is able to be used . . . as a laughable screen"[2] in front of absolutely novel dangers, and the courage to denounce the confused conjunction of interests, lack of awareness, and negligences in the face of this "taboo subject." It is significant, revelatory of a new awareness, that a technician himself denounces not only the technical carelessness and inadequacies of security systems, but also protests the entry of a major risk into a rationality dominated up to now by the calculation of costs, the exigencies of the workplace, economic traditions, etc.[3] Analyzing the catastrophes of Flixborough, Toronto, Three Mile Island, and several others, Lagadec shows that these events are only the alarms that, unfortunately, reveal and announce a new type of risk: these risks are far-reaching in the number of human communities affected, and they are radical due to the irreversible and even partially invisible (such as the dioxin at Seveso)[4] biological alterations involved. The qualitative leap is such that there is no longer "either a territorial or temporal upper limit," which assures a barrier against potential megacatastrophes. Yet, one still maintains, rather artificially, a boundary between the civil and the military, thanks to which the technical civilization is somehow almost innocent of Hiroshima and other acts of war ("necessity" having been the law); thanks also to this artificial boundary, one more or less closes one's eyes to the military control of space and the ocean depths, as well as to the implication of the military in technology—the serious business of the "special services." Lagadec senses the fragility of that boundary, yet at the same time he appears aware of the fact that the "solutions" he proposes (to strengthen controls, "to ferret out the irrational," to circulate information, to democratically open the processes of decision) risk being only palliatives or slogans easily co-opted—and neutralized—by political demagogues (and/or those in the media): "At the moment of history when scientific rationality flourishes in an industrial and social form, this same rationality explodes with the danger of chaos, the possibility of a breakdown where reason is absent . . . What would happen if the unbearable occurred at the heart of productive activity, not directly as the result of the military? What industrial policy would exist after the first civilian holocaust? What kind of crisis would this precipitate for technology, democracy, reason, and social functioning?"[5]

Such questions must be taken up and elaborated philosophically: a skeptical and "defensive" attitude will not suffice, although it is inevitable—and understandable—that the technician and the politician are not able to renovate their language or radicalize their vision overnight. As attached as we are to democracy, we must learn to recognize that the increase of power implies constitutive risks that the best democracy in the world will not be able to eliminate. The incalculable is not that easy to isolate or circumscribe; if a radically new phase of potentialization has begun, the old conceptual means—even the most "reasonable"—will not suffice. The scientific or parascientific literature completely admits this inadequacy when it is a question of confronting the prodigious changes of scale or schema that the most recent discoveries of astrophysics impose upon us: quasars from millions of light-years away, pulsars turning on themselves at dizzying speeds, solar winds, black holes, etc. But it is necessary that this cosmological *decentering* not mask the *recen-*

tering of power that techno-science simultaneously brings about in absolutely new and unbelievable proportions.[6] The cosmic incalculable (confined to theory) could conceal the planetary incalculable (the closed field of praxis), a separation that is rather convenient for the rulers of the world: religion and science fiction nourishing the imagination of the masses, while at the same time definite procedures are set in place to control their actual behavior.

With the crossing of such thresholds, the danger is everywhere—in the effective risks of combustion, the loss of control over stockpiled forces, and, more subtly, in the paralysis or the confusion of thought. The magnetic needle spins frantically when approaching one of the earth's poles. Likewise, thought, surrounded by the intensity of risks, by too much information, and by the complexity of problems, might be tempted to renounce unity, coherence, responsibility; it might be tempted to surrender to the formidable pressure of the incalculable.

Yes, there is no world more bizarre than the one in which we live: one-dimensional yet splintering, more and more centrally planned, but also anarchical. At which extremity shall we try to understand it: from the developed or the underdeveloped side, or else from the marginal? Is its "truth" California and Japan in their industrial and "post-industrial" performances, or is it Calcutta and its overflowing misery, or is it the law of the jungle in Uganda? Is it the universe of concentration camps and weapon factories, or the clandestine activities that abound everywhere?

We grasp the limit of every phenomenology in the face of this explosion. How do we *describe* and reduce to the essential that which never stops moving *in differentiating itself*? How do we describe a metastasizing state that might not even have the "systematic" unpredictability of Brownian movement? If the articles and accounts of journalists give us so little assistance, it is for this reason: contemporary reality is advancing vertiginously toward the *indescribable*.

The historian was always eager to encompass the *entire* epoch, provided he was willing to try; there was a time when he could also hope to open avenues for posterity: Thucydides and Tacitus succeeded in doing this. Imagine today the best reporter doubling as the most formidable analyst: will this remarkable person overflow with confidence at offering the keys to the future? We have excellent observers of the world: this is the century of journalism and the media. But a surplus of intelligence and an increased knowledge of the unknown will not suffice to dispel our perplexity in the face of the disturbances of the present and our anxiety before the future.

It is hardly difficult to observe—indeed it seems tempting to suggest—that this world is a world of power. To observe or to suggest? On the surface, power does not merit more than minor attention: it circulates, harmful or not, like a current that has greater or lesser voltage. In the world today, in an increasingly banal form, some degree of *power* is always being released in widespread effects. Because of it, one weeps, one laughs, one dies, one eats, one explains, one puts on weight. These discharges of power are just as irrational (or arational) as the eruption of a skin rash.

Is it enough only to record these differences of tensions, often stormy and at times not easily seen, as much as possible? Or, dressing up with the tawdry finery of Sense, add a touch of mockery (cynical or Dionysian) to these mechanistic discharges? It is perhaps not the least bit modern to always suppose a latent sense somewhere behind, to the side of, or beneath the phenomenon. It is perhaps not very modern *to go on and on about* devalorization, functionalization, or neutralization. To surrender language itself (and the keys to the City named Philosophy) to the blast of the fire that consumes everything? *Rightly or wrongly*, this is to use and abuse the *will* to be "resolutely modern." Rimbaud hardly had need of a "justification" other than that of his ardent passion, the poem. We can hope for nothing other for his unknown heirs.

But this text does not pretend to be the place of illuminations. Its "madness" is other: to take the sense of a word and see what ensues, right up to the senselessness of power which discharges itself *in this way* (*comme ça*). Inversely, it is to gather the *"this"* (*ça*) of the discharge of power and to see whether it structures itself, whether its threads bind together: the power-effect, the sense/nonsense of power.

This inquiry, therefore, will not be limited to interrogating the growth of power and the loss of sense, the rise of the irrational, as if one were independent of the other. We will follow rationalization in its connection with potentialization, because—as C. Castoriadis writes in a little-known text—"The unconscious illusion of the *virtual omnipotence* of technique, the illusion that has dominated the modern epoch, rests upon another idea, concealed and not discussed: the idea of *power*."[7] The most important incalculable is without doubt not to be found elsewhere than in this "event," which continually confirms itself before our very eyes: the alliance between rationality and power (which, originally mythic, was made more and more real, until it overcame the rest—in the form of the present Complex of Power). Hence, before analyzing and describing it in the following chapters, it is first necessary to locate this Power, at once a forceful idea and an impressive actor. But it is equally necessary to anticipate without delay a little of its genealogy, to designate the place of its engenderment: rationalization. By presenting in this way the presupposition of this inquiry—that rationality potentializes to the point of being joined in an unparalleled degree to power—we will also take account of the dangers. Beyond all the calculations that our world so reveres, there is a more originary law that will claim its rights, the Law of Reversal, by which the rational discovers within its own power an irrationality that mocks it.[8]

Definitions: power, Power, domination

The concept of power cannot be used effectively without being defined. Nothing vanishes as easily as a reserve of power for which the conditions of conservation and of intervention have not been carefully considered. Physically, economically, militarily, politically, power is always threatened by a flight into "abstraction" such that power constantly struggles to maintain and redefine itself in relation to that which it is not.[9] But this truth, illustrated first by the living organism whose power could be circumscribed in Bichat's terms as "the network of forces opposing death," is already applicable to logic: every concept—including power—is threatened by misunderstanding and senselessness, that is to say, impotence. This explains the care Aristotle took in defining the definition—by a proximate genus and the specific difference—and to establish precisely logic as *organon*. This also explains the care taken by Kant, from the first paragraph of the *Prolegomena*, to articulate precisely the three criteria for truly scientific knowledge (the object, the origin, and the mode).

Nevertheless, it is not my goal to establish a general theory of power. Every great ontology from Aristotle to Hegel, from St. Thomas Aquinas to Nietzsche does this, and thus assumes an economic project in the most fundamental sense. Spinoza gave such a project its somewhat canonical formulation in the Axiom of Book IV of *Ethics*: "There is no individual thing in nature (*rerum natura*), than which there is not another more powerful and strong. Whatsoever thing be given, there is something stronger whereby it can be destroyed." It is remarkable that this general, yet phenomenologically evident, truth should come so late in *Ethics*, in the section that concerns the relation of singular things between themselves and at the threshold of a theory of "human bondage or strength of the emotions." Is not God, an absolutely infinite being consisting of an infinity of attributes, at the same time power taken to the absolute? According to this hypothesis, his power of destruction would be equally absolute. Spinoza does not proceed in this way; this method would consist of defining absolute power on the basis of a lesser being and "subjecting God to destiny." The authentically Spinozistic formulation of the power of God is given in proposition thirty-four in Book I: "The power of God is his very essence." Hence, there is nothing in common between the power of God or of Nature (which is necessary reason, eternal life) and the finite, determined power of a singular being, unless it is not the very thought of power that allows us to understand this gap, that is to say, the true idea whose positivity is to be found only in the divine essence.

Since the time that reason, "one and the same," was originally postulated, no thinker (with the exception of Hegel) has formulated the circle that encloses the power of the rational more perfectly than Spinoza. Such is the greatness and the limit of metaphysical rationalism: it comprehends power starting from the power of the rational, it recognizes only the rational as truly powerful; but it must take the inverse, reciprocal (and nevertheless perilous) course: it must *justify* the existence of all power on the basis of rationality. The rationality of the real is always double. The interminable debate over the Hegelian dialectic undoubtedly teaches that no one truly knows which side wins in the game called "whoever loses wins"; rationality wishes to be exhaustive and sovereign, to have the effectiveness to comprehend and justify itself.

No academic homage necessitates this detour through metaphysics. No more than my renunciation of a general theory of power is a work of modesty. All research rests upon a presupposition: mine ought to become evident in this attempt at definition. My presupposition is completely different from that of ontologies of power (to present it as the inverse or the contrary would be again to subordinate it to these ontologies): it results from the *phenomenological* confrontation of rationality with its *destiny* of power. Destiny denotes the irreducible exteriority of a *partage* that is not inscribed in any rational necessity, even though it concerns the latter. That rationality is such a *partage*,[10] and further, that this *partage* reveals itself objectively in a world where power is at stake, can be established and analyzed by gathering the phenomena and not through an a priori ontological affirmation. Nevertheless, from the very fact that this gathering of phenomena organizes itself in a philosophical discourse, and investigates (in a historical-worldwide sense as Hegel would say) the origin of the currently dominant phase of rationality, it cannot be a matter of producing an exhaustive definition of Power (in the specific sense that particularly interests us), nor at the same time, does it relieve us of the long work of inquiry and identification that will constitute the subject of this work. There is power and there is Power. This is not an eternal truth, but a phenomenological presupposition that will be tested against the empirical. The same is true for rationality. Nothing guarantees any longer that it is ontologically "one and the same" in its essence and in its effectuations.

What is feasible and even highly desirable at the outset of this inquiry is the determination of a conceptual delimitation that allows us to begin to differentiate Power from power in general—and also from domination.

The capital letter used provisionally with Power ought not lead us to believe that it concerns a metaphysical entity, even if the *origin* of the actual Complex of Power is identified genealogically as metaphysical. As Jean Ladrière indicates, a formal system "is always more powerful as it is rich in means of expression and demonstration."[11] Semantically, the Power that I seek to describe and to analyze is closer to this evaluation of power (due to the potential for demonstration and explication) than to any rational theology. As I understand it, there is not an in-itself of Power; it is known first of all only through its effects. A further series of examples will clarify this by introducing the notion of "the effect of power."

If it is still too soon to establish that the current phase of the deployment of power involves an increasingly radical focusing on the effect of power as such, it is without doubt time to delimit the semantic link—reinforced by History—between "power" and "effect of power". What would be a power without effects? If the original meaning of *dynamis* is not attached to an effectuation, if even our sensation of physical force increases proportionately with the threatening or impressive putting into reserve of that force, this is not the case in the technical world where power is systematically articulated (especially mathematically) according to its already realized or possible effects. The physical definition of power is, then, significant. The history of the experimental theory of motors reveals that over the course of the nineteenth century there was an increasingly pronounced semantic osmosis between the "useful effect" and the "power" of a machine.[12] The power of a steam engine, first determined roughly, became "effective power," that is to say, its utility measured by the time it takes to brake the engine. Work itself, in the physical sense, came to be defined as "the fact of producing a useful effect by its own activity."[13] Indeed, for modern physics, but only in the second half of the last century, power is literally the measure of the quantity of work produced in a given unit of time. Physical determinations will be completed by other physical measures (for example, the measuring of the intensity of electrical current), but also in our hyperquantified world these determinations will be made by measures of power, which is to say, they will be measured with extreme precision by the effects of power: in the economic order (production potential, GNP, etc.) as well as in potential destruction (the calculation of "dead forces," and "live forces" of battling armies, the quantification of overkill, etc.). However, the impressive compilation of all these batteries of measurement does not imply that the Complex of Power as such is *only* measurable, that it can be understood *only* in calculated terms.

We must now distinguish Power from domination. Max Weber can help here, having made a very clear distinction between *die Macht* (power) and *die Herrschaft* (domination) in these terms:

> Power (*Macht*) is the probability that one actor within a social relationship will be in a position to carry out his own will despite resistance, regardless of the basis on which this probability rests.
> Domination (*Herrschaft*) is the probability that a command with a given specific content will be obeyed by a given group of persons.[14]

Domination is more socially determined than power (unless one reduces it, as von Neumann does, to the "intransitive" notion of "superiority."[15]) Weber even specifies that "the concept of *power* is sociologically amorphous."[16] Although diffuse, power is defined by Weber according to a sociological perspective, that is, in terms of a will and at the heart of social relations. Such is not the point of view of this study: we cannot thematically treat the sociological problems of domination— or even power—in such terms. A clarification, however, proves indispensable due to the innumerable intersections among power, Power, and domination. It is also indispensable given the ideological connotations that sociopolitical domination comes to convey in the exercise of power.

Quite contrary to what is believed, domination in its generality, which is in some way transcendental, is not any more scandalous than the appearance and recognition of differences. Every difference is, or can be, a source of domination. The mastering of the senses, the understanding of a situation, the possession of a weapon or of a piece of information—these are so many advantages that are perhaps specific, but are nevertheless not negligible, since they permit one to make a decision, to save a life, to open new perspectives. Differentiation is vital. Where are we all equal if not in death? But to discern and to legitimize differences, however minimal, is this not therefore to justify domination?

However, the passage from the establishment of difference to a system of domination does not follow automatically. It is one thing to accept differences, another to congeal them and to extend them systematically. The hierarchical system does not allow differences to disclose themselves; it uses them, playing them to its exclusive advantage as images or relays of a Leader or Commander-in-Chief. The recognition of inevitable differences does not necessarily lead to the fixing of rigid hierarchies, authoritative economies, or organized terror. Indeed, in our developed societies, organization has not eliminated contingencies; instead it has formed systems of restrictive conditions, complex configurations of the distribution and control of power. Luhmann has recently demonstrated the possibilities of a sociological theory that combines the systematic formalization of Parsons with a general theory of communication (that is to say, more precisely, of the organization and selection of symbolic messages).[17] Insofar as our perspective is not sociological, but more fundamental, I can only refer to this theory here, reserving the right to return to it later with a critical analysis.[18]

Philosophical wonder, therefore, does not have to surrender before the recognition of minimal domination, or before the conceptualization of its most complex forms. When one passes, however, from this still general examination to ideological fixations and justifications, wonder must then yield to critique.

In an unconditional ideology dedicated to domination (Nazism, for example), one is not satisfied with recognizing hierarchies (as in the naturalist conservatism of Aristotle and St. Thomas Aquinas); in such an ideology, one reinforces the hierarchies and exalts them; one creates a system and a mystique with them. The lessons of life and the precepts of biological science are solicited and brought to bear not only on the social question, but also on the role of vital struggle and aggression.

In the Nazi perspective (too often implicitly shared by the general public), states of domination are vigorously explained and justified by the universal *instinct* of aggression. What is this instinct? According to Konrad Lorenz, this instinct testifies to the universality of aggression, but only as long as one does not understand by this a generalized will to power nor the aim at systematic extermination of rivals. What Lorenz presents as an absolute certitude is rather the inverse of Spengler's presupposition of a will to power: "*Never* does this type of "combat" between those who eat and those who are eaten end with the killing of the hunted by the hunter. It *always* establishes a state of equilibrium, perfectly supported by each member of the species."[19] We must remember, however, that in the animal kingdom, aggression essentially manifests itself within each species in the form of a *competition* that, from an evolutionary perspective, will be called selective pressure. Within these limits, the most violent form of aggression is *critical reaction*, a desperate defense motivated by fear.

Lorenz does not minimize the dangers of intraspecies aggression that are also found in human beings. But he shows that at different levels of its evolution, life invented compensatory rites designed to channel aggressive energy, preventing it from being excessively destructive. Thus, the ceremony of triumph performed by the grey goose is a "rite of appeasement";[20] many human gestures or expressions (such as laughing) are equivalent to such rites.

Biologism has its limits. One may be surprised that I find it necessary to attack its Nazi variants, given that its monstrosity has already revealed itself. However, as George Steiner shows (if one knows how to decipher his novel *The Portage to San Cristobal of A.H.*),[21] it is not clear that Nazism has been *morally* conquered to the extent—a fact that Lewis Mumford has already deplored[22]—that the victors have taken up the politics of power and have extended the priority of the military-industrial complex. If the spirit of Nazism has spread like a gas and has contaminated us, it is perhaps because it gave life to specific characteristics already present in this Power that I address here: the social refraction of an objectified Complex.

The complicity between ideologies of domination—be they acknowledged or not—and techno-scientism is perhaps the most decisive feature of the current state of the world. Moribund or gasping for breath, these ideologies find an unexpected reprieve in techno-scientism; they find here an enormous appeal to power, supported by material ameliorations or fascinating gadgets with a specific type of imagination (science fiction, for example, but also industrialism and technicism in their various aspects). I raise the question that has already been posed to Nazism: what would become of Marxism if this techno-scientific development fails to appear? There is no doubt as to the answer.

The necessary critique of ideologies is, therefore, to a large extent, a rear-guard struggle. I will not detain us excessively with it, nor with a critique of sociopolitical domination. But this discussion will at least allow me to clarify the question that manifests itself in the space principally delimited by science and technology. Domination can no longer be conceived solely in its transcendental groundwork nor from the simple point of view of social relations—important though they may be. There are in the insane race of technocratization, if not an essence, then at least some common characteristics that explain the convergence between techno-science and domination that we observe.

Malraux said that we are the first atheistic civilization. I must add: "and reduced to the willing of power." (Does this latter explain the former?) However, it is this modern sense—or nonsense—of domination that ought to be henceforth questioned and from a very specific perspective: the link of this domination to rationality, which is to say, to the specific forms of logical-scientific-technical articulation that make this Complex of Power possible in its absolute novelty, and in its vertiginous unpredictability. This novelty is manifest in the *irreversible* character of the "phenomenon," in a certain number of points of no return that have been crossed. The unforeseeable is due to the immeasurable growth of the risks taken, and to the specificity of the new phase of potentialization presently under way.

Through these detours and thanks to these differentiations, one no doubt understands that the specific question taken up here is posed neither within the limits of a general theory of power nor in terms of a social theory of domination. This is to say that rationality does not intercede in the course of this project in order to reduce aberrations, inequalities, nuisances, etc., as if the philosophical project had to make itself practical and "useful" before even having tested its conditions of intelligibility. Rationality itself is the primary focus of this inquiry: at once subject and object, method and institution, and perhaps even more: the critical distance in relationship to a process of rationalization that has become irreversible and inevitable—like a destiny.

The thread of the question: rationalization

We have seen that the question of power apparently interested Max Weber only from a sociological point of view: the study of the social and economic ties of dependence in the modern Western world (especially in contrast with more traditional conceptions of hierarchy, in particular those in India). However, when it concerns the development of capitalism or the formation of an institutionalized state, a single question guides the observation of power in the Western sense: "To what chain of events do we ascribe the appearance, in Western civilization and in Western civilization alone, of cultural phenomena which ... have taken on a *universal* meaning and value?"[23] Even though Max Weber does not *directly* pose (not in our terms) the question of the bond between rationality and power, the astonishment that animates his monumental investigation has to do with the unique character and specific development of the West. One word summarizes this development, especially in its modern phase: *rationalization*.

This theme is certainly well-known. One believes that everything has been said in evoking and celebrating it, even though the difficulties have only then begun. Rationalization can thus be reduced to something evident that one salutes in passing; it can also be used to the exclusive ends of self-justification, this "self-rationalization of Western rationalism," as Castoriadis calls it.[24] Finally, it can be the basis of a social theory that is as complete and exhaustive as possible—such is the spirit which animates the renewal of this thesis by Habermas.[25]

I do not believe that the real importance of Weber's research can be reduced to these various points of view. When Weber isolates the specific characteristics of Western *science*, of rationalized *art*, of the *organization* of society and the state, of *capitalist exploitation* of economic and labor possibilities, he is considering a singular phenomenon, however macroscopic and complex: the substitution of an archetype (not only epistemological, in Kuhn's sense, but global—philosophical and social) for the religious archetypes of traditional societies, whose last support in the West was the Middle Ages. It is, therefore, neither the quantitative increase of power nor the multiplication of technical means that interests Weber; rather, it is the philosophical mutation, this rationalization whose underside is the disenchantment of the world (*die Entzauberung der Welt*). Weber addresses the heart of the matter when he asks: "Does this process of disenchantment, realized over the course of the millennia of Western civilization, and, more generally, this progress in which science participates both as a part and as its driving force, have a significance which surpasses this pure practice and this pure technique?"[26] Citing Tolstoy, but also including Nietzsche and his assessment of nihilism, Weber responds negatively: no overarching significance orients this immense power. Neither life nor death has any sense in this civilization of progress, which is always subordinated to new progress. The distressing paradox of this *Zweckrationalität* is that it continually establishes ends, even though it fundamentally lacks a highest end (because it no longer recognizes any "value" outside of science and technology). How is it that Western humanity is thus constrained to will the absence of sense by an excess of sense? Therein lies the *incalculable*: this enigma of a rationality becoming an end in itself and ensnared by its own power. We see that "disenchantment" has nothing to do with a purely subjective and contingent sentiment; it is the inevitable flip side of a process of rationalization that acts in such a manner that neither things nor beings can any longer engender power *by themselves*. Henceforth, we know above all that "we can master everything by *prediction*."[27] All essential power is *mediated* by the knowledge of scientific laws and technical or practical rules. Weber insists on this specific character of the new power of the rational. This power is itself largely potential because rationalization is founded upon this primary belief (where a priori knowledge replaces all religion and faith): "We *are able*, provided *only that we will it*, to prove to ourselves that there exists no mysterious and unforeseen power which interferes in the course of life."[28] It suffices that the power of the rational be potential, solitary, and unique, in order that all other powers be eradicated. By pointing to this "great principle," Weber goes infinitely further than scientific or technical historicism; whereas the latter exposes the facts according to a linear conception of progress, as if rationalization were only the sum of discoveries and inventions that survive along humanity's journey, Weber again shows the spiritual and social unity without which science could not impose itself. The thesis affirming the link between the Protestant ethic and the spirit of capitalism is too well-known to require a detailed account. However, it is also the case that it is too superficially known to be appreciated under its most interesting light: neither merely nor even principally as a theory of "the sociology of religion" that contradicts the priority accorded by Marxism to social struggle and material factors in the relations of production,[29] but primarily as the deepening of an authentic philosophical wonder confronting the fundamental mutation of our modern Western world.

There is not just one way to be a philosopher, and the most philosophical is not necessarily the most effective in demonstrating the spirit of a system. Weber neither explains nor justifies rationalization starting from a metaphysics of History; he *shows* this enigma by revealing its generation in dispositions and institutions, scientific conceptions and technical apparatuses. Placing the emphasis on the role of the "rational conduct of life" in the constitution of our technical society, he is not less than philosophical in that—starting from the analysis of a considerable accumulation of facts (interpreted in relation to diversified correlations)[30]—he stimulates reflection on the double character of this rationalization, as destiny and as process. As Karl Löwith remarks in his comments on Weber, rationality has become our "destiny",[31] a thought that is taken up again and again in Weber's writings,[32] and that is seen in the closing statement of the *Protestant Ethic*: "The Puritan *willed* to be a hard-working person—and we are *forced* to be so."[33] Weber's reflection is organized around an incalculable event: rationality has become an unavoidable constraint, and *in this sense*, all-powerful and omnipresent. The first lesson goes well beyond the establishing of a "fact." The second essential characteristic is apparently in contradiction with the first: a process develops that is overt and dynamic. How can it be at the same time a destiny? Weber notes that scientific progress "prolongs itself infinitely,"[34] and that the rationalization that emerges from science (and that is its "driving force")[35] is not constrained, in principle, by any limit. This dynamic of rationalization, supported by human freedom, masks, by its orientation toward the future, the absence of choice that is embedded in its original impulse. According to Husserl, the infinite opening of rationality effaces its destinal origin. It is not understood this way by Weber, for whom destiny and process are co-present: our *partage* is freedom. Hölderlin already had a premonition of this; Weber discerns it, less prophetically, more scientifically, in the very structure of *modern* capitalism, which is nothing more than the rational organization of *free* work.[36] Weber himself rationally assumes this disenchantment with rationalization: the stalwart acceptance of the "destiny of our epoch".[37]

Rationally? After having shown that "this notion of rationalization brings together the method, the research, and the philosophy of Weber," Raymond Aron notes this great paradox: "Against this rationalization of existence, Weber perceived refuge only in the irrational, total freedom that he proclaimed."[38] In the end, the guarantee of freedom appeared to him to be the "irrational determination of values,"[39] the heroic but fragile affirmation of the supreme priority of *Wertrationalität* over *Zweckrationalität*. Is this an ultimate leap into the irrational? Not at all. Raymond Aron notes that the unconditional affirmation of freedom is not made in Weber at the expense of "legitimately rationalized spheres." Is not the hero who is capable of elevating himself above the logic of means and ends, the logic of "realist" responsibility, proof of a superior rationality, precisely that of value?

This difficulty remains formal if its position continues to be included within a tautological definition or within a logic identified with rationality. In meditating on rationality, however, Weber went much further; he saw that rationalization is open to the possible at the point where it is *also* exposed to its own reversal. The rational produces the irrational; the process of rationalization leads to that which Löwith calls a "specific irrationality."[40] It is only by paying the price of first elucidating this

Reversal that the ethical paradox can be explained. For this same price, rationalization will guide us towards its ultimate aim, its own incalculable risk.

The rational, the irrational, and the incalculable

The ultimate confrontation of rationalization with irrationality remains—for Weber—an apparently arbitrary enigma or "choice," so long as one does not understand that this outcome is the result of the very same structure of rationality considered in terms of its ends. Rationality can unfold and develop only by exposing itself to a risk, necessarily having realized sooner or later that the end itself becomes the means. If all action is amenable to the means-end relation, it could be *successively* means and ends (or the inverse). For example, financial gain, first of all subordinated to the necessity of ensuring a decent life, easily becomes the ultimate priority. The slippage is even easier when an activity—the daily exercise of an artisan—has been "rationalized," that is to say, methodically refashioned according to a given aim. The process of rationalization forms a whole: the decisive event occurs when rationality substitutes itself for religions and magic; then the slippage between means and ends becomes inevitable and incessant. We grasp that examples are, of course, not lacking, particularly in the dialectic that watches over all modern political action, which is very quickly imprisoned in the means that it establishes and thus is caught in its own game. Therefore the Party, the instrument of freedom, becomes an end in itself, the apparatus to which all must be sacrificed. Thus the Puritan notion of "duty," issuing from the religious notion of "vocation," is transformed into an end in itself, the "iron cage" where modern humanity is imprisoned, even though it no longer has faith. But Weber arrives at an even more radical philosophical position; he concludes that there is an inescapable reversal of all rationalization into irrationality. This reversal is the result of the promotion and "automatization" of general "conditions" that are systematically allowed to organize human action. As Karl Löwith writes, echoing Weber, "the rational and total organization of conditions of life produces, from itself, the arbitrary and irrational rule of organization."[41]

Even if Weber does not succeed in philosophically isolating the "inexpressible criteria"[42] of this reversal of the rational into the irrational, even if his analysis of rationalization was still incomplete (as Habermas has recently shown),[43] his insights have nevertheless proven invaluable—for two principle reasons. In the first place, his disillusioned analysis of the reversal of rationalization much more faithfully explains contemporary reality than does the rational optimism of a dialectical philosophy: whereas in Hegel and Marx the dialectic always operates finally to the profit of a rational reconciliation (because it is the most general law of rationality), the dialectic in Weber is henceforth entrusted with a higher truth to which rationality itself must pay tribute. The Reversal is no longer a rational, dialectical law; it is a law against which rationality comes to learn of its own limits. All dialectic is not rational; the inverse is true. Thus Weber goes further than Marx, because the latter, having dialectically analyzed capitalism and having demonstrated the reversal of bourgeois freedom into the alienation of workers, shelters a new dialectical finality (the abolition of the exploitation of humans by humans), which strategically exempts the "instances of freedom" (philosophically: rationality; politically: the Proletariat or the Party) from total submission to the law of Reversal, that is, to the profound truth of the dialectic. Therefore, in the name of the "scientific application of science to industry," he dispenses with the analysis of the contradictions and aberrations resulting from an intensive techno-scientific development. Likewise, Leninism and its quasi-unconditional justification of the Party will always be prepared to take up residence in a political space ruled by an ad hoc dialectic. Contrary to this, Weber's lucidity appealed to the inexorable law of Reversal and signaled a beneficial return to the actual. Listen to Weber on the eventual suppression of private capitalism: "Supposing it succeeds, what will be its practical significance? A shaking of the steel cage of modern industrial work? No! Rather, it would signify that the organization of nationalized enterprises which are more or less "socialized" would become more bureaucratic."[44]

In the second place, the consciousness of the ambiguity of rationalization allows for preserving the chances for an apparently irrational individual freedom in the face of the widespread rationality of organizations, but at the same time submitting this irrationality to a higher rationality of another order. It appears far too paradoxical to claim the superior rights of heroes in the era of *managers* and *leaders*; but we have just seen that Weber is led to this apparently irrational claim by the same irrationality of a *Zweckrationalität* pushed to its limits. It is precisely at the time when efficiency claims to be the sole truth that the "impossible" heroes make the impossible possible—and bear witness to *Wertrationalität*. An anticipatory announcement of the incalculable role of dissidence in societies that have been, in principle, designed by planners, ruled by administrators, and controlled by police! Acting as the revenge of responsibility on fragmentation, this reversal of the Reversal has as its guarantor in the world of the mind the "intellectual integrity" of the scholar.

In drawing on Weber's analyses, I do not contend that they explain everything, nor that they have fully thought out the status of the rational: it remains for us to question the Kantian conceptual limits (essentially the difference between *Zweckrationalität* and *Wertrationalität*) that mark the horizon of Weber's thought. In any case, in Weber as in all metaphysics, the irrational returns to the rational and vice versa. Just as the "irrationality" of $\sqrt{2}$ can be posed only in relation to a mathematical *logos* that began with mastering the arithmetic of whole numbers, so, too, does the "irrationality" of contemporary bureaucratic society have sense (or non-sense) only in reference to a project, determined, ruled, and disclosed from an articulation of the social bond. But while failing to delimit the metaphysical limits of rationality *as such* in this play of return between rationality and irrationality, it does appear that Weber is able to help define the degree, or the limit, at which rationality must own up to its characteristics and declare itself (or see itself denounced) as "reversed."[45]

One point must now be clarified; it concerns the status of the irrational. One must not confuse formal irrationality (identified within a well-defined rational procedure, and done for its sake) with a *second irrationality* that results from a process of rationalization. A more attentive examination of the status of irrationality in itself and "in context" must, in effect, permit us to clarify the status of Reversal and to mark the limits of Weber's setting free of the irrational.

Irrationality is not in itself an objection "against" the rational. In order to prove the contrary, is it not necessary only to recall the greatest achievement of Greek mathematics: the very concept of "irrationality"—whether in arithmetic or geometry? Philosophy, with Plato, goes even further: the ontological meditation of the *Sophist* raises the very possibility of error. The irrationality of π or of the relation of the diagonal to the side of the square proves to be less of a philosophical "scandal" than the irrationality of "semblance" found in the sophistic reasoning that presents the false for the true, with every appearance of demonstrative rigor. Aristotle, for his part, classifies, in *Posterior Analytics*, the typical cases of logical error, and includes among the legitimate modes of demonstration the "reduction to the absurd," which, although indirect, permits the demonstration of a proposition by the refutation of its contradictory.[46] More generally, there is no formal or scientific rational mastery without a knowledge of irrationalities, irreducible aberrations, and constitutive limits. Infinitesimal calculus, statistical calculus, the axiomatization of the undecidable (or, for example, the impossibility of demonstrating that a given sequence was produced randomly), such are the stages, among others, of mathematical progress, symbolizing the integration by science of irrationalities that were even yesterday considered scandalous. How true it is, as Spinoza has noted, that rationality dies when it mistakes itself for harmony!

The irrational, as it is identified by the logical or scientific spirit, has a type of exemplary role: it is the very life of the Relational. Does the same apply to the "irrationality" of the process of rationalization? Let us first note that the conditions under which the problem is posed have changed: the region of the deployment of the rational is no longer purely formal or epistemological, but rather social, economic, ethical. Rationalization designates precisely this extension of rational models to the *totality* of life (even if this extension is produced in a differentiated manner, by separating, for example, the private from the public). Next, in formally analyzing the "irrationality" in question, we must recognize that it is only relative, and that it still remains highly rational. I take up only one example, but it is extremely significant: bureaucratization. Every organization, every State, has its archives. But when this "memory" develops excessively to the point of becoming an end in itself, when the circulation of directives, of documents, and the checking of files become autonomous aims, then "irrationality" has triumphed in practice. It is evident that this irrationality in no way compromises the logical or the legitimate functioning of rational and scientific research. It is a matter of something else, namely, the articulation of a practical field, and the profound disharmony between means and ends. The term "irrationality" can therefore deceive, leading to the accusation or to the very swift indictment of reason in cases where its formal quality is not implicated. This is, in my opinion, the error made all too often by those who too quickly follow a certain type of rationalism (Marcuse, for example). Weber does not fall into this trap. As Löwith shows, what is irrational for Weber is "the fact of universal rationalization,"[47] or, as I suggested earlier, the total rationalization of life.

To unmask the "irrationality" of total rationalization, therefore, cannot be a formal endeavor. Nothing can be thus "demonstrated" either *for* or *against* the rationality of the rational. However, this critique is inevitable and beneficial. Why? Because the purely rational justification, or rather, the purely rationalistic, justification of these rationalizations is totally abusive. Surely, the American strategy in Vietnam was rationalized. Certainly, nuclear deterrence obeys a logic, that of game theory, and benefits from information systems methodically extended to domains that, not long ago, did not lend themselves to any control. And undoubtedly, the introduction of the computer into the school or even into the household can more efficiently facilitate the learning of languages or skills, as well as make everyday life easier and more comfortable.

These are many examples of rationalization, which, according to Weber, or conforming to his analyses, ineluctably reverse themselves into "irrationality." The "rationalization" of the Vietnam war denied or neglected too many actual conditions of the struggle: the terrain, the history, and especially that elusive element—the *spirit* of a people. The logic of deterrence operates at a global level, apparently with greater success, but it does not prevent local conflicts, it rests on a disastrous technological escalation of the stakes, and most especially, it does not absolutely eliminate the supreme risk of universal explosion. The extreme rationalization of teaching and everyday life also results in a complete paradox: why save time on the one hand, if, on the other hand, the freed time is empty, devoid of goals and significance?

In all these examples, which could be multiplied, the call for reason to justify total rationalization is just as unfounded as the inverse attitude, that which accuses this rationalization of being irrational. One can just as well argue *for* as *against*, ad infinitum—as Hegel remarks in "Logic of the Essence," regarding the search for *Grund*.[48] Formal rationality is not compromised, but its ontological foundation becomes uncertain.

This clarification of the question both supports and indicts Weber's account. It supports the account when he critiques this rationalist overcompensation, which is obviously excessive yet nevertheless so widespread that it obliterates the specificity of rational deployment in the practical field. It is not sufficient that a reason be good in itself, in order for it to conform to my freedom. In recalling the practical priority of the legality of the *Wertrationalität*, Weber exercises an incontestable, critical role,

as Kant did in his time. And thus appears, at least for now, at this point of the inquiry, the meaning of the law of Reversal: it is an essentially critical reversal of a
rationality having to undergo the task of judging itself irrational in the light of a
harsh analysis of means, ends, or situations.

Nevertheless, Weber is also wrong, or at least reveals his limits. In acknowledging this, I also concede *my* limits: must I try to outdo his argumentation, having recourse only to a language that is itself a reasoning about rationality? This question
is fundamental, and its difficulty ought not paralyze us. The limit of Weberian
thought has already been established: to return the rational to the irrational and
vice versa, in a perpetual play. When faced with the development of power, the risk
of nuclear war, the attempts at genetic manipulations, the total "control" of consciousness, does it suffice to protest: "This is irrational!"? This outcry only returns
to reason, only appeals to the cause of the "illness," as if, in order to treat it, it were
necessary to increase the dosage that provoked the crisis in the first place. In this
way, the critical analysis risks, in the course of its development, a return to the rational as the only recourse, without having sufficiently specified it, and more impor-
tantly, without having seen that the rational by itself is impotent (even if it is the
results of its own power that are presented to it).

The object of my perplexities in this text is not the power of the rational *in
abstracto*, nor is it this or that effect of power that is particularly outstanding: my
object is the very effectuation of Power—whose degree of rationality is, at this time,
problematic. Never more than today has there been such total calculation, and at
such an unprecedented level of precision and exhaustiveness. Yet, never more than
today has it been so difficult to gain a sense of life. Power itself is endlessly adjusted,
computed, reevaluated—as a potential energy, as the economic and military power
of nations, as the quantification of production, work, and consumption. Time itself
is taken into account by worldwide economic planning and investment. However, as
Castoriadis objects: "No *rational calculus* exists that can show that a temporal horizon of five years is (for society) any more or less *rational* than a temporal horizon
of one hundred years. The decision would have to be made on the basis of considerations other than 'economic'."[49]

These "other considerations" are the *incalculable* postulates of development.
"The crisis of development is obviously also the crisis of these 'postulates' and of
their corresponding imaginary significations."[50] What does Power postulate? This
is, indeed, the question that is increasingly raised, in spite of itself, by the technical
world; it is the question to which *ratio* as calculus is constitutively incapable of responding.

The incalculable that is the subject of our investigation is not simply some sort
of provisional *x* that can be reduced or established. Rather, it is the very movement
which Heidegger names *das Unberechenbare* (the incalculable). Starting from the
representation of the world as object or picture (*Bild*), the predominance of science
and technology is established; its essential configuration is the rule of quantification.

Everything, in principle, can be represented with exactitude, that is to say, it can be calculated. "But as soon as the gigantic (*das Riesige*) in planning, calculating, adjusting and securing leaps beyond the quantitative and becomes its unique quality, then what is gigantic and what can seemingly always be calculated completely becomes, precisely through this, Incalculable."[51] What calls for explication in this very dense thought is less the notion of the "gigantic" than its leap "beyond the quantitative." By "gigantic," Heidegger obviously means the omnipresent—if not exclusive—role of the quantitative, but not from the merely formal point of view: the fact that the approach to being, its study, and its use are no longer subject to any limitation. According to the terminology of the present study, it is a question of designating the modern phase of the potentialization of the rational; however, in this very phase will appear the ultimate moment (which will be called Phase IV) when the Gigantic "becomes a specific quality." How does the Incalculable reveal itself? It does not do so in any particular sector nor in the abstraction of the purely quantitative. Can the shift from quantity to quality be isolated at a given moment? The response is apparently temporal ("As soon as . . . "), but in actuality it is historical. The Incalculable has been the "invisible shadow" of the calculable since the dawn of modern times, since the establishment of the method as a project of the mathematical mastery of nature and of being in general. This decisive trait of our epoch reveals itself even more clearly as soon as one grasps the unity of our epoch, and as soon as the scientifico-technological exploitation in all its forms reveals its epochal character.

To grasp the Incalculable within the general rule of calculus is never a magical operation, but it is the revelation of that which is *eventful* in the epoch. "Do we see the flash of Being in the essence of modern technology?" Heidegger asks.[52] When the universal rule of computation appears as the Incalculable, it is indeed—beyond all representation—Being itself in its contemporary dispensation that begins to be differentiated in its withdrawal.

By differing paths, we have arrived at this crossing where the very sense of that which marks the present epoch, Power, is in question. The Incalculable is nothing other than the unconditioned advancement of Power, which is continuously measured and reevaluated. To submit everything to "economic imperatives," to respect nothing other than scientific-technical "success" (itself continuously submitting to competition), to mobilize the masses for an enterprise of unlimited "development" and of "expansion" (and without any highest aim to be definitively assigned): in this "project" the characteristics of what is perhaps the final phase of History are gathered. Is it so difficult to detect the Incalculable sheltered within this? Why does the dominant world language refuse to take this step? Why must the extraordinary (the splitting of the atom, the manipulation of the living cell, the computerization of language), appear to go without saying?

If the answers were locked in some drawer of the office of universal Reason, it would suffice to procure its layout, the number of the office in question, the combination of the lock. But, as Hegel said, we are dealing with reason in History; it is necessary to add in a less Hegelian mode: reason assigned to humanity as ontological *partage*, reason delivered to its own destiny of power. We know, therefore, the work required by the power of the rational must be combined with a phenomenology of the effects of Power, which will patiently prepare for this understanding of the Incalculable: rationality confronted with its enigmas.

2

Power in Its Effects

> I use the word Attraction only to explain an effect that I discovered in nature, a certain and indubitable effect from an unknown principle, a quality inherent in matter whose cause will be found, if they can indeed find it, by persons more clever than myself.
>
> —Newton, according to Voltaire[1]

ACCORDING TO VOLTAIRE, Newton clearly announces that modern science renounces the search for hidden causes. What Newton isolates is nothing other than an *effect* (a word that physics constantly uses since Newton), but an effect that is "certain and indubitable." This last is the essential phrase: to give an account in mathematical terms of regularly observed phenomena in nature. Thus *certitude*, in the modern Cartesian sense, is attained; its contemporary translation would be objective truth. Whatever the terminology (to which the scientist does not aim to fall prey, not wanting "to take the chaff of words for the grain of things" as Leibniz says), obviously universal attraction as "scientific fact" is no longer derived from the search for the efficient cause. The effect, in Newton's sense, is no longer the effect of *one* cause, but instead, is the *effectuation* of a general law. Newton writes that he "could not yet deduce from the phenomena the reason for the properties of gravity."[2] We know the theoretical and philosophical conclusions that August Comte has drawn from this methodological revolution, which contemporary science does not reconsider, even as it enriches, partially corrects, and assuredly renders more complex the understanding of the laws of nature. In any case, the interpretative limit reintroduced by "theory" in the contemporary sense is obviously not a return to the ancient and medieval search for causes.

Why recall the obvious? To answer this question, we must address, through an approach which is first of all negative, the methodological setting of this inquiry. In what sense do we speak of Power as effect or of "effects of power"? Is it legitimate to isolate, even provisionally, this concept of Power? I will attempt in the following pages to *begin* to respond to these questions.

Some effects of power

A bomb explodes, a ship is launched, a satellite is set in orbit, a radio broadcasts: these are some effects of power. Seen from Sirius, our planet appears as the site of a gigantic effect of power—techno-science. The methodological difficulty leaps out at us: where does the effect of power begin or end? Can a definition ever be exhaustive?

The very concept of "the effect of power" implies a discrepancy, at times considerable—and at once logical and chronological—between causes and their results. The entanglement of causal series is even more complex as techno-science develops. The "borrowings" from one domain to another, the shocks of technology responding to science, the effects of a discovery upon the course of "progress" are so numerous that it seems impossible to list them. Simondon has shown that technological progress does not follow a linear direction, but obeys a jagged evolution. Thus, two elements (the tubular stove and Stephenson's slide), originating from a group of eighteenth-century artisans, are adapted in the invention of the locomotive, a new "individual" that is going to contribute powerfully to the development of a new ensemble of the nineteenth century—the thermodynamic phase of the industrial complex.[3]

This first approach reveals the *perspectival* character—glimpsed by Nietzsche—of every determination of the relations of power. While there is no power in itself, but only power at the level of a differentiated and determined relation, all the same there is an effect of power *known as such*, only under the axis of effectuation that furnishes the scale of measure or the limit of all measurements: the destructive power of a weapon, the population employed by the industrial sector, the quantity of information exchanged, the amount of money in circulation, etc.

An initial negative consequence develops with regard to the interest of an eventual phenomenology of the effects of power. In one sense this "phenomenology" already exists, in a rudimentary state, in the manner in which American television presents, without any order or sense of priority, a barrage of shock-waves whose psychological result (saturation) has already been noted and studied. The television screen is the contemporary frame of immediate representation (although secondary, preserved and conditioned): a sort of informational frozen food. Although it is fascinating due to the neutrality of its perspective, the free-floating quality of its "imaginary variations," and the overabundance of its images and sounds, this rudimentary reserve of a phenomenology of the technical world is also extremely deceiving because it lacks that very thing that defines a phenomenology: the perspective of an *eidos*. In our terms, this is to say that one searches there in vain for a *perspective of effectuation* (and, at the same time, an evaluative perspective). Everything is there, but one does not know what to make of this fluctuation of raw effects that are simply registered and quickly forgotten, this moving picture with neither verifiable "entry" nor "departure" points, this screen which captures reality but glosses over it, capturing all of these "events"—but permitting no retention of them. At the same time, this limit imposed on our knowledge gives an indication of the world that is the concern of knowledge: in the same gesture, brute phenomenality (at the "basic" level of television) *diffuses* power and renders it taboo. A true phenomenology would be completely different: it would presuppose a methodological aim, a principle of evaluation.

But precisely at this point the second consequence appears: insofar as this world is in constant mutation, its axes of power are continually redistributed. We currently see this just as much in industrial strategy as in international politics. Is the aim of power today (which is to say tomorrow) to have at its disposal a cheap labor force or to control a highly sophisticated technology? Does it seek to own a stock of hydrogen bombs or intercontinental missiles? The cards are constantly being reshuffled: we have seen this during the 1970s and 1980s in the area of energy; we have perceived it in the military domain (for example, in the Falkland Islands conflict). But this difficulty has nothing redhibitory at our level. This study is not dependent on the current state of affairs. The study of the effectuation of power ought to be neither purely phenomenological nor simply strategic; rather, it leads to a genealogy of Power. This means that, confronted with the axes of the effectuation of power, I will engage neither in their description (which might become "journalistic") nor in their economic evaluation (which is not my subject). Instead, I propose to demonstrate the strands of their *essential* conditions of possibility. For example, what is it that fundamentally makes possible the thermodynamic phase of industrial development? What is it that fundamentally makes possible a passage to the "post-industrial" (if this concept makes sense), etc.? We must, therefore, invert the phenomenological perspective in order to understand the condition of possibility of all these fluctuations of power. If this rudimentary phenomenology is so fascinating, it is because in the process it *simulates* a boundless reality. If a philosophical genealogy is so demanding, it is because it must truly keep *in view* its sources within power itself. From these difficult observations a first lesson emerges: at least we know that it is useless for us to remain passive spectators in the face of this current explosion of the effects of power.

Of Power

—You have said: Power. In capitalizing this notion, you have removed it from all efficacy. You seem to refer to a metaphysical entity, to a substantial unity defined in itself and for itself. This is a prescientific procedure. It was not until Galileo and Newton that the notion of force became determinable and measurable within the framework of the principle of inertia. The desubstantialization first performed by mechanics was then extended to other levels of reality: even if in various places there remain imponderables and margins of indetermination, political power itself rests upon an economic, technical and military potential which is measurable and largely relies on statistics, polls, and calculated prediction. In the modern world, power never appears without precise differentiation or quantitative measure. Without these discriminations, there is no power. Your own power, hypothesized, becomes fiction.

—Science has already used fictions, at least in its theories: the demons of Laplace or Maxwell. But here Power is the least "fictitious" of the substantives; it is the denotation of the most real of our realities, the scientifico-technological complex in its internal dynamic.

Hic et nunc are the co-ordinates of Power. It is as determined as possible: its time is our time; its area of operation is planetary and already spatial. It is—more deeply still—determined in its principles. Neither fiction nor purely emipirical-material (not even in the sense of Sartre's practico-inert), it has none of the former faces of domination: ownership of the earth, direct constraint of bodies, possession of gold or pure financial speculation (in spite of its ties, in its modern version, with Capital). Unprecedented, this Power will reveal its scope only if the most facile denials of its existence are removed. Science insists: "I am only the agent of Progress. I am myself at the direct service of Reason." Technology insists: "I am only the result, ever more perfected, of scientific development." Now that the bandages of rational idealism have been removed, we can see the scientist's nervousness, the good conscience of the technician facing this living reality of the present (forging the future): Power.

If any contemporary thinker can contribute to this project, it is Heidegger. In setting aside the traditional definitions of technology (instrumental or anthropological), in determining the unity of the technological phenomenon and its source starting from the Greek *techne* understood as "mode of deployment," Heidegger allows for understanding that "the essence of technology has nothing technological about it,"[4] that modern technology accomplishes metaphysics and defines the ultimate epoch of the West.

Why then not take his qualification of technology as *Gestell*? In the first place, *Gestell* has been the object of numerous misunderstandings: its clarification is first in order. Second, Power offers the "advantage" of being a denomination which is as phenomenological as possible (considering the very limits of phenomenological inspection in this matter): it acknowledges the massive presence of the "phenomenon" without *immediately* making presuppositions about its genealogy. More decisively, the approach from the perspective of Power presupposes neither the history of Being nor a fortiori the completion of this history, even if the present path leads to the edge of the ontological enigma.

The misunderstanding that surrounds the Heideggerian interpretation, which runs as deep as its insightfulness, leads to the interrogation of the strange denial which affects Power as soon as one addresses its provenance and its destination, its sense or its lack of sense.

If power in general has its spontaneous finality in the augmentation of its field of application and of its intensity, contemporary Power does not escape this rule: it does not reveal its own essence. It is philosophical thought that uncovers it. Several factors explain this state of affairs. First, there is a structural reason: the operative is dynamic, not genealogical. Technological or practical successes do not invite a meditation on the origin or on the essential; it is not even certain that failures lead us to this meditation; Power spreads itself, but conceals its essence. Heidegger showed this (concerning the *Gestell*), and hence predicted that his own interpretation could only be masked and misunderstood by the dominant language-world. A second factor also comes into play: Power as a techno-scientific complex is relatively new. The large ultramodern megamachines (state projects or supranational "conglomerates") are no more than a few decades old: humanity has been mobilized prior to having had the time to take full measure of these events. Thirdly, this estimation has been rendered even more difficult by the diverse ideological positions that have pretended to justify immediately, and at any cost, the expansion of science and technology in the name of Progress or the Proletariat (the latter having appropriated the former). In its first great stage, science found in positivism an ad hoc ideology that reconciled dynamism and order, and stimulated research in the name of a rational harmony which is always presupposed. Marxism, for its part, inheriting its founder's enthusiasm for science and technology, has continuously tried to make its sociopolitical practice and scientific theory converge (as well as technical efficiency) by constantly (and very adroitly) reinforcing the political messianic message with the objectivity or efficacy of science and technology. Thus Power closes itself to fundamental questions and continues to avoid them.

Philosophical thought, I have said, must seize the essence of Power. But can this endeavor be reduced to a definition? Power is neither an idea nor a thing such that it would suffice to exhaust its senses in a definition. We have already identified it as the techno-scientific complex. Viewing power as a complex, concrete mega-ensemble (which is not to say merely material), we can apply the Aristotelian caution in matters of definition: the notion of *energeia*, Aristotle writes, "can perhaps be elucidated by *epagogue*, with the help of some particular examples, not attempting to define everything, but contenting ourselves with analogies."[5] Since the analogy of proportion is too narrow, there remain the combined resources of clarifying distinctions and of a progressive induction. Power is a complex in the sense that it is indeed an immense worldwide "Kombinat" which includes and integrates the technical system. Its complexity, however, is even greater when we include (to a large extent) science in a twofold manner: as institutional network and as a body of both theoretical and technical potentialities.

In spite of this impressive complexity and development, the ultra-empirical zones of power remain outside my delimitation of Power: politics in particular.

Why? Precisely because I believe, with Heidegger, Ellul and several others, that the main current of power passes less and less through politics proper and more and more through new inductors that are fundamentally technical or techno-scientific, although often hidden, disguised or manipulated by ideologies or politicians. This coronation of techno-science as Power has come to challenge the so-called primacy of politics. This challenge is justified on many levels. First, it is philosophically essential, perhaps even vital, not to get caught up in words, but to face this Complex of Power which has taken humanity hostage. Second, we will see in what follows that the decisive importance of the political will concern the survival of the political as such in the "era of technology."[6] Third, the political situation is only one aspect of the entire apparatus of the phenomenological *epoche* that allows us to proceed from the effects of power to their constituting power.

Three authors have recently evoked Power in a manner worthy of our attention: Ladrière, Serres, Ellul. For Ladrière, we know that it is a question of a type of "super-domain or unique super-structure," which science and technology *tend* to constitute both conceptually and practically.[7] For Serres, it is "the rational alliance of theoretical reason, practical reason and of calculable, predictable, goal oriented reason ... the most powerful and productive triangle that has ever taken place. It is a motor. The abominable motor of contemporary history."[8] Ellul, less a genealogist, identifies Power as Technology which autonomizes itself, becoming a System.[9]

It is not surprising that Power is surrounded by terms that do not exactly mean the same thing. Neither an idea, nor an isolated phenomenon, nor a "discipline," nor a fixed structure of reality, Power is also not reduced to the sum of its effects. One could be tempted to relegate it to a minimal or formal unity, a possibility that we must reject. Its time and place, however, have been established: they are our time and place. It is undoubtedly Mumford who has most effectively understood Power in its phenomenality: neither purely "ideal" nor purely mechanistic, but a dynamic articulation of politics, economics, technology, and the military, the Complex of Power has effected the transformation of the planet and the conquest of the cosmos. But this new nucleus of world Power is not a fixed and definitive moment; it is a process that constantly readjusts its objectives according to the availability of raw materials, struggles for influence, and the economic "conjective." It is, therefore, necessary to grasp the constitutive elements that form a network that is always moving and "performing" in an average way, in relation to which we, as observers, cannot pretend to abstract ourselves completely.

The pertinent recourse to the notion of Power is, thus, not made through an ideal genesis seeking only to display a definition. The concept of power is intermediary, whose meaning appears only at a macroscopic scale and under the steady light of a phenomenology that describes and identifies the effects of power, as well as a genealogy that reveals which type of rationality engenders these effects.

From phenomenology to diacritical genealogy

If we proceed from the effects of power to Power itself, identifying all the while the effects of power, are we only conducting a search for causes in a classical sense, that is to say in a prescientific sense? To this objection, the following response is given: for the type of inquiry carried out here, the logic of efficient causality cannot be applied as such to a specific case; it is indeed impossible to verify the truth of the adage *cessante causa cessat effectus*, since the observer does not have the power to stop the phenomenon or to experimentally modify its cause. The scale of the observed "phenomenon," as well as its constant state of becoming, has in advance rendered difficult a study that integrates the totality of variable and imponderable elements. Certainly economists have rather recently articulated a polyvalent model of the world economy according to an "input-output" analysis.[10] However, beyond the fact that the model is subjected to constant revision according to new "input" or data, it offers only an economic and quantitative approach to a situation that is much more complex, wherein thought itself (and not only in the form of "psychological factors") plays an essential role.

If the problem of the scientific character of our approach is a particularly urgent one, it is first of all by virtue of our interest in rationality, which connects us— by its very logic—to this problem. It is, secondly, because the power of science is particularly at stake: one cannot truly speak about science from a place that is absolutely exterior to science (in this century of information overload, it is necessary to battle two "impertinences" that are antithetical, yet, at times, accomplices: "background noise" as well as ignorance). Finally, it is not ruled out that the scientific intervenes at a level that is completely comparable to that which phenomenology has known since Husserl's first efforts, namely, the search for *sui generis* rigor in the study of phenomena, which by their very nature are shielded from exact science. To what degree did Husserl succeed in raising the phenomenology of conscious states to a "scientific" level? Certainly at the very least his phenomenology renewed the question, all the while remaining (in spite of the *Crisis*) on the threshold of a truly fruitful expansion of this type of inquiry into the world-historical development of contemporary humanity. We recall the example of Husserl only to make the point that the recourse to phenomenology is at once inevitable and yet surpassable. It is inevitable because the "thing itself" that we have in view does not give itself beyond actual phenomena. It is surpassable because of the limits of every eidetic analysis and even of every phenomenology in the face of the indescribable, the unthought that confronts us throughout our destiny of power.

The time has come to explain why my inquiry begins with a phenomenology of technology in order to transform itself into a critical genealogy that reconstitutes the world-development of a certain type of rationality. This is not simply because it is necessary to view the phenomena (the effects of power) before explaining them.

It is essentially because when phenomenology is first of all confronted with power in its actual results, it raises difficulties that it is not capable of resolving—precisely because of the type of rationality that it practices. Remaining "rationalist through and through," Husserlian phenomenology shows itself incapable of taking, vis-à-vis rationality, the *minimal distance* that seems to be required in order to avoid reproducing and reactivating the ultimate potentializations of the rational. Husserl, presupposing the purity and "innocence" of rationality, rejects the destiny of power, placing it on the side of contingency and irrationality. He prohibits (a new illustration of the turning of rationalism into the irrational) a diacritical analysis of the destiny of power starting from rational potentialization. If this paralyzing prohibition is lifted, a new situation emerges: rationality no longer has to be defended or attacked, but merely—a task that is not easy—recovered in its destinal oscillation, which the very word *power* enigmatically conceals. In the production of dormant possibilities, whose effectuation, long suspended, is activated in an accelerated, vertiginous process, rationality is neither guilty nor innocent; it above all is not reducible to the intelligence of pure forms: it *generates* itself ever anew. This surplus is power in its innumerable effects. In thinking this dimension, in recovering this domain from the senseless condition of the world, thought itself increases considerably; it returns to the source of its own power, which is also the source of all potentialization.

Many of Heidegger's readers fail to understand the necessity of the immense detour that he imposes on phenomenology, toward the unthought of metaphysics. But it is along this astonishing deviation that I shall follow him most closely; it is the lesson that I shall retain most decidedly from his itinerary, and I intend to apply it in understanding the contemporary technical world.

An example of this deviation is furnished by the following passage from *Holzwege*, taken from the Heideggerian commentary on the Introduction to *The Phenomenology of Spirit*: " . . . the modern technological world with its absolute claim is nothing else than natural consciousness which, in its opining and representative way, accomplishes unconditional productivity—and assuring itself—of every being, in the objectification of each and every thing."[11]

In order to understand this text in its essential deviation, we must grasp that, according to Heidegger, "natural consciousness" has become *metaphysics in its modernity*. This point is confirmed later on, in the commentary on paragraph six, just before the enigmatic text that was just quoted: "What Hegel calls natural consciousness is in no way congruent with sensible consciousness." To which Heidegger adds the following comment, which is less obvious: "Natural knowledge lives in all the figures of the spirit; each one of them lives in its own way (i.e., in natural knowledge), even precisely the figure of absolute knowledge which advents (*ereignet*) in the form of absolute metaphysics—an event which is perceptible to few thinkers and only from time to time."[12]

The important point here is not that there is a movement from consciousness to knowledge, but that a statement is made on the ontological status of the natural. Heidegger defines the natural in the same terms as metaphysics: "Natural consciousness is, as far as this is concerned, nothing more than the representation of beingness in general in an indeterminate fashion. . . . "[13] Thus, if in the passage quoted from *Holzwege*, we replace "technological world" with "metaphysics," everything suddenly seems clear: "*Metaphysics* in its unconditional claims is nothing other than natural consciousness. . . . " Again we find Heidegger's well-known interpretation: worldwide technology is the accomplishment of metaphysics.[14]

A considerable shift is at work in these few lines of Heidegger. He begins with a statement that every Hegelian is able to accept in order to arrive at a thesis that is not at all believable from the Hegelian point of view: he starts from the noncoincidence between natural consciousness and sensible consciousness in order to arrive at the coincidence (or at least the correspondence) between natural knowledge and absolute knowledge insofar as the latter is *absolute metaphysics*. Heidegger accomplishes this trajectory by passing through a middle proposition that a Hegelian would dispute. "Natural knowledge lives in all the figures of spirit," a middle term that prepares us to accept the conclusion: therefore natural knowledge lives also in absolute knowledge.

The Heideggerian "procedure" is quite clever: it consists in pushing Hegelian knowledge as far as possible from its own point of view (by admitting that there is an immanent naturalness to all knowledge),[15] in order to show that this natural knowledge is nothing other than metaphysics itself (carried to its completion by Hegelian knowledge). This certainly explains the caution in the sentence that precedes the text of the *Holzwege* (*nur wenigen Denkenden zuweilen sichtbar*),[16] in which Heidegger hints that we are now involved with a truth that escaped Hegel himself. This is a dizzying ellipse through which the treatment of self-overcoming applied by Hegel to natural consciousness is, in turn, applied to Hegel, thanks to the overdetermination of the *absoluten* (which from now on means both scientifically true in the Hegelian sense and historically accomplished in the Heideggerian sense).

The equivalence, that had at first surprised us, between the *technological world* and *natural consciousness* can now be explained: it is a matter of the technological world, but understood according to its primordially metaphysical essence. Heidegger adds a useful qualification: the technological world "in its unconditional claim" (*unbedingten Anspruch*). This claim reveals the Cartesian and Modern exigency of a *fundamentum inconcussum* that metaphysically finds its Archimedes' principle in the *cogito*, its epistemological development in the method, and its "world-development" in the exact sciences as well as in their industrial applications (a genealogy that justifies, perhaps in rough outline, the link stressed several times by Heidegger between absolute metaphysics and positivism).

By exposing the claim of the technological world, reduced to the mere equivalent of "natural consciousness," Heidegger shows his cards, presenting his interpretation of metaphysics: natural consciousness is metaphysical consciousness, not so much in its most sublime claims as in the *effective results* of its exclusive orientation toward the beingness of beings; it is an *ontical consciousness* that from now on is interested in Being according to the modern mode of productivity. We can now discern two aspects of this unconditional self-assertion: ontologically, it is the *objectification* of all beings, which are destined to be represented as objects for a subject; ontically, it is the cycle of productivity: between demand (*Bestell*) and supply (*Bestand*). For Heidegger, this production is as unconditional and totalizing as its metaphysical principle: the epoch of will-to-power is also that of the devastating domination of the earth.

It seems, in the end, that the passage quoted from *Holzwege* has now been clarified, but *at the price of a global interpretation of metaphysics*. From Heidegger's point of view, neither a critique of natural consciousness by means of a theory of knowledge, nor an approach to the contemporary world by means of an autonomous phenomenology—be it critical—suffices. The Heideggerian deviation contains its own motive: *it is only by thinking the unthought of metaphysics that one thinks the extent of modern technology*—which has nothing of the purely technological. The allusion to the technological world in the commentary on the Hegelian text concerning the experience of natural consciousness is not, therefore, gratuitous: it reflects the profound logic of the Heideggerian choice. It indicates that we too must also face a choice; this is precisely what I am doing here in my attempt to show more clearly the limits of the phenomenological method when questions involving the very essence of technology as Power are pointedly raised. The Heideggerian detour stands in defiance of the principle of economy, which explicitly or not, rules over modern science and technology. But does contemporary thought itself become rigid as it becomes all the more reckless? In order to go to the foundation of the phenomenon of technology, it is necessary to discern in it, following Heidegger, what is neither visible at first sight, nor even grasped cidetically. In this study, I intend to benefit from those analyses that permit us to hold Power at a distance.

If phenomenology proves indispensable on the two levels of gathering phenomena and providing a methodical inquiry into their essential characteristics, it is because it will be accompanied by critique. Most of the recent and significant studies of the technological world—we cite Marcuse, Ellul, Baudrillard—are as much critiques of this world as they are descriptions. Is it possible to examine closely technoscience in its expansion in the same way that one establishes the ideal genesis of geometric forms or even as one identifies the fundamental structure of the imagination? *One-Dimensional Man* denounces (at the same time that it describes) the extension of the technical mode of rationality to all domains of life. *The Technical System* puts us on guard against the closure of technology upon itself, wherein autonomization follows, from phase to phase, from sector to sector. By retracing the steps of the "desubstantialization" of merchandise in the latest forms of the market economy, *The Consumer Society* exposes its forms of conditioning, its advertising and "media support." Are these critiques simply indulgences in these essays? I would prefer to discern in them an emerging awareness of the urgencies imposed by the manifold expansion of the effects of Power. Moreover, the analyses are not always unfavorable: Gehlen's instrumentalist anthropology is not able to refrain from admiring the ascent of humankind toward a surplus of power; the progressive technocrat or Marxist permeates the discourse with the thought that for the most part technological society holds about itself. Simondon is undoubtedly the writer whose impartiality is synonymous with properly scientific objectivity, even when he denounces the dangers of technocracy.

On the other hand, my attempt wants to be *diacritical*. What is meant by this? I mean that the critique functions in all its senses, placing itself between phenomenology and genealogy, playing the one with the other, the one against the other. I mean also that the internal critique at times turns against itself in a sometimes healthy way. Power does not have to be treated with tact, but its critique will not become an end in itself. The ruse of thinking: it will not be a question of prolonging the technological or scientific attitude of neutrality, but of taking some *critical distance*. *Diakrinein* means to see better, to discern at close proximity the phenomenon in its essence, as well as in its deformations, its dislocations, and the eventual aberrations which are exceptions to its rules. It is to repotentialize Power, to turn from aggravated effectuations toward possibilities yet held in reserve: for a new *place* of the rational.

3

The Ruses of Technicism

Thus modern technology can cry with Mirabeau:
"Impossible! Never say this foolish word to me."

—Karl Marx[1]

Technology and power

IT IS INSUFFICIENT to claim that we are in the epoch of technology: a declaration that is formally correct, but one which risks exposing a deficiency of thought. Such a general statement is made *for the want of something better*; it is an acknowledgment of the most common denominator, which still lacks specificity. Moreover, it is certainly not necessary to accumulate facts and arguments around the nucleus of this study: the link between the increase of the power of a nation and technicization. The hierarchy of nations in the order of economic power is supported by the delicate euphemism of "development." At the top, in terms of its GNP, are the United States, Switzerland, Sweden; at the bottom are Mali, Bangladesh, and an impressive number of others.[2] That the rise of GNP is inversely proportional to the population is neither surprising nor significant at the level that we are examining here. Likewise, the particular case of the former Soviet Union—with its hyperdeveloped military sector, an island on a continent of underdevelopment in the areas of agriculture, civil rights, etc.—does not directly concern us here. The general motto, from the largest to the smallest, from the most capitalistic to the most collectivistic, is "Let us become more technical." Certainly, there is a distance between a wish and its fulfillment, especially for the poorest countries. But the dynamic is thus defined, and nothing appears to have escaped it—even the Iran of Khomeini, at least in military matters.

Concerning technology, Jacques Ellul already said twenty-five years ago, "There is nothing equivalent to its power in the world."[3] A related declaration, albeit more prudently stated, was made by Ladrière, who points out that the technological *logos* "becomes an exterior power" that tends "to impose its own law" on humanity.[4] This acknowledgment masks an enormous ambiguity: does it point only to the form of technological domination, or, more generally, to its dominating content? For Ellul and for me, less so for Ladrière, the two are inseparable. But Simondon clearly distinguishes between an "autocratic philosophy of technology," "which considers the technological network as a place where one uses machines in order to obtain power,"[5] and a technological philosophy in the true sense, which would develop the study and meaning of these interconnections and regulations. Is technology saved by communication (and vice versa)? One can sense our skepticism on this point. Simondon's position, although widely shared and respected, appears to us rather as the indication of a constant *denial* of technological activity in regard to the effects of power that it incontestably induces through its informational networks.

Technology has woven connections with power that are as enigmatic as they are tight-knit, judging from the *denial* that technology offers to its analysis. To return to Ladrière's definition, "To obtain the desired effect with the maximum of efficiency": What could be more justified and more neutral at the same time? Where are we to discover power in this definition of "the methodological problem proper"? It does not appear more directly or distinctly than in the traditional definition of technology as the deployment of "appropriate means in view of an end." At best, Ladrière has modernized the formula by insisting upon efficiency. But technology in itself, constitutively, always hides behind its incompetence in addressing the ends, turning its hyperconcentration on the means. Just as it is not for the executioner to decide whom he will execute, but only how he will operate the machine with which he has been entrusted—complicated though this might be in the case of *The Penal Colony*—so each technician awaits for someone to entrust to him a task for which he is competent, acknowledging only a single responsibility that is defined and isolated, releasing him in advance from the choices, judgments and practices that result from his work. To be sure, there are technicians who dominate: military, militants, and specialists in every field. But technical neutrality is projected on the aggressiveness of each of these activities in order that the denial in question can function. The person in the military will say, "I am only a soldier, I carry out political orders. And besides, my task is increasingly technical." And that person will be "right." The militant and the specialist will also make humble references in the direction of the Party or the Organization. And they are never wrong.

These denials have their truth (not absolutely, but phenomenologically), in the sense that they constitutively define, *nolens volens*, the "normal" technological project. Does this constant denial, however, tell the whole truth about technology, especially technology in its most contemporary forms? To accept without any critical distance these denials as they present themselves would lead us to an endorsement or reinforcement of a generalized technicism whose contradictions and limits I would like to expose by unmasking this technological denial, as well as by acknowledging its inevitable character.

Granting the technician the "right to denial" is the least that one can do (if the technician is nothing but a technician and asks for nothing more). But perhaps it also leads—if he or she ever took up the question—to a kind of psychoanalysis which risks compromising the delicate equilibrium. Because to deny is to acknowledge; it is at least to acknowledge that one can envision this dreaded possibility (to be implicated in the ends, a reversal of the Kantian nightmare), and therefore to have this dizzying thought: "Beyond my own speciality, I can be implicated by the consequences of my actions." It would be at least to examine the eventuality of having a responsibility as a human being (note that technology, in this very game, lends itself to all types of distinctions, having engendered casuistry, the technique of soul-searching and "justification"). It is as if technological activity, massive and self-confident, were conscious of this "danger," sheltering its good conscience behind what we call *hypertechnological silence*.

This first analysis already allows for an assessment: technology as technology does not want power for the sake of power. Technology as such wants nothing other than its auto-effectuation. Let me emphasize a word that has been neglected until now in Ladrière's formula: "To obtain the *desired* effect with the maximum efficiency." The will is very much implicated, but how? Ladrière responds that "the objectives . . . are in the end dictated by the systems of value which command action."[6] Here it is a matter of an affirmation of a principle which has not yet proven to explain *adequately* the actual functioning of the technical world. Hence, our perplexity.

—"If the *desired* effect is power, I obviously have my specialists," technology could reply, if one were to risk allowing it to speak.

This prosopopoeia is necessarily very short and the rhetoric ought not make us lose sight of the central question that has begun to emerge, around which our inquiry is being formulated: What is of concern in this contemporary complex of Power? Is it simply technology? If the process implies *more*, if it is necessary to qualify it more specifically, in what ways is it new and perhaps even unique?

The method followed in this chapter will be, as previously articulated, *diacritical*: it concerns itself with the understanding of techno-science in its specificity (and its effectuations of power) by a series of corrections of incomplete or insufficient positions. Successively, this will involve the technicism of "techno-discourse," the anthropological conception of Gehlen, Marx's controversial legacy, and the most recent transformations of technology in the Marxist movement (the scientific and technical revolution as progressive myth).

Technicism and techno-discourse

If it were a matter of uncovering an explicit philosophical position, the task would be easy. Too often philosophers give themselves paper-tiger adversaries, all the more easy to dismiss because no one claims to hold the positions presented. In the present case, it is not a particular doctrine that is at stake, but rather "the atmosphere of the times." Today, humanity is devoted to technicism; no other choice has really been offered. This encompassing technicism is composed of a large margin of passivity; it allows the operation of "circuits" of technological networks; it accepts the "functioning" of preconditioned frameworks; it welcomes gadgets and innovations which move "in the direction of progress"; it wraps it all in a silence that is indifferent to everything that is not "technologically viable."

Spread universally and like a mist in our mental atmosphere, technicism raises an inevitable question: Is technology thinkable? The denial of technicism justifies itself formally by a sort of paradox of networks transposed to the technological problem. This denial could be formulated as the following: Is the network of all technological networks itself technological? If so, if the discourse on technology is itself technological, it presents only a technological procedure, but does not really focus *on* technology itself. If not, if it is no longer itself technological, it can still focus *on* technology, but it no longer views it as the network of all technological networks. In the first case, one will have perhaps a systematic worldwide planning, a comparative study of technologies, a general theory of models, but one will not have thought about technology in its global aspect nor above all in its essence. In the second case, one will have a philosophical interpretation which runs the risk of being challenged by specialists as inadequate, because it is too general, metaphysical, etc.

We are apparently far from the technology of "natural consciousness." And yet, without having read Aristotle, "natural consciousness" is vaguely aware that technological production is irreducibly different from theory, that it is necessary to practice it in order to master it, and that it "moves in the realm of the contingent."[7] At the same time, it is spontaneously respectful of the diversity of qualifications: "to each his own job." This acknowledgment of specializations assures efficiency, but also a kind of solidarity in the division of labor.

Technicism, however, seeks at any cost to "homogenize" technological attitudes by rooting them in a fundamental "instinct" of human beings; in doing so it tends to minimize as much as possible the differences between ordinary technology and techno-science.

It is well known that the reemergence of natural consciousness in technology did not escape Heidegger. Even if the statement cited earlier[8] has metaphysical implications that we cannot yet discuss at this level, it undoubtedly illustrates the principal force in the technological world (as technology, nothing more). Its foundation is in Being; its stamp is placed on things by taking charge of their management and transformation. What fascinates natural consciousness and stimulates its attraction to technicism is that it constantly concerns itself with effects and results, even if it benefits only indirectly or imperfectly. "Productivity" is the key word of this thinking, which is endlessly striving for efficiency. By this term, Heidegger did not think primarily about the growth of the norms and output of production, but rather about the fact that modern technology essentially seeks to be *productive of effects* (as we have seen with Ladrière). If it reaches this goal, which is its proper end, the powerlessness of the discourse and the obscurity surrounding the ultimate end of the techno-scientific process are of little importance. How could one not hope for a world of miracles from this systematically stimulated efficiency, an ever greater confidence in objects, the mirage of complete happiness, "absolute knowledge" according to the standards of natural consciousness? Myth comes to the aid of the prosaic: a powerful future which will relieve the monotony of an almost chaotic present. However, technology itself promises nothing more than that which it can produce—effects corresponding to a specific demand, with respect to a given situation (materials, financial, resources, human capital, etc.). Though all of this is "reasonable," limited, and less than exciting, "it produces results"; hence, a better future is invoked, which nourishes hope. Thus, personal mastery and collective power will be summoned and united by the process of technicalization. While inevitable phantoms, "second thoughts," will arise unbidden, making an affective contribution, nevertheless intellectual stimulation totally organizes the world.

This operation takes place as if by itself, having no need of being "articulated." Technicism is, by instinct, silent, once the instructions have been given and the orders passed along—a silent instinct earned to some degree by the tacit acquiescence of machines. In the most technological society, in an absolutely closed technical system, only codes would break the silence, as orders would be received by a perfectly disciplined army. A hyper-technical society *would function and would remain silent*.

But why speak in the conditional? To a great degree this society has already been realized and continues to develop. To be sure, we are not yet in Huxley's or Skinner's "brave new world," nor even totally in Marcuse's one-dimensional world. Yet year after year, under our very eyes, we see political directives, the reality of the marketplace, and the "general atmosphere" that direct the young toward technical jobs, rejecting the "humanities." One hears, "It is necessary to adapt to global competition." Or again, "We live in an era of technology, let's go with the times." That this kind of advice or imperative is formulated shows that technology has not yet reached its dream. More radically still, the existence of a declared technicism proves that the "perfection" of technology is not yet accomplished.[9] A totally technological society would have no need of technicism. To cite some different examples: The "Japan Computer Usage Development Institute" in Tokyo has studied the possibility of planning and administering a city of 20,000 inhabitants through computerization.[10] But this remains a well-delineated *project*, and, even if the Japanese are easily "conditioned," their technology remains a very coded discourse wherein patriotic elements and psychological investments are not negligible, thereby exceeding the "purely technical." This also applies in the same way for Stakhanovism and Taylorism. The first, Stakhanovism, indeed corresponded to a phase of massive and "heavy" industrialization in Eastern countries, but technologically it made no sense (given that one cannot base mass production, planned over many years, on peak performances); the dominating factor, therefore, was ideological and imaginary, bordering on paranoid delirium. It was a compensatory discourse, already a techno-discourse. The case of Taylorism is different. It was a necessary and essentially technical phase of rationalization; however, it contained an idealistic connotation also falling into techno-discourse. Taylor presents his "management" of production as a "true science," but praises Anglo-Saxon virtues and the ideal of efficiency in quasi-lyrical terms.[11]

Finally, technicism that formulates and thematizes itself in propaganda is already "reactive"; techno-discourse compensates for the inevitable cracks that exist in the most well-oiled social machines, neutralizing the protests of workers who do not follow the rhythm of the machinery or do not assimilate the logic of "restructuring." A supplementary factor opposes itself to the "perfection" of technology, making recourse to techno-discourse inevitable: the internal conflict within technology itself. One type of technology pushes aside another, but the changes are not without conflicts because innumerable elements are involved: political, social, financial, strategic, etc. There is not (yet) a sovereign technological authority that would decide at what moment which industry in which country ought to be stopped, which ought to be launched, on what timetable, etc. All of these developments follow paths which, for the most part, escape technology itself: for example, in spite of the "division of labor" that Michel Ghertman analyzes in the form of "atomistic decision making," which intervenes in the management of multinational societies, Ghertman recognizes the indetermination and the unpredictability of many decisions, that, in the end, are more political than purely technological.[12] Beyond the political, or at its margins, contingent factors come to challenge technicism's claim to complete mastery: the "laws" of the marketplace (which are laws in name only), bureaucratic viscosity, sabotages, etc.

Obviously, technicism takes many complex forms, whose ideological "consistency" must be examined in a more detailed and subtle manner than that required of the mainly "reactive" components arising from conflicts engendered by mutations in technological society. From this point of view, as Habermas has understood, technology and science have become ideology.[13] It is not sufficient to denounce this ideology. It is necessary to grasp, on the one hand, that it probably will last much longer than other ideologies, because it is more recent, becoming even the equivalent of a religion. Skinner's *Walden II* is the prosaic version of this dream of the total control of existence, wherein "behavioral engineering" will have resolved every problem.[14] On the other hand, just as it is insufficient to denounce technocracy in order to lay the foundation for a politics, the critique of technological ideology remains ineffective if it reverts to a Marxist point of view of class struggle, or if it remains at the level of an anthropological conception of technology, as is the case with Habermas.

Technicism as techno-discourse

The idea of techno-discourse, often glossed over in this discussion, needs to be elucidated in order to critique the anthropological conception of technology, which is the implicit or explicit "philosophy" of technicism.

I call techno-discourse a discourse which is neither strictly technological nor "autonomous." It is a parasitic language inextricably woven into technology, contributing to its diffusion or, for the lack of a better expression, making almost impossible (like Muzak) any radical analysis or any questioning of contemporary technological phenomena in its specificity. Every technology has its vocabulary, its codes, its "list" of activities, its problems, and its operative "scenarios." Such is not the case with techno-discourse; it is neither strictly scientific, nor philosophic, poetic, etc. Some examples will allow us to locate this parasite. Techno-discourse: a large part of the "functionalization" of language used by the audiovisual medium. Techno-discourse: advertising. Techno-discourse: technocratic "thought." Techno-discourse: informational excitement à la Servan-Schreiber. Techno-discourse: all of the political-ideological-audiovisual "blabber" in the discussion of global competition, productivity, etc.

If these discourses abound, is it not because they have a function to fulfill in (and for) the technological universe? They perform, without any doubt, social and even technical functions (the latter referring back to the former). It suffices to think for a moment what the technical world of the West would be without publicity, or the East without propaganda. From Tchakhotine to Baudrillard, these functions have already been articulated. Reflecting the technicization of society, these techno-discourses stimulate it and mold it in turn. They play the role of "informational relays" which perfect and accelerate planetary technicization.[15] However, the reality and intensity of these functions do not dispense with a truly philosophical critique.

To this end, we must first criticize an anthropological type of techno-discourse in order to show how this prevents access to the understanding of techno-science. It prevents it for the simple reason that "techno-discursivity" *itself* seems to be the result of techno-science and not of traditional technologies. As Ellul has shown, "the technical system is a real universe which constitutes *itself* in a symbolic system";[16] it is an autosymbolization that tends to recodify the totality of the real into "sugarcoated" manipulated information.[17] This is an absolutely new phenomenon; it exists neither in traditional societies of diverse and dispersed craftsmanship, nor in the first "megamachine" states (Babylonia, Egypt, Rome), nor even in the industrial societies of the nineteenth century. Why does the anthropological type of techno-science mask such a considerable rupture?

The anthropological conception of Gehlen

In the face of technology, Gehlen takes a position that one often finds among our leaders and engineers: we have for too long indicted technology through reductionist representations or even caricatures of its realities. Technology cannot be reduced to mechanization, and it does not bring about dehumanized automation. Rather, let those of us in western Europe be inspired by the American or Soviet models; technology can and should enjoy an immense popularity, if one stops judging it from concepts handed down from the theoretical-contemplative tradition, and furthermore, if the technological mode appears as an accomplishment of the needs and aspirations of humanity. Reacting against Nietzsche and Spengler, Gehlen claims that technology should be given a value that would allow it to attain an "equality of right" with the other cultural domains.[18]

According to Gehlen, the method for the study of technology is also positive. For Gehlen, everything should be done to smooth the tensions between humans and technology; likewise, he seeks every solution to establish a continuity between primitive and contemporary technology. Human beings, condemned to action because of the lack of specialized organs, must intelligently transform natural objects that existed before them. There are three levels in the constitution of this *ersatz* technology: complement, reinforcement, alleviation. The hammer extends the arm, reinforcing it; the motor replaces the hammer, freeing the arms. The airplane combines these three advantages: it replaces the wings we lack, it carries their power beyond any organic possibilities, and it spares our energy, permitting us to travel great distances without much effort.[19]

There is nothing surprising in this anthropological conception of technology. As Gehlen himself admits, we already find it in Scheler, Bergson, Pradines, Ortega y Gasset. This conception is anthropological at the level of definition as well as at the level of the model it utilizes. At the level of definition, there is a co-extensiveness between man and technology; man has been a technician since the beginning; technology is a "mirror" of human beings (or, in other words, the totality of human "resources"); through technology, humans project and increase their "artificial nature." At the level of the model, it is the organism that holds the key to technology, since organic deficiencies will dictate which tasks need to be performed. Technology will be the body of the human being, enlarged and adapted in proportion to the potentialities of this active, dynamic, creative being.[20]

Furthermore, the continuity between humanity and technology is reinforced by an *instinctual theory*. In acting technologically, human beings do not use only their intelligence in order to accomplish certain conscious ends. They also follow deep, unconscious drives. The desire to master the environment, the aspiration to security and predictability are as old as humans themselves. Magic, a supernatural technology, has responded to these desires throughout the millennia. From agricultural rites to astrology, this "drive" has never ceased to affirm itself. The fascination with machines is even stronger within us in that it echoes those automatic rhythms inscribed on our organism (for example, the heartbeat), and we want therefore to reinscribe, objectify, and project them in inanimate nature.[21]

This definition and genesis of technology within terms and on a horizon which are purely anthropological do not have to be discussed here. This is due less to the lack of originality of this theory than to the following: my essential inquiry does not bear on technology in general, but on the *specificity* of modern technology and its enormous release of power at a level that is perhaps no longer human. But it is extremely interesting for me to establish how Gehlen's anthropological technicism answers this question. It does so by erasing the difficulties encountered and the dangers created by the *actual* development of technology in the twentieth century. Moreover, at the level of explanation, it is not critical of the emphasis that modernity has placed on rupture. This is a seductive and practical optimism, a rather clever theoretical conciliation, whose weakness I am nevertheless going to expose.

Regarding the further developments of technology, Gehlen gives a quasi-idyllic view. Thanks to the extension and perfection of the system of regulation, cybernetics will be able to further the harmony between man and nature (by emphasizing the "isomorphism" between psychological systems and artificial structures). "With progressive technology, man imports into inanimate matter a principle of organization that is already functioning at many points in the interior of the organism."[22] Likewise, the systematizing of communication will allow a fruitful cooperation between technology and the sciences (as well as among the sciences).[23] The future appears to belong, therefore, to cybernetics, the science of the sciences, the supreme regulator, which is also called upon to transform the human sciences.

This vision seems fragile, primarily for the reasons that have already been suggested: one cannot constantly substitute a *program* for an account; one cannot in the name of tendencies or final ends refuse to take into consideration certain facts. Neither the negative effects of progress, the huge massacres of the world wars, nor the ruptures of balance and their harmful effects seem to exist for Gehlen. It is difficult to believe that these lines could have been written in this cruel and bloody century, in which the survival of humanity has been put in jeopardy by the unbelievable perfection of the technologies of extermination. In one sense, how one would hope that Gehlen was right, and not Heidegger and his apocalyptic view in the "End of Metaphysics." One sees here that a truly equitable assessment of technology is difficult; it concerns a domain where emotional and even religious investments are inevitable (Gehlen's theory agrees with this). Given all the evidence, such a vision of technological developments is at least as "reactive" on the optimistic and pro-technical side, as Heidegger's is on the pessimistic and anti-technical sense (even if Heidegger's thought on the question is not exhausted by this characterization). But the weakness of the argument is also in the conception of man that Gehlen presupposes: it is an architectonic and "organizational" conception in which the unpredictability of passions and of freedom does not seem to exist, and where there is also a lack of any other ideal than integration within the environment and balanced mastery. A prosaic thematization—made in good conscience—of the technological world.

Concerning power—is this a coincidence?—Gehlen addresses it only in order to inscribe it in a timeless theory of magic: by a phenomenon of "resonance," human beings want to discover themselves everywhere in nature and to master once and for all its resources. It is because of this, and in order to present a credible facade, that this anthropology of technicism must minimize the fracture that scientific-technological modernity introduces into the relationship between man and his environment, because if the mathematization of physics in the sixteenth century revealed its revolutionary impact, it would become difficult to pretend that its result (modern technology) is only a "superstructure" of traditional technology.[24] Gehlen does concede, however, that a *qualitative* mutation took place with the advent of modern technology. He is, moreover, correct in asserting that this mutation does not consist principally in the passage from the instrumental to the machine, but in a change that affects, in the age of classical modernity, the *Naturwissenschaften*, that is to say, essentially physics. At this point, when he touches the nerve, Gehlen remains strangely and significantly silent; he gives no explanation at all for why science suddenly became so essential for technology itself, nor for the consequences which are going to result for technology. A decisive mutation takes place: he recognizes that it does not come principally from the evolution of instruments and technologies; he points out the role of science, but he goes no further, as if—the nature of technology having been determined once and for all (from the Stone Age to the atomic age)— nothing new could happen, except a closer "cooperation" between these well-defined domains, for the greater glory of humankind—which is itself well-defined. The reason for this silence quickly appears: it is in the interpretation that Gehlen gives of the modern scientific revolution. He characterizes it as "analytic-experimental," and he maintains that, in putting an end to the previous mixture of "speculations and random observations," science moves closer to technological praxis. The new experimental devices are comparable to "simple machines," and the "logic of experimentation" isolates the processes that humans will have grasped in order to apply them.[25] This is a surprising argument, which reintroduces technology as the cause of the "modern revolution" just when this had been put aside. What is the logic of this reasoning? It is in the analogy between the "logic of representation" and technology. But from where does this new logic spring? This remains quite mysterious. And yet, the inadequacy of a purely "experimental" interpretation of the modern scientific revolution is clear. For now[26] it suffices to cite Alexander Koyré on this point: "It is thought, pure and unadulterated thought, and not experience or sense perception which is the basis for Galileo's new science."[27]

Finally, concerning the most contemporary aspects of the problem, Gehlen is compelled to recognize that technology is no longer "pure," but that the current situation is characterized by the "functional connection" of industry, technology, and *Naturwissenschaft*. Once again we are on the threshold of important insights— perhaps this new complex is no longer measurable on a completely human scale because of its immeasurable scale and the redistribution of the powers that it deploys? Gehlen makes sure not to ask such a question, because, for him, the connection is "functional," which seems true only if this functionality is referred to the "self-augmentation" of modern technology.[28] For Gehlen, this is not the case: technology itself is always presumed to be human through and through (a thin veil being thrown over those inhuman, suprahuman, or "ahuman" aspects). While he recognizes, therefore, the connection of science, technology, and industry, it leads to nothing other than the reaffirmation of a massive pragmatism, because the science that concerns him is only experimental science—a thesis we believe is false.

Marx or the missed opportunity

More frequently than an anthropological technicism as explicit as Gehlen's, one finds his equivalent prevalent among our contemporaries. One can ask, in this regard, whether Marxism is not the most widespread form, exemplary insofar as its fundamental beliefs are shared by many of those who oppose it—but who nevertheless continue to take up its presuppositions.

The great mobilizing themes of Marxism-Leninism are without doubt more technical than purely anthropological. They are of a technicism that would have perhaps shocked Marx himself, who accused the bourgeoisie of having organized the idolatry of work.[29] Even when the technological and productive obsession did not take the extreme form it did under Stalin, it retained an important place in the communist mythology, as if the projection of machines to the forefront of the conciousness of reality (going so far as to put them on postage stamps) could magically accelerate production and the progress of the march toward "real socialism." To linger with the mythology is hardly of interest within the limits of this study; it is important, however, to understand the presuppositions, to grasp whether—and why—this situation has been made possible by Marx himself.

Marx's work has been, to this day, so scrutinized that attributing a "reflection on technology" to the great Karl would not seem to be the most scandalous enterprise. However, I do not believe that this thesis is tenable.[30] What is striking about Marx, if one boldly pushes the inquiry, is rather the absence of a reflection on technology *as such*. Castoriadis rightly remarks that "the word technology is not frequently used by Marx," but that "this is nevertheless aimed at when it is a question of *work*, *industry*, and *productive forces*."[31] To claim that Marx did not *think* the modern technological phenomenon in its depth and specificity is not at all equivalent to rejecting the totality of his analyses, which nevertheless appear to me to be limited by the presuppositions (or prejudices) that have become the current exchange in this century of triumphant technicism.

"It is in praxis that man ought to demonstrate truth, that is to say reality, power, the precision of his thought." This central phrase of the second of the "Theses on Feurbach" reveals the first presupposition: a metaphysical humanism that none of the "epistemological incisions" of the world are able to erase. This humanism, when it is not purely and simply denied, risks being trivialized: the usual corrective of an obscuring idealism. The priority of praxis, asserted here by Marx, has nothing trivial about it, nor does it merely echo common sense; it manifests the spirit of a conquering modernity where man constructs himself and must prove to himself the power of his own thinking. Praxis is the relation of man to the human. How could it be otherwise when "the entire so-called history of the world is nothing other than the production of man through human work"?[32] Marx, therefore, recognizes the importance of technology, but *only after he has put in place the anthropological-practical system that will give an account of History.* "Technology lays bare the mode of action of man vis-à-vis nature, the productive process of his material life, and therefore, the origin of social relations and the intellectual ideas or concepts that result from it."[33] If there are those who are inclined to credit Marx with having placed technology within a total social and historical process, I will respond to them that one must also consider the metaphysical price that has been paid in order to accomplish this operation.

The second, correlative presupposition is the technicism of Marx in his definition of the human being, as well as in his conception of the historical process. He takes up Benjamin Franklin's definition of the human being as "*a tool-making animal*,"[34] establishing a correspondence between the means of labor and the level of development of the society. The formulation is famous: "The hand mill will give you a society with a sovereign; the steam mill a society with an industrial capitalist."[35] However, this abrupt formulation should not cause us to attribute to Marx a technicism that is in some way literal and mechanistic, as if each technical progress carried with it—term-for-term—a radical modification of the social structure. Technology is a "productive force": taking it into account is necessary in the global strategy of a critical enterprise seeking to disentangle the "material base" from the historical-social process.[36] Whether it is a strength or a weakness in Marx's theory, this technicism is fraught with ambiguity, because the technology which conditions social superstructures is considered as a material factor, even though it is—to a large extent—the product of knowledge and of intellectual projects.[37]

Marx's revolutionary perspective, his systematic desire to shift the focus from human phenomena to their *specific* conditions of appearance constitute trump cards for understanding the specificity of modern technology. Indeed, the precise analysis of the movement from manufacturing to industry, and the exploitation of man by machine places Marx on the edge of a double understanding: technology as techno-science, and as technological power transcending the human. Concerning the first point, by calling attention to the importance of the "application of science to industry,"[38] Marx actually assigns to science the role of a productive force, at the risk of once again compromising the "purity" of his own materialism. This productive power of science, however, was not detached from the general anthropological-practical schema that we made reference to; insofar as it is one aspect of social activity, it was therefore hardly recognized in its specificity, but rather appeared as "taken hostage" by praxis: "*the development of science* as an enhancement of both the theoretical and the practical is only one aspect, a manifestation of *the development of the productive forces of man*, that is to say, of wealth."[39] Even if after 1844 human praxis is no longer understood as the pure and simple self-development of human beings, the practical project remained doubly subordinate to humans, on the one hand because of their needs,[40] and on the other because of their will to master nature (it is human mythology which "tames, dominates and shapes the forces of nature").[41] However strongly his analyses critique mechanization—this is my second point— Marx was amenable to considering both the enormity of the forces at work and the minor impact they made on humans. As he shows in chapter 15 of Book One of *Das Kapital*, "Mechanization and Large-Scale Industry," Marx has a remarkable grasp of the most recent improvements in industrial equipment. He thus makes a distinction between the "cooperation of several homogeneous machines" (for example, the fabricating of textiles, "formed by uniting a vast array of power looms") and the emergence of "systems of machines," assembly lines that combine and articulate segmented operations (one machine to card, another to twist, yet another to spin, etc.).[42] Noting this new equipment as well as "cyclopean machines dedicated to the construction of the first motors," Marx already describes what Jacques Ellul will call the "auto-increase" of technology. "Large-scale industry was . . . forced to adapt its characteristic means of production, the machine itself, in order to produce other machines."[43] This extraordinary process surpassed humans on several levels: the human as a driving force, becoming weak; the human as technological agent, becoming less precise, less adaptable than the machine (for example, in the case of machines that drill, machines that plane, etc.); finally, the human as a labor force (the machine "requiring" unskilled labor at that point: women and children).[44] Mechanization therefore becomes a formidable instrument of exploitation, since it is the human being who must adapt to the rhythms of the machine and its "logic" of production and profit, not the reverse. "The movement and activity of the means of labor became the machine that has independently erected itself in front of the worker."[45]

As we know, Marx assigns the monstrosity of this process to capitalism alone, even though he manages in passing to take note of a "purely technological" factor in the division of labor.[46] Faced with capitalism, Marx has a double attitude: he is moralistic in condemning cynicism and the revolting excesses of inhumanity; he is cold and "scientific" in analyzing capitalism's functioning and its laws (for example, the inverse ratio between the productivity of a machine and the value assigned to its product).[47] One cannot be content, in this regard, with separating the scientific results from the rest of the work, as do some current economists—the new attorneys at the tribunal of intellectual history. Marx's scientific work has been constantly determined by his political choices. Would an enthusiastic partisan of capitalism have made these two errors—the theory of pauperization and the tendential decrease of the profit margin? Undoubtedly, that person would have made other errors, and would not have gone as far as Marx in his critique of large-scale industrial capitalism. However, the contrast is there: it exists between the extraordinary ingenuity used to demonstrate the logic of capitalism from within, and the good conscience of the proposed "solution": to change the social mode of the "exploitation" of technology.[48] Surely this paradox is even more decisive than that which burdens Marx's "materialism"; his revolutionary progressivism (another name for humanism) allows him to entertain the illusion that a mastery of social practices will permit the reappropriation of technology. Yet, this is the same lucid observer who explained how historical processes follow a rigorous necessity, that the development of the modern Western world was an absolutely new and specific conjunction of "productive forces" changing the "relations of production," and who observed to what extent this imposing process would escape the will of capitalists as well as proletarians. On the one hand, the lucidity of a researcher; on the other, the ardor of the militant. The two coexist in Karl Marx, Promethean militant, brilliant and banal at the same time. Should we abandon the past to the researcher, the future to the militant by proclaiming that the dictates of the political will are "scientific"? This has been practiced too often already, and one can see the results: the frenzied paces of production becoming "progressive," and strikes prohibited because they are "reactionary."

Marx's theoretical advances are therefore compromised by his revolutionary anthropologism. Marx believes that the rational domination of nature is good in principle, and that it will suffice simply to invert the social order of that mastery in order to end exploitation, diverting the surplus of work toward the workers, making the "proletariat" play his new role of "universal class," dissolving class lines. It is, however, naive to presume that technology (or techno-science) will experience an essential metamorphosis with this changing of hands. As if it were sufficient to decree that one has appropriated Power in order to eliminate its massive and harmful effects! The socialist techno-science is no less polluting than the bourgeoisie; its excessive armament, no less fraught with danger. If the socialization of Prometheus liberates him from his evils, how would his indecisiveness be possible? But history has already given a negative and often tragic verdict. Technological power is more revolutionary than any revolution; it comes from above, no one can know where it is going. Marx allows us to *touch* its novelty and breadth: no more, no less.

False recognitions of Power: The Scientific and Technological Revolution

Today it is common to "rescue" Marx's thought from the contradictions of Marxism. Although some deny the Master's humanism and others his scientism, there is historical evidence of the convergence between progressivism, humanism, and techno-scientism in the thought and action inherited from Marx. Whatever the exegetic discussions might otherwise be, Marx had hoped for "the development of all human forces,"[49] and Marxist regimes—excluding the case of Cambodia's Pol Pot and, to some extent, Maoism—have interpreted this optimistic humanism in a way favorable to the unconditional development of techno-science. Revolution viewed itself as sustained and dispersed by Progress (the Soviets plus electrification) until the more recent day when the technological advance of Western countries and the bureaucratic conservatism of the so-called socialist countries were led to pose the problem in a different light: is not the true revolution of the twentieth century technological? Has not science become the productive force par excellence? The question we are asking is the following: Is the effort to adapt the concept of the revolution sufficient? Can this concept—even if altered—allow us to discern the new characteristics of contemporary Power? I am not questioning the application of this concept to the political, social, and even epistemological upheavals of modernity; rather, I am asking whether this concept limits intelligibility and clouds the horizon, as if it alone were able to decide the future and it alone were able to give any measure of the unprecedented.

We have already seen that it is not sufficient to acknowledge the importance of technology in order to understand it. This has certainly been the case with the so-called Scientific and Technological Revolution, which renders the most decisive reality of our times banal. Just as in the last century it was both inevitable and yet superficial to celebrate with Renan, Littré and many others the "future of science," it seems inescapable that the progress of scientism is assured in the twentieth century by a "techno-scientism" whose forms and variations are multiple, both inside and outside of Marxism. We find significant evidence of this in the work elaborated by Richta and his collaborators, *Civilization at the Crossroads*,[50] which at first glance presents some interesting characteristics. A critical work situated before the "Prague spring," this text breaks with the rhetoric of the Marxist vulgate, acknowledging the decisive importance of technology and especially of science as "productive forces," hence permitting one to surmount instrumentalism on the one hand and political obsession on the other. Should this not satisfy us? It is undeniable that one sees here a laudable effort to draw closer to the "new relationship" that follows the industrial revolution: Richta recognizes that the theme of the class struggle has lost its immediacy,[51] even as it increasingly affirms—in the wake of automatization, "chemistrization," and cybernetics—"the priority of science over technology, and of technology over direct production."[52]

Nevertheless, at the same time, this example demonstrates that nothing is more difficult than understanding "that which is", when that awareness—as informed and coherent as it is—can hide behind an expedient concept, the Scientific and Technological Revolution, which becomes in fact a means for avoiding *thinking* technological power in all its aspects—both manifest or reserved.

The Scientific and Technological Revolution is truly *celebrated* by Richta, whose thesis could be summarized by the following: we are leaving the industrial age in order to enter into a new era where automation substitutes itself for mechanism; science becomes a direct *productive force*. "Presently, science and its applications in technology, organization, qualification, etc. replace simple and fragmented work which up until now was the basis of production."[53] The transformation of production by science, in particular by cybernetics, permits the progressive abandonment of the old economic rationality; the regulation of automatization puts in place a new model of growth, which is no longer extensive, but intensive (one will work less, but better). "*Ratio* . . . returns to man as the *rationality of human development*."[54]

The essential flaw of this thesis is that it claims to be scientific, yet reveals itself to be taken in by an ideological position. The alignment with the Party in the Leninist sense is replaced by an alignment with the Scientific and Technological Revolution; everything else is accused of "romanticism." The "cybernetization" and the "chemistrization" or the "nuclearization" [sic][55] of life have become panaceas. This enthusiasm is already outdated (we know that current world economic conditions are proving to be far less favorable than those of the sixties), but it also deserves to be distrusted because of its gross generalizations. It presupposes a qualitative difference between "industrial" and "post-industrial" without demonstrating it—a situation very popular in the United States during the same years; one remains silent or systematically minimizes the danger of the "consequences" of scientifico-technological power, whether from over-armament or pollution.[56] Retaining only the positive or optimistic aspects of the Scientific and Technological Revolution, Richta has an easy explanation for the alienation of work and the sufferings caused by harmful effects: it is a matter of that which survives or results from imperfect technology.[57] He proposes even more technology: everything will then be marvelous. Thus, the technology of consumption will allow humans to avoid serving "the world of objects."[58] This is a techno-discourse whose naiveté—real or imagined—hardly appears believable to us, the privileged society of consumers, who have learned to decode new fascinations, and who understand that an abundance of goods undoubtedly liberates us from the "world of objects" less than does scarcity.

Actually our problem is not whether Richta was completely faithful to Marx. It is, however, the problem of Benjamin Coriat, whose small book *Science, technique et capital*[59] represents still another case of a false understanding of Power, albeit for different reasons. Coriat does not deny that science is a productive force and that it extends its "sphere of influence to ever-increasing sectors of production."[60] But he does question whether it is a *direct* productive force that would allow new forms of production to escape, through the virtues of the Scientific and Technological Revolution, the capitalistic law of value. We do not doubt that Coriat is quite correct in contesting Richta's interpretation of the Marxist theory of relative surplus value,[61] but this serves only to confirm two convictions. The work of Richta is a circumstantial techno-discourse that performed a hasty marriage between Marxism and "post-industrial" phraseology; second, Marx himself hardly delved into the discussion of the power of science, completely preoccupied as he was with articulating his "readings" of modernity as the expansion of capitalism.[62] That the expansion of mechanization was connected to the exigencies of capitalism was obvious to Marx. But this hardly takes us any further, unless one has decided, once and for all, that the truth is contained in *Capital*, particularly in the fourth section, which concerns the "technological application of science."[63] Such is the case with the Althusserian Coriat. I do not share his view.

The orthodox repetition of the dogma of class struggle, of the unconditional priority of the law of surplus value and, finally, of the necessary initiative of the masses does not take us any further than Richta or the Master himself. The question of the significance of rationality, modern or ultramodern, remains constantly obscured by the omnipotence of the law of value.[64] Can we explain everything from the sixteenth century to the present by the quest for profit? In order to respond to this question, one must confront the difficult problem of the relationship between potentialization and capitalization. In principle, Marxism ought to begin from a concrete analysis of given situations. Coriat, like Althusser, does the opposite; considerable areas of empirical reality (the intensive militarization of "socialist" countries, the *precise* situation in China) are ignored; the autovalorization of Capital is applied automatically to the case of war,[65] as well as to that of contemporary research.[66] There is a lack of any critical distance—as with Richta—in order to consider the new rationality of *optimal efficiency* in its positive aspects (and in its dangers). This reigns as true in the "socialist" positions of Taylorism and Fordism as in the new modes of information-systems or media control, which the Party in power is able to appropriate for purposes of propaganda and "disinformation."

The recourse to philosophy, when it is restricted to opposing materialism and idealism, is a caricature. Technicism benefits enormously from this: every question is neutralized, a critical genealogy of Power made impossible. To be sure, technology is not "neutral," but to think of it only in terms of its social or economic use is a unilateral step. It is to condemn ourselves to not seeing that the Scientific and Technological Revolution is only the most recent manifestation of a process that is much older, more fundamental: the *potentialization* of knowledge as power, without any limit or end assigned to this growth of power. It is also to misunderstand the importance of the technicalization of language, which cannot be reduced to the production of supplementary techno-discourses, but represents at its core a "taking control" of the symbolic by the functional.

4

Jonas in the System

My dear friends, let us know ourselves, and how it standeth with us.
We are men cast on land, as Jonas was out of the whale's belly,
when we were as buried in the deep; and now we are on land,
we are but between death and life.
—Francis Bacon[1]

Is there a technological System?

"POWER, ENEMY OF the worker,"[2] "power . . . foreign" to individuals:[3] Marx did not
minimize the effects of the economic-industrial development of the West. Rather,
in proclaiming that "this power, so mysterious for German theorists, will be abolished by the reversal of the current social order,"[4] he overestimated the scientific
character of his own endeavor. One hundred forty years after the drafting of *German Ideology*, it is difficult to claim that History has not rendered a verdict on this
point. The mercantilistic state of the mid-nineteenth century has been inverted in
one part of the world, profoundly transformed in the rest of the world. In any case,
History has not followed the simplistic model of Marx and Engels: the abolition of
domination by the reversal of capitalism, a conviction that *The German Ideology*
claims to be "empirically founded."[5] At the close of the twentieth century, it is undoubtedly possible to have a view of technological power that is free from Marx's
messianism as well as from his technological humanism, but which is nevertheless
enriched by the scientific knowledge resulting from his research. In his two books
on technology, Jacques Ellul is totally disillusioned by humanist-progressive illusions, whether Marxist or anthropologico-technological. His understanding of technology as a System represents one of the most lucid efforts made thus far to measure
Power.

Given that I must restrict myself to the essential points, it is not here a question
of undertaking an elaborate analysis of these two works, which are filled with judicious remarks; neither is it a question of taking an inventory of the differences between *The Technological Society* (1954) and *The Technological System* (1977). At any
rate, these differences are not major. The first book, more carefully written than the

second, already goes a long way. Twenty-five years of reading and reflection are evident in the second book, as Ellul became aware of the importance of information systems, problems in communication, and the role of self-referencing (*feedback*) regulation in cybernetics. His reading of Parsons[6] allowed him to identify technology as a System, in a sense that is precise but not formalized. What is Ellul's essential contribution if one reads his two books on technology together?

A first distinction proves to be absolutely fundamental: that between technological operation and technological System. The first "has always existed throughout history"[7] under the form of an instrumental relation (which I will identify as the first phase of the potentialization of the rational); the second has been foreshadowed by the passage from craftsmanship to mechanization, because if it is true that the machine "uses man" in order that he serve it[8] (Marx had already established this), then the System realizes itself only in modern technology that clears the way for a qualitatively new stage. Whatever the mode of approaching modern technology in its specificity would be, we must not measure the new beginning from the old: we do not move to a technological System by a combination or a sum of technological operations or machines. How can we, therefore, isolate that which constitutes the System as such? "The system is an ensemble of elements, interrelating in such a way that any evolution in one element provokes an evolution in the entire complex, and, conversely, each modification in the complex has a reciprocal effect on each element."[9] This definition, in itself, is nothing more than extremely classical; the innovation is to grasp technology as an enormous concrete complex without confusing this system with either society, the economy, or even with subsystems of equipment. To recognize in technology as such a "specific organization . . . relatively independent of man"[10] is to cross over a threshold in front of which many have shied, either because of ideology (as with Marxists) or because of scientific caution (the traditional reaction of specialists to technological problems).[11] Ellul does not go so far as to sever this system from "other systems or realities",[12] and he himself concedes a "margin of error".[13] Nevertheless, the affirmation of this specificity calls for the deployment of an appropriate analysis of the constitutive characteristics of technology as System.

What are these characteristics? Four of them are grasped *statically* ("the characteristics of the technological phenomenon"); two, *dynamically* ("the characteristics of technological progress"). This distinction could appear artificial; but as to the characteristics themselves, it seems that Ellul would agree with me: what is essential—beyond the subtleties of presentation—is to accept and understand that *there is a technological System*, that is to say, a self-organization and self-augmentation of technological power marked by a growing independence in its relation to human decisions.

This automatization of technological power is increasingly recognized, but clashes with extremely strong ideological resistances. A significant example of this has been given by Philippe Roqueplo. Making perfectly correct remarks characterizing contemporary technology (its systematic linkage with science, its quasi-"transcendent" power, the cumulative and almost irreversible character of its "progress," its technological "self-ideologizing"),[14] he sees very well and clearly declares that it is necessary to give an account of the dynamic that operates like a destiny,[15] taking possession of the "norms of rationality"[16] as well as of reality itself (whose operational nature becomes the sole criterion).[17] He even makes the self-augmentation of scientifico-technology his *hypothesis of reference*.[18] Everything presented thus far is quite acceptable. But this hypothesis is very quickly criticized, (as if a hypothesis could be in any way refuted a priori without having been truly tested), and critiqued in terms that literally reverse it. We quote the text: "*The thesis of the autonomy of scientifico-technology is false* BECAUSE IT IS NECESSARY THAT IT BE FALSE. The production of society by itself forbids accepting any FATUM TECHNOLOGICUM, and it is precisely this which is in question in the DECISION to institute a social control of technology: it belongs to such a decision to put into place the apparatus that will permit it to prove its own possibility."[19]

It is indisputable that Roqueplo readily assumes his own begging of the question: he himself uses that expression. But is it enough to proclaim the necessity of "voluntarism" to justify a similar theoretical *coup de force*? The core of the argument is the "self-production of society" (having already presupposed this, as the author himself acknowledges, in his earlier critique which gives a first indication of this same reasoning). Insofar as society produces itself, it is society that is the sole spring of autonomy. If techno-science plays this role, every politico-social project of democratic "taking control" would collapse in its very possibility. Therefore, it is not admissible; it is necessary to prove the (contrary) movement at once!

One could be content with brushing it aside: "This procedure is not scientific." And one would be correct. All in all, the "refutation" of Ellul's thesis reduces itself to the enumeration of questions, *none* of which is truly substantiated.[20] Was there not a "decisive" human intervention with the atom bomb? Yet, human interventions count for little in the face of the very logic of the process and with the appearance of a new phase of potentialization.

Despite all this, I would like to respond briefly to Roqueplo on his own terms: do you serve your democratic ideal by masking reality to this extent? Voluntarism certainly has claim to nobility, especially when it is manifested by individual heroism. However, collectively, is it not the producer of the catastrophes of the twentieth century? Hitler and Stalin were the champions of voluntarism; they both showed the most complete lack of deference to techno-scientific autonomy. Admittedly, techno-science returned their contempt. The democratic ideal has nothing to gain by taking up worn out slogans. It amounts to an electoral or media parade, an avatar in the manipulation of the masses, an alibi for intellectuals who prefer their clear militant conscience over a rigorous examination of their understanding. In fact, at the bottom of this voluntarism, there is *faith* in the self-production of human beings and society. I have shown in the discussion of Marx the merits and, especially, the fragility of this position.[21] Furthermore, this thesis is profoundly metaphysical; it leads to a praxis which is purely and simply ideological. Moreover, it cannot claim that any of these accounts rest on "that which is": a techno-scientific discourse that is autonomous from truly social and political criteria. While I agree that there is a "social dynamic" which must be accounted for, I categorically refuse to transform this "social dynamic" (which, incidentally, is not sufficiently determined) into the key for everything else, turning it into a dogma. If History is a process of "domination" (to use Althusser's phrase), then the question is to know what is really dominant. The self-production of society is only a *secondary effect* (in large part ideological) of a process which is not created by society itself. In addition, we will see later that the thought of rational *partage* avoids making technology a pure and simple *fatum*.[22]

This polemical excursus touches a sensitive point: the question of the autonomy of the technological system is not at all academic; it has become urgent: our very present is infused with it.

Autonomy, unity, universality, totalization: these four characteristics of the technological system can be isolated statically, but nevertheless are profoundly interdependent. To be sure, autonomy is not absolute,[23] but it can be decoded at many levels: it is already present in the "concretization" of the object and in the technological environment, as Simondon has shown, as well as in the abandonment of instrumentality (available, present-at-hand, multiple) for the profit of programs which are self-defining and which bring about decisions by virtue of their own weight, and finally, in the absence of taking into account human,[24] ethical, or religious[25] concerns. The unity of the phenomenon is obvious: it is the very character of development that it is an "immense system of uninterrupted correlations,[26] from which follow constant "restructurings," the globalization of the economy and the intensification of exchanges, as well as the impossibility of a state to rise to an honorable rank in the System by focusing on a single specialization (this hyperspecialization would make it even more dependent on the rest). Geographic universality is evident. Qualitative universality as the extension of technology to every domain of life presupposes a generalized "mental appropriation" of materials, but also of technological codes and values, thereby leading to a psychological homogenization, thanks to the conditioning of mass consumption. Totalization brings about the closure of the System upon itself: it is regrettable that Ellul devotes so few pages to this, merely critiquing theoretical totalizations that *anticipate* the closure of the system, accelerate the installation of technological man, and undermine any spiritual resistance to technicalization.[27]

In his first book, Ellul did not separate the exposition of technological "progress" from that of the characteristics of the "phenomenon": automation and self-augmentation were presented as the two primary characteristics of modern technology. He was correct. It is artificial to separate the evocation of its dynamism from the phenomenon. Indeed, it seems difficult to conceive of the technological System in a simple state of "survival": from its origins to the present day, its renewal and the perspective of its perfection have been co-substantial. There is not the technological phenomenon *plus* an attribute ("progress") which qualifies it: under its modern form, technology is constitutively "progressive." Self-augmentation is therefore a metaphor: "By self-augmentation, I understand the fact that everything happens *as if* the technological system increased by a force which is internal, intrinsic and without *decisive* intervention of man."[28] This is a more prudent formulation than that given in 1954, although Ellul himself notes the frequent revival of this idea by other authors. This autonomous or quasi-autonomous growth is explained by the irreversibility of technology,[29] as well as by what Gabor calls "the fundamental law of technological civilization: *Anything that can be done, will be done.*"[30] The pressure of a possible increase by a series of actualizations and convergent inventions,[31] and finally, by the fact that "technology has harmful effects which only technology itself can solve."[32] This last is the ultimate and subtle form of self-augmentation.

As to automatism, Ellul does not seem to understand this in the sense of mechanical automatism that Simondon shows to have a technological character that is rather underdeveloped. Technology as system is more supple, according to Ellul, even if it yet lacks *feedback*. Nevertheless, "all obstacles must yield in the face of technological possibility. Such is the principle of automatism."[33] This will come into play through a kind of internal arbitrator in the technological system. Thus the choice between two technologies for the same operation will be made on the basis of efficiency.[34] We see here what Ellul understands by automatism: an automation of choice,[35] an automatism which is very often effective, at times merely desired (however, expressing by this a kind of irrepressible aspiration to the technological world), when it is a matter of perfecting the capacity of human adaptation to the point of perfect malleability.[36]

Concerning all these aspects of contemporary technology, in all his critiques of every form of contemporary technology, Ellul appears very often to be penetrating and beneficially acerbic. This is all the more reason not to hesitate in frankly presenting my critique of both his method and conclusions. The method is more rhetorical than scientific; the mode of exposition appears frequently as a discussion; the polemic tends to strain conceptual patience. It results in uncertainties and redundancies. Some examples: Why isolate the automatism of technology (in a sense, moreover, that is very loose), and in so doing make it an autonomous principle, when one recognizes that "this already results from the autonomy of technology"?[37] Why mix, almost constantly, the descriptions of the *consequences* of technicalization with the analyses of the *principles* of the System?

His conclusions call for a more complete and sustained discussion: they concern the power of the technological System, the problem of the correction of its dysfunction, and the amount of hope that humanity can hold onto. Ellul concludes with the thesis of the total power of the technological System. In 1954, pointing out the multiple and difficult effects of controlling DDT on the food supply, he writes, "The human being is delivered helpless, in respect to life's most important and trivial affairs, to a power that is *in no way* under his control; there can be no question today of his controlling the milk he drinks or the bread he eats, any more than of his controlling his government."[38] And very recently, "Nothing today has sufficient power to force technological projects to modify themselves."[39] This statement concerning the powerlessness of people, and, in particular, the individual is obviously the corollary of the principle characteristics of the technological System: autonomy, self-augmentation, totalization. Since humanism proves illusory, making it necessary to go to the limits of a radical reassessment of the facts, there is no hope for anything more than a correction of the System by itself, in the form of cybernetic *feedback*. Integral technicalization has its logic: the remedy for technological evils can only be technology. However the bitter, but undeniable, conclusion that Ellul reaches is clear: the System is not capable of correcting itself.[40]

Technological power is far from limiting itself to the material or mechanistic domain. It also has an enormous power of psychological integration: non-conformism sees itself recouped by the system, "creativity" itself becomes a factor of adaptation.[41] Technological civilization's tolerance of zones of uncertainty has already been pointed to as a way of regaining confidence in the society we call post-industrial: islands of freedom are henceforth open for leisure and creativity, which are going to expand themselves to the degree that their progress will permit society to be more prosperous and tolerant. Ellul opposes this idea. He connects what he calls these "zones of indifference" in the System to an increase in organization, not the inverse. To his mind, it is an increased adaptation which allows for the diversification of the system; it is not possible to take the allowance of autonomy in some subsystems for a real challenge to Technology. Technology appears more and more flexible because it perfects its efficiency.

In his desire for logic and a radical stand, Ellul does not seem to have escaped this alternative: either Technology is all powerful, ceaselessly refining its integration of resistances and problems, or, if the System *as such* is not capable of self-correction, then its power blocks itself, inverts itself, and becomes impotent. The second eventuality is not clearly explained. Concerning self-correction, Ellul writes, "if from all evidence we establish the negative effects of technology, the errors, irrationalities and dysfunctions, still we have not observed any method, automatic or semi-automatic, of rectification."[42] This is certainly true, but at the global and essential level: the System is incapable of correcting the "defects" that *constitute* it. On the other hand, we see it continuously working to repair or to rectify the defects, proceedings, and processes whose inadequacies have been demonstrated in practice (anti-pollution devices, recycling, antidotes or anti-allergenics in medicine, new legal recourses, etc.).

Therefore, one should not expect a global auto-correction that would be *fundamental*. Technology can correct only technological errors. No massive retroactive measures can recoup those principal errors, those which arise from the very nature of the System. As much as Ellul is correct in showing the automatism that is increasingly dynamic and subtle in technology, this is also the place to ask whether he proposes an idea that is too mechanically dogmatic concerning our entrapment in the technological System as well as the future of technology. Perhaps himself a victim of the surrounding ideology, he pushes the fact of our powerlessness to the extreme consequence of a technicalization of our freedom: it is still technicism to transfer thought itself to the model of *feedback*.

Certainly, the question is one of extraordinary scope and difficulty. Ellul has moments of perplexity: "It is difficult to respond."[43] And more thematically: "It is thus possible that man, unsuccessful in controlling, orienting, and utilizing Technology reasonably, may in his turn become a restraint and cause a recession."[44] Who can pretend to escape Ellul's contradictions, his oscillation between extremes? On the one hand, "It is perfectly vain to pretend that one can check this evolution, much less control and orient it"[45]—but why then even write *about* technology? On the other hand, "It is in establishing voluntary limits that man establishes his humanity ... And I believe that nothing is as fundamental as this problem of voluntary limits."[46] Thus, everything is not definitively sealed by the System.

Ellul recognizes Power in its technological figure. But in the end, his method and his conclusions expose him to objections which can be summarized as follows. When he moves from the phenomenal to the essential, he does not make a sufficient distinction between the problem of technological self-corrections and what is beyond technology proper. He brushes over more than he treats the question of the relationship between science and modern technology. He hardly recognizes the reversals of "for and against" that are produced when technology becomes total, because he does not take the critical distance necessary for attaching the technological question to its rational origin and to the structures of rationality.

Technology in language

Despite the preceding objections, I have responded affirmatively with Ellul to the question, "Is there a technological System?" Even if Technology is not a system in the strict sense of a theory of systems, even if I cannot yet produce a model of Power such as that for the global economy, it nevertheless remains impossible to account for planetary evolution without acknowledging the *automatization, unification, universalization*, and *totalization* of the techno-scientific phenomenon as the essential motor of Power. The last "stage" of this is technology's seizing control of consciousness and language itself, through which it perhaps places its lock on all of reality, down to its symbolic aspects.

In contemporary reality, technology—which I understand not in the partial sense of the technology of this or that, but in Ellul's global and encompassing sense—is universal from two points of view: constitutively and in its extension.[47]

Constitutively, there can be no modern technology without the support of scientific rationality which is itself universal by vocation. In its extension, there can be no development without the construction and diffusion of models, in the industrial sense, with all that this demands in terms of the unification of attitudes, projects, and vocabularies. "It is necessary to normalize everything in order to universalize everything," said a specialist,[48] presenting as an imperative that which appears to be rather a fate already powerfully set in motion. The normalization of a gun or a gadget is not the most decisive moment. The radical revolution operates at the level of language: the conversion of the maximum amount of quantitative and qualitative data into information-processing systems, as well as the adoption of a cybernetic "philosophy" covering the totality of the social and material reality, and stockpiles the world as an immense informational fortification.

This constitutive universality is first affirmed, according to Simondon, in the teaching of the *Encyclopedia*, "Celebration of the Federation of Technologies," which discovered their solidarity and constituted an homogeneous field which presupposed an "internal resonance."[49] If this teaching was already, two centuries ago, "doubly universal," first through the public to which it addressed itself and then through the objective and dynamic information that it dispensed, what does this say about contemporary technology, relayed by innumerable channels wherein the audiovisual medium plays no small role? The expansion of technology rests, effectively, at once on the expansion of the objective space of knowledges as well as on an indefinite opening of the "public space" in order to stimulate adaptive behaviors.

Universality, in its extension, is also double: at once the annexation of new sectors to the technological field and a worldwide geographic expansion. Numerous works have already been devoted to the first aspect: from the environment to art, from sports to cultural affairs, from leisure to propaganda, technology is queen, empress, despot. The second aspect is all too easily observed in China as it is being "opened," and in the fact that every region on our planet is crisscrossed by planes, plowed by some machine, or "blasted" by a transistor. We live in a time in which there is no government that does not sponsor a plan for development, and in which our globe sees itself surrounded by satellites, relays of telecommunications and, soon, telematic "banks."

"We do not need to understand each other in order to carry out the most important tasks of our times." This statement, made by Jacques Ellul more than twenty years ago, still holds true.[50] The massive reality of technological consensus could not be formulated more elliptically or successfully. Smiling faces and handshakes characterize our presidential candidates. The issue is too important to debate: leave that to a few old hippies or to the ecologists, eternal adolescents; contracts are signed and indifference settles on the gulags and torture chambers. Babel without God: one constructs (and destroys) in the *hypertechnological silence* that surrounds the essential issues (the objectives, the effects, the sense or lack of sense of this frantic technicalization).

According to Ellul—we will return to this critique later—it is precisely because technology as a "universal language" establishes itself in the "unformulated fraternity" among technicians and even among all people,[51] that it becomes futile to look for another global link, be it religion, ideology, or language. But the main point that Ellul clarifies less well, even in his remarkable last book, than does Lewis Mumford in his *Myth and Machine* is this: To what degree does the reality of technological universality go hand in hand with a *mythical power*?

Considering the movement which is already under way and the massive mobilization which has already occurred, one could think that technicalization will continue to develop itself automatically. Although no proof can be produced, everything is happening as if the contrary were true: technology maintains its own myth; and if it does so with such intensity, it is apparently because it needs it.

At this point the adoption of technological priority is so widespread that one was surprised to see, for example, President Boumediene being crowned with accolades for having launched the industrialization of Algeria: what other political choice was offered? It is, however, necessary to have saints for a new sect, because it seems that in the "race for development" every moment and every atom of energy counts. Far from comprising an inevitable necessity, the dominant ideology constantly mobilizes its troops. Apparently, even a developed country like France is not developed enough. The obsession with endlessly setting records, the fear of being "outdistanced" becomes acute. The celebration of "development" has all the obsessional characteristics of a cult, the difference being that its *sursum corda* lacks the pomp of another era, and the heart is absent.

Mumford shows in a convincing manner that the Megamachine needs a superhuman consecration for its power to attain maximal capacity. Five thousand years ago, on an already impressive scale, kingdoms of divine right set into motion tens of thousands of human beings, reduced to playing the role of "moving parts"—like cogs in a machine, according to Reuleaux—in order to construct the great wall of China and the Pyramids. Today, the Pentagon of Power does not seem impressive enough with its bureaucracy, army, munitions, and bomb silos: it must undertake the conquest of the heavens and the cosmos. Better yet these exploits must reverberate in the four corners of the world. Thus we see a convergence between technology and the most impressive symbols of an ancestral myth—Power, as well as a new myth: the Future. Technology is celebrated as Power that comes towards us with the irresistible force of the Future.

But the mythical power of technology is also reinforced as a supreme privilege wherever we find universality transformed into *transparency*. Baudrillard showed how people in consumer societies must banish all reserve; in order to participate in communication, almost as functional as their furniture, they must do everything so that the "current flows" and so that nothing opaque can be erected as a screen in front of the flux of unities of production and consumption. Modern human beings are beings of public relations. Television is their chosen place. Telecommunications already permit them to consult the database from their homes. "Power, universality, and accessibility are the advantages which stand out against the telecommunications horizon," say Nora and Minc[52] in a study certainly more worthy of attention than the empty theorizing of Toronto's "electronic magi," a study in which the part of myth (with its demagogic currency) is not to be underestimated.

Technological universality doubles itself in the myth of the *transparency* of the communication that it claims to have founded. The neutral and well-determined concretion of technological problems, the discontinuous segmentation of machines and equipment, and the frustrating character of industrial and bureaucratic work are hidden in a luminous superstructure that is fluid and multilayered, the source of a compensatory communication (leisure as opposed to work), but also the place of an ever-increasing displacement towards the ideal of transparent availability.

In producing such a mythical radiance, technology reinforces the power of its influence. Not content with requiring us to use its services, it invests in cultural models. We name "techno-discourses" the "linguistic networks," which, at very diverse cultural and ideological levels, come to activate and stimulate the power of technology in celebrating its virtues and almost its cult. The Nora-Minc report appears to be part of these "techno-discourses," even if, in this context, myth is played out more subtly, this time for the benefit of the intellectual elite.

This study is exceptional, not only because of the information concerning the future of telecommunications—which it summarizes with remarkable clarity—and not only because of its brilliance, but also because of its political-cultural situation, which makes it the almost ideal object of the sociology of literature. The result of a command by the Prince, it addresses itself—through him—to every member of the political elite; it speaks to political power on the destiny of Power—in the name of the technological future. Insofar as information is power,[53] the study announces to Power what its *potestas* of tomorrow will contain. But this power wants to be enlightened, inventive, and considerate. The study addresses itself to an intelligent Prince who knew to have the most illustrious ministers of Finance sign a public statement, designed to enlighten the elite and, if possible, the less exhausted fringes of the masses.

To enlighten means here to reassure opinion, to attempt to show that the computerization of society does not necessarily lead to "brave new worlds," which is to say to an increasingly fatal conditioning of behavior and attitudes. Given that one has fewer proofs to offer, one can take, with increasing assurance and assertiveness, the opposite point of view from the pessimism that is eternally "late." To be sure, one will first repeat the eternal banality, so convenient, that technology is "neutral" and the future is a "thing of choice."[54] Then one will offer assurances that the society of tomorrow will be aleatory, multifaceted[55]—exactly the opposite of a planned gulag. What arguments are given to support this? They are that the new age of information-systems is less cumbersome and centralized than the old, that the emerging technologies have an extraordinary flexibility of use, and, in particular, that they permit a decentralization and diversification that has been hitherto unknown. However, if we consult, for example, page 53 of the study (impatient as we are to finally learn what the *autonomy* attained through computerization will be), we find a small and not very meaty bone to gnaw on: "Not connected, computers depend on autonomous information where the user is the sole master." This is a curious mastery about which nothing is said: the actual content of autonomy is not specified.[56] The authors at once reaffirm that the "new information-systems" escape "fatality," then envision the case of the autonomy of a production facility (an autonomy, it seems, that is relative, fragile, and intermittent), as well as that of the decentralization of banking networks. About autonomy itself and, in particular, the reasons that are going to push the user to disconnect, about the real limits of this autonomy, nothing is said. Elsewhere, however, even in adjacent passages, the reader is warned of the considerable dangers of concentration and authoritarianism that stalk the system. Now the dangers appear with a solidity much more indisputable than this famous, phantom-like "autonomy." Among these dangers (and in spite of the miniaturization or the "intelligence" of the terminals), the two principal dangers are size and preeminence, which, moreover, are already to a great degree actualized. The first is the concentration of means; the second proves to be much more fundamental, this time at the philosophical level; it is the codification of language, its becoming "modular" and basic. At this juncture, the essential questions come together: "Where will computerized communication end when households begin to be equipped with computers? ... What will become of traditional writing when computerized language, unpolished but sufficient to express the essential messages of daily life, will be offered to us?"[57]

Two readings of this study are therefore possible. One is very careful, authorizing only a very measured optimism, or let us rather say, an enlightened pessimism. A stroboscopic reading is also possible, which, in its haste, retains only the "titles" of the text. The latter will be how the great majority reads. Cleverly, the study will have nearly fulfilled its goal of describing the dynamic of power without having totally sacrificed intellectual honesty for a "happy few," a minority who are not the least bit dangerous.

The dynamization of progress: it is in fact to this end that social *transparency* is extolled, but it is also what is seen as "dangerous" in the eyes of the "pessimists."[58] Who would not, in principle, be in agreement with a claim as "innocent" as this: "The goal should be to support freedom in all its transparency, rather than to preserve the privilege of some in and through obscurity"?[59] A very tempting Manichean vision of the division between shadow and light! At the same time, it is a recognition of the mythical choice of a society of communication wherein transparency would be universal and justice would be distributed like light. We rediscover the myth and its inevitable power constitutively linked to modernity. This link is acknowledged on the preceding page:[60] The "allergies to modernity" are presented as emotional reactions (this is stated) and therefore (this is suggested) obscure to the "general public" (must we once again recall that one addresses the *enlightened* elite, much closer to the ideal of transparency?). Let us throw down the gauntlet: Nora and Minc, in extolling a new social transparency, do not admit, to hear them tell it, of any passion. Is pure objectivity the only choice on behalf of optimism and transparency? It would be contradictory to claim this, since it was acknowledged that there are problems, great dangers, possible choices; it would be equally difficult for the authors to deny that they have minimized the guarantees offered to freedoms by a certain obscurity, insofar as this is always in the name of an ideal of universal transparency, itself produced by the irresistible thrust of the "meaning" of History (the world-development of the scientific-technological interpretation becoming the "transformation" of the world, to borrow Marx's terms, if not his thought). At least here the myth of universal transparency has the advantage of offering itself openly.

From the *fact* of technological universalization to the *myth* of universally diffused transparency, the lock has not yet been completely fastened—what is lacking is the consideration of the properly linguistic character of technology.

Ellul, while acknowledging the "unformulated fraternity" that is established among technicians, also diagnoses a progression of misunderstandings, a disorientation of social forms and moral norms, the reduction of the social body to a collection of individuals.[61] And McLuhan himself[62] is alarmed to see the *media* lead toward a "loss of identity" of the person within the group, and predicts that the United States "will rapidly become a third-world country." To these misunderstandings we must also add religious or secular fanaticism and the ardor of nationalism.

Inasmuch as technology establishes and develops a certain kind of communication, inasmuch as it installs and reinforces the domination of its higher and lower priests, specialists, businesspeople, and PR specialists, its Universalist movement is not univocal. What it universalizes is especially, but not uniquely, its own processes and procedures. The psycho-ethno-cultural destruction that it effects assures its ascendancy, freeing *libidos*, passions, random violence. Is it necessary to push technicalization even further, so that the technological system encloses the whole of society and merges with it, in order to realize the great Saint-Simonian utopia of universal communication through technology? We are not yet at that point. Even if it were the case, this society would be completely different from a Republic of Letters or a community inhabited by the Pentecostal spirit; it is probable that it would be even more compartmentalized, excessive specialization being compensated, at the macro-social level, by a hyperconditioning: in short, there would be an extreme increase in the distortions of communication, which Marcuse has already shown in his critique of one-dimensionality.

Natural languages, vital treasures, have not yet been destroyed; rather they are shaped, even increasingly stifled by a computerized language that is fashioned from its reserves and networks, from its banks and potential data. The efficacy obtained is evidently tied to the rationalization of programs and to the automatization of operations; however, in reference to properly linguistic problems, it is necessary to see that information-systems produce a large economy of phonetic difficulties, because its "language" is "spoken" only by a machine. The material support being—for example—a magnetic tape, we are now in the presence of a language which no longer has the corporeality that weighs upon "spoken language," with its fragile, infinitely diverse, and suggestive opaqueness—but which, from the operative point of view, is also heavy, slow, and unpredictable.

We must go to the heart of the problem to understand why the universalization of technology creates the appearance of a true and integral communication, while destroying its living roots. What is a language? The question resonates in the ambiguity of the French term: between the extremes of language and that of code. That technology is no longer *stricto sensu* a language (assuming that one defines it as "a system of signs directed at communication") is admissible at the level of definitions, but in no way excludes the many imbrications between technology and language. Before examining the latter, let us mark—for the sake of clarity—the difference between linguistic sign and technological object.

Even if one defines language as a "system of signs," the technological object cannot in any way be endowed with the supple qualities of the linguistic signifier. The linguistic signifier, due to its "arbitrary" character, in effect allows for the flexibility of multiple semantic combinations, almost at will: witness the phoneme *in* contained in the French words *pain, brin, pin*, etc. In an impeccable analysis[63] Simondon has shown that the technological object, to the contrary, does not maintain an "arbitrary" relation with the material, and therefore and especially tends toward "concretization," that is to say, toward a functional auto-reference, at the heart of which the parts are more and more interdependent. An eloquent example, in this regard, is the comparison between the first gas motors, noisy and inefficient (in which each element went its own way), and the motors of today which are almost perfectly "integrated."

Of course, the problem is complicated in our consumer society because, as Baudrillard has shown, the entire network of nontechnical connotations (advertising, in particular) refers to the properly functional coherence of technological objects: "The connotation of the object . . . burdens and perceptively alters the technological structures."[64] Thus, the tensions experienced in consumer practices interfere with the stability of the "technological system"; however, this loss of stability, which encumbers the "system of objects," cannot be strictly assimilated into the movement that speech [*parole*] infuses into the richness of language [*langue*].

Having conceded this difference, let us return to the imbrications between technology and language, this time, however, at a more essential level than that of our initial remarks. Many threads connect technology and language—from at least three points of view. First, modern technology presupposes a constant elaboration

and application of mathematical and informational languages.[65] Second, all technological procedures, especially today, do not use speech so much as they use an exchange of signs of agreement, of information, which presuppose a minimally communal project: there is no technology without the beginning of *consensus*—even if
it is as superficial and fragile as the cracked cement of "detente." Simondon notes
that the machine is a "detached human gesture."[66] One could say more generally
that technological activity is a fragmented and materialized form of language. Finally, once technology constitutes itself in this immense dynamic system, where
each forward step is set against the capital of the work and inventiveness of the previous generation, the living language (this time without excluding speech [*parole*],
the practice of language [*langue*]) continues to denote the class of technological objects, procedures, and experiences which are constituted: technology becomes the
object of everyday language, while at the same time at the level of the transformations of projects, mental attitudes, and vocabulary it is to a great degree its driving
subject.

Despite these incontestable and lasting ties between technology and language,
recent research is in agreement in noting how the process of technicalization, in its
massive effects, *desymbolizes* human activity. Baudrillard: "The objective public system constitutes . . . less a language, because there is no living syntax, than a system
of significations—it has the poverty and efficacy of a code."[67] Ellul: "The technological system is a real universe which constitutes *itself* as a symbolic system . . .
Symbolization is integrated into the technological system."[68] This integration is
recognizable at all levels of actual experience, but perhaps most strikingly and
poignantly in the loss of symbolic investments tied to the traditional "gestures" of
work.[69] If it is true that the technological system symbolizes itself, it does this by
carrying its own power to a greater degree of complexity—that is to say, to an abstraction for the ordinary user. Clearly, what the process of technology thus threatens is the relation to language [*langue*] as such, but it threatens also—a significant
solidarity—the richness of embodied life and its gestures.

What has just been established at the level of effects can again be established
at a more sophisticated level by demonstrating the movement from consumption to
production, more precisely, from the consumption of objects to the production of a
code. Norbert Wiener is of help here. Cybernetics, in the strict sense, is the "science
of the command and transmission of messages by men and by machines."[70] To confine ourselves to this would be to wander wisely among other scientifico-technological disciplines, but to no avail and, moreover, without achieving unanimity on this
point in the scientific community. Nevertheless, a reflection on the etymology of the
term *cybernetic*, but especially a reflection on the function of the "command" in
contemporary society and on the vision of the world proposed by Wiener, should
take into account cybernetics in a much more fundamental manner: precisely in the
sense that Heidegger understands it in the metaphysics of the atomic age.[71] Indeed,
when Wiener defines information in order to expose his cybernetic conception of

the world, he echoes and goes beyond the techniques established by Shannon in the field of telecommunications; it is the living organism in its relations with its environment and, finally, all of human society that are the domains to which this new method of intelligibility applies itself, pronouncing itself on the essence of language and taking possession of it: "*Information* is a name used to designate the continuity of that which is exchanged with the exterior world to the degree that we adapt ourselves there, and apply to ourselves the results of our adaptation . . . To live effectively is to live with adequate information. Thus communication and regulation concern the essential part of the inner life of man, even as they concern his life in society."[72]

Heidegger echoes this, in a spirit that is obviously completely different from Wiener, but which makes a statement just as radical with respect to the mutation of language in its relationship to the world: " . . . the representation of human language as an instrument of information always gains the upper hand. For the determination of language as information first of all creates the sufficient grounds [*zureichenden Grund*] for the construction of thinking machines and for the building of frameworks for large calculations. Yet while information in-forms, that is, apprises, it at the same time forms, that is to say, arranges and sets straight."[73] The domination of the scientifico-technological sphere does not manifest itself in principle nor principally by an "eating away" of the traditional "humanistic" language: it is the very essence of language which is overturned. If there has been a seizing of power by the scientifico-technological idea—who could deny this fact?—it is undoubtedly at the heart of *Logos* that it operates, and not at its periphery.

Certainly, one will say; but does not all of this show, at the very level of the (philosophical) production of language as information, that technology has become, if not a language in its actual materiality (it was correct, in this regard, to challenge the *assimilation* between technology and language), then at least the region where the new production of language will take place? When Wiener analyzes the exchanges between an electric power plant and the outside world, he does so in terms of information, which is to say in terms of language: the opening and closing of switches, of generators, etc., "perhaps considered as a *proper language*, having its own system of behavioral probability linked to its own history."[74]

The ultimate in technicalization is less an investment of language by technology than an investment of technology *in* language. To think of technology as the dominant "world language" is not—to repeat—to assimilate technology and language term-for-term, it is rather to recognize a new historical conjunction at the heart of which the instrumentalization of language is the decisive agent of technicalization.

The inversion realized by the cybernetic "seizing of power" is obvious: whereas, until now, in the history of humanity, symbolization has always preceded and even surpassed codification, the latter now tends to become the rule of equilibrium and to encroach upon the symbolic resources of language. A *language* with a living, mysterious, multiple and unpredictable relation to the world is what the techno-scientific *language-code* can never replace. This second language of code is rooted in significations that are more subtle and more fragile; but the dangerous paradox of our world consists in founding language upon its phantom, sacrificing the delicate richness of symbolization to the certain, but unilateral order of Organization.

A language rooted in the obscure and moving abyss of bodies, provoked by singular utterances and unspeakable pauses, a language which is the accomplice of silence, the guardian of secrets, a language which accompanies the most cherished rites of a group from birth to death, a language which—in immemorial myths as well as through the gesture of the writer—carries within it what Merleau-Ponty calls "the miracle of expressivity," I ask whether such a language can continue to live and develop *if* standardized space, manipulable at will, of television, other audiovisual means, and soon of information-systems, becomes the exclusive linguistic and cultural medium.

If I think, with Mumford and several others, that humanity runs the risk of experiencing an unprecedented cultural collapse, perhaps even a loss of its identity, it is hardly because I am reproaching the audiovisual media and its retinue for not allowing a universal communication, precisely because *it is this universal transparency which is the myth of modernity.* Technology achieves perfectly its immanent "goal": the installation of a universally accessible space where information is exchanged at will, (whose counterpart is distributed, at the linguistic level, by the mediating indifference of money). Thus the "phantom of pure language"[75] finds itself pushed to the extreme and realized: the myth of a communication without mystery, the "algorithm" pursued by Western metaphysics.

Confronting this mythical transparency where it infiltrates our contemporary drift, I am not so naive as to believe that ready-made solutions exist, or even—as Ellul thinks—that technology could itself produce the antidotes to correct the dysfunctions which are part of it. What is placed in question is beyond "unforeseen events": it is the essence of "technology"; it is perhaps humanity itself, on the verge of losing the "most precious of its possessions," its living language. No technical means will prevent this "veering off course" which marks thought itself.

One will perhaps find this conclusion at once too pessimistic and too partial. Indeed, if one does not lose sight of the necessary distinctions between the audiovisual media, computerization, and a possible "cybernization," one will equally take into account specific complex reactions, not all of which move in the direction of an erosion and eradication of symbolization. The audiovisual medium is not uniformly accepted. After the first wave had passed, reactions of familiarization and rejection intervened; in addition, the creation of local radio stations, the development of video ("consumed" and produced), of CB radios, and of the use of tapes (pre-recorded, but also autonomously recorded) short-circuited a centralized dispatching of information. At the second level of computerization, one will object that the "markets" left to conquer are still enormous and that a huge educational and apprenticeship effort is left to be deployed, whose task it is to accomplish the computerization of developing societies, then to establish this model worldwide (educational programs via satellite are still only in their beginning stages). From the point of view of the specialists themselves, an engineer recently confided that there subsists a "tinkering" (without doubt ineradicable) in the work of programming, not to mention the imagination—if not poetry—in the development of computer labs.[76] Finally, at the third level (cybernetics), one must not confuse incontestable technological successes (security systems, alarms, etc.) with the problem of a "cybernetization" of society, which remains for now at the level of science fiction.

This series of objections does not take up the question at the level where it had been previously posed. If it is inevitable that divergences appear in the assessment of the effects of technological language, and even if one concedes that significant nuances could be introduced into the rather somber picture that has been painted, two fundamental points cannot be denied: the domain of language is massively infused by technicalization, and that radically changes the immemorial relation of man to symbolization.

Not only is it not surprising that important gaps exist between technological projects and their realization, it is indeed a constitutive trait of technological progress. Technology always sets new objectives: the inevitable distance between projects and results is *potentially* filled in by mythification (the techno-discourse which anticipates and stimulates technical expansion). Partial disconnections and tinkerings slip into this gap, playing an inverse role (but as an accomplice?) from mythification. This still ambivalent dynamic would disappear only if it could realize the complete "lock" of Technology on the whole of reality, a hypothesis conceived and nourished more by the "adversaries" of technicalization than by technologists themselves. The latter do not dream of a *Brave New World* as much as they dream about technologies which are ever better—to infinity. There is only a "danger" (or a "salvation") because the future remains open.

The domain of language is not one zone of reality *alongside* others, as technological representatives would have us believe. Everything takes place there. Everything, that is to say, free access of the human being to its possibilities and, among those, its self-determination.

It remains to understand how and why that which we have until now called sense, consciousness, freedom, risks being swept away by an all-powerful technology, not limited in its project, however, always limited in its effectuations. Nevertheless, it is not from this contradiction that the true *resistance* to Power arises, but from an Antinomy whose constitution, which is no longer technological, proves however, to be rational.

Jonas returns: towards an antinomy of the power of the rational

Yahweh spoke to the fish who vomited Jonas onto the shore.
—Jonas, 11, II

Jonas's trial is the prelude to his salvation, as it was the sanction of his rebellion. Imprisoned in the belly of the monster, Jonas beseeches his God, and it is he who saves him. There is no battle with the whale.

"Only a god can save us," Heidegger says in his posthumous interview.[77] Is this an incongruous rapprochement? One will judge this in light of the question that we would now like to outline. If, at the end of this first attempt, the technological phenomenon no longer has a beneficent innocence nor especially the neutrality that it still enjoys among most of our contemporaries, if it appears to accomplish an age-old project (however recent, even relative to the history of the West) of the practical mastery of nature, if this scientifico-technological complex is in continual expansion without there being any knowledge of where it will stop, then is it necessary to invert the perspective and stimulate a pluralistic resistance to this technological "imperialism"? This first investigation will require us to pose the problem of the *limits* of technological power, perhaps even question whether the representation of technology should be placed under the sign of Power.

What is striking when we attempt to pose this problem of the limits of technological power in all its complexity is that we run up against a tension (limitation? lack of limit?) which is impossible to resolve either in purely factual terms or even in terms of an eidetic phenomenology. The most extensive effect of power is still finite (to take an extreme example, the explosion of our planet, a negligible event on the cosmic scale); and technology deploys procedures which are—by definition—*mediations* (be they as rapid as the speed of light). Technological power is therefore limited. But the inverse is equally true: technological power is without limits in that its limits are continually expanding with no region of reality secure from its advance (from the transformation of the natural environment to the manipulations allowed by biotechnologies, from the new means of psychical control to generalized informationization, etc.); this lack of limitation at the interior of technological power is conditioned and reinforced by Research and the always possible perfectibility of processes and procedures.

A complete inventory of scientifico-technological power would be impressive but would also be never-ending—considering the speed and number of discoveries, inventions, and innovations.[78] Such is not our task. Should philosophy not do better: anticipate progress by arming itself with research on the future and—why not?—science fiction? Because technology is dynamic, a study confined to the *present* seems to be quite obtuse: it is the technology of tomorrow and the future that is of interest, whether threatening or fascinating. Is Gilbert Hottois not therefore correct in requiring us to consider the technological conquests which present themselves, some partially realized (the constitution of a "technocosmos" and of "technocracies" without anything in common with the traditional environment and the usual perceptions of human beings), others on the imaginary fringes of Research and Development, at the limit of science fiction (for example, *Genetic Engineering*, the constitution of "Cyborg," astro-genetics, etc.)?[79] Here we have, as far as I know, a very useful, even indispensable work.

Perhaps boundless in its potential expansion, technology will remain limited by temporal-spatial conditions of deployment and, fundamentally, by the physics of the cosmos. But even if the boundaries of what could be conceived and realized (or not) were indefinitely displaced, the problem remains the same: even if a gap should always exist between technological power and absolute omnipotence, technology, at our level, has already attained its aims of omnipotence. Confronted with the incalculable, we have identified the Power which is able to destroy the planet, manipulate life, control consciousness, and condition language.

The result is that the accumulation of surprising facts and astonishing potentialities no longer has any significance from the moment that one *makes a fundamental assessment* of the technological phenomenon as a specific Power in the process of expanding, transcendent to all prior human realizations (which remain relatively anthropomorphic), leading to a surplus of power that does not seem to be oriented toward any finality.[80]

The antinomy, at once factual and constitutive, between the limitation and the limitlessness of technological power, is, in fact, *for us*, more apparent than real: it resolves itself phenomenologically in an indefinite expansion, a perpetual narration of the frontiers of the possible, an *effective lack of limitation* (whose limits in principle become simply conditions).

But this lack of limitation is not yet the true limitlessness of technological power, just as the phenomenological tension which was just analyzed does not lead philosophy to the heart of the *formal antinomy* which shatters the power of the rational in the era of technology.

The limitlessness of technological power becomes a formal principle when it is connected to the *internal logic* of scientifico-technological development considered not from a solely empirical angle, but as a law governing the conduct of the "agents of progress," sweeping away every moral law. "Everything that is technologically feasible ought to be realized". This rule, which I will for the sake of ease call *Gabor's rule*,[81] constantly recognized *de facto* by the writers who have studied the technological progress of the West, remains empirical and relative as long as one considers it as the very principle of the functional, which resolves the *phenomenological antinomy* between limitation and non-limitation. The "technologically feasible" is not just anything, but obeys physical limits as well as a selection of possibilities. It "ought to be realized": this does not denote a formal, absolute duty, but a probability akin to certitude, according to which the technician will be tempted to add conditions—profitability, facility of circumstances, etc.

Gabor's rule becomes formal, a principle acting as a *thesis* in a true antinomy of the power of the rational, from the moment when it confronts, as a counterprinciple, the very principle of morality such as Kant himself definitely gave in his description of the rational (the self-determination of the will as the condition of practical rationality). One will forgive me for presenting this antinomy in a formulation which is so apparently scholastic; it is so formulated by a desire for clarity, not as parody. And if I slip in a joke about a Jonas who takes Hans for his given name, it is because I hope that a bit of irony in this analysis will not jeopardize the fundamental seriousness of the debate.

The antinomy of the power of the rational in the era of Power can be thus formulated:

Thesis: Everything that is technologically feasible ought to be realized, whether that realization be judged morally good or condemnable.

Technology itself cannot, by definition, ask itself moral questions. It is "neutral" in the sense that it does not appeal to ethical considerations (either Good or Bad) in the conception and effectuation of its operations. The only factor that is considered is the "feasibility" of the operation itself, as concerns its formal conditions and the materials necessary for realization (procedures, materials, labor, etc.).

Antithesis: Nothing which is technologically feasible should be realized if it gravely imperils or suppresses the ethical capacity of humanity.

By "the ethical capacity of humanity" we understand the constant possibility to decide whether an operation (or an action) is bad or good according to conscience, which is to say the free determination of the will as practical rationality.

If a technological operation gravely compromises or suppresses this capacity (by physical or chemical constraint, or any other form of coercion), it can only be considered morally and even *absolutely* wrong (because it not only commits an act which is relatively condemnable, but attacks the very roots of morality—freedom).

Consequently, no technological operation ought to be accepted (or "condoned") if it attacks the very principle of morality, namely humanity in its freedom.

Remark I: Hans Jonas proposes this fundamental ethical law: "Act in such a way that the consequences of your action are compatible with the permanence of an authentically human life on earth".[82] This imperative conforms to the ethical *thesis* (presented here as an antithesis to the unconditional affirmation of "Gabor's rule") and to the spirit of Kant's practical work. It reformulates the Kantian categorical imperative at the heart of a new problematic of the *conservation* of life. Indeed, in Kant's time, it was not a question (nor even conceivable) that life and the human psyche would be technologically manipulable to the point of being gravely or fundamentally (and massively) compromised. Nevertheless, this displacement (which restricts the application of the moral law to "essential crime"—without considering other actions) does not in the least alter the spirit of ethical "resistance." It is a radical perversion—more insidious than any specific evil—that humanity must now face. And in this very perversion, the incontestable character of the moral law in its essential principle is illustrated.

Remark II: The Law of Jonas does not concern an individual subject in its daily life, but humanity as such (or a responsible individual representing humanity), confronted with an aggressive extension of the reign of Technology (linking science to "mercenary arts"). In the new Technological/Ethical antinomy, the place that was occupied by *nature* (according to Kant) is taken by Technology (becoming in the proper sense the *determining* factor), and the impulse of the anti-ethical motivation passes from hedonism or perversity to efficiency. The greatest "enemy" of the moral imperative today is no longer the desires or the malice of a person, but the efficiency of humanity (itself technicalized).

Remark III: From a Kantian point of view, one could, however, raise the following objection to the Law of Jonas: the reformulation of the categorical imperative is accomplished by virtue of *empirical modifications* with respect to human life. The restriction of the sphere of application of the aforementioned law to operations that endanger the permanence of an authentic human life on earth appears as the *last line of defense* of the ethical (but is, in fact, a casuistic concession to every technological abuse). If it becomes minor to treat the other as a means, except in the limit cases where the essence of humanity is threatened, will it not be tempting to perpetually push away the limits of the tolerable? For example, was Truman's decision to bomb Hiroshima and Nagasaki incompatible with "the permanence of an authentically human life"? Can we not ask the same of Hitler's gas chambers, etc.? In these cases, the Law of Jonas would not only be useless (by virtue of the methodological operation called Occam's razor), but perhaps even harmful (because it is a source of confusion concerning the essential).

No conciliation between the logic of the development of Power and the moral Law is possible, because there is no common measure which governs their respective norms. This antinomy rationally ordains our impotence and distress before Power—which we have seen imposes itself *as a destiny*.

Principally, there has been nothing new in this regard since Kant. It is on the empirical level that the tension between our impotence and Power seems (with the potential for the destruction—or mutation—of life on earth) about to cross a qualitative threshold such that this desperate situation requires some sort of an *urgent plan* for humanity. Environmental politics, moratoriums on nuclear matter, legislative "safeguards" (which still oppose complete manipulation of genetic material) therefore ought not to be treated with disdain. Indispensable intermediaries of ethical exigency, translating that exigency into the dominant world-language of Power, they have value, however, only as makeshift measures. It is this which we would want to recall in critiquing the "Law of Jonas": one must not believe that a "strategic retreat" into the conservation of life in any way changes the *formal* nature of the question.

To remain with this formal antinomy would be, however, to infringe on my initial project, which was to follow rationality in its effectivity and to explore its possibles. In search of possibles that perhaps exceed the effective and effectuating power of the rational, it is still necessary to examine this power. In replacing phenomenology with a philosophical genealogy, we will see whether the power of the rational is able—through its internal differentiations—to help us better confront its fundamental antinomy.

PART II

What Power for the Rational?

> At the threshold of a great new country, without title *nor* device, at the green threshold of a great bronze country without dedication or date,
> Lifting a finger of flesh in the rush of the wind, I question, O Power! And you, take note that my inquiry is not commonplace.

—Saint-John Perse, *Winds*

AFTER HAVING CONFRONTED Power in its brutality, I noted the necessity of critiquing in it the masks of technology, and of more clearly discerning its specificity. Now it is necessary to go further: diacritical phenomenology must pass the baton to a genealogy of technologically powerful rationality. Is Power that of the *rational*? And how is this possible? I will attempt to respond to these questions by first understanding the phases of potentialization.

Clearly, this investigation cannot avoid the question of the contemporary status of science. If Power essentially expresses a phase of rationality, is science entwined with the later? Science does not seem, in its original impulse, to interrogate *power for the sake of power*; however, how can we deny that it *finds* it more and more with each of its activities? Engaged in a *destiny of power*, which at times it still shields itself from investigating, is science nonetheless capable of being proposed to humanity as a new Alliance? And whence comes the ardor of some who anticipate this Alliance, diverting their eyes from Power?

No longer nourished by illusions about soteriologies, if they are scientists, the answer will only serve to intensify the efforts of thinking, reoriented here according to three paths: absolute power, critical resources, and the path of the ontological secret. And it is in the test of a dialogue that each will make the most of his or her chances.

5

The Phases of Potentialization

Ambiguity of power, dynamic of potentialization

BEFORE EMBARKING ON the concept of potentialization itself, we must recognize a terminological ambiguity that would be easy to minimize, but whose richness and interest will be discovered bit by bit. There is power [*puissance*], and there is power [*puissance*]. One is not saying exactly the same thing when one thinks power as that which *effectuates* [*s'effectue*] and as that which holds itself in *reserve* [*se réserve*] (in order to effectuate or not). Usage, however, unites these two senses. Is it merely a linguistic deception arbitrarily inherited from the Greeks, or the sign of an essential connection?

What does one mean by *power* [*puissance*] in the expression "Science for power"? If power is taken only in the sense of the possible, one then faces an excessive generality. It is at once incontestable that science brings about the possible, yet untrue that it restricts itself to this. One would not see, in this case, what would distinguish it from total invention or total creation—including the order of fiction. If power is meant only as a coercive type of domination, then the expression is still too general. We have seen that there is domination only by differentiation. It is therefore not power in general that would constitute in that assumption "the imperialism of science": it is a well-determined sense of power that must be exhibited.

The two meanings—the reserve of the possible, the effectuation of the act—are co-present in the expression "Science for power," as well as in potentalization such as it occurs here. The two meanings communicate, nevertheless, thanks to a determined sense of the possible that the philosophical tradition receives from Aristotle: *the virtual*. If science potentializes, it is not in an immense and vague universe of possibilities; it starts from an *energeia* that has already excluded by its own position many other possibilities, and which, nonetheless, by its *theoria* opens new *perspectives*, which is also to say new technical and practical possibilities. The act, for Aristotle, is anterior to power. "It is not to possess the faculty of sight that animals see but it is in order to see. . . . "[1] The virtual manifests this ontological primacy of the act: it is weighted with its effectuation. The virtual is, in a large sense, "that which is already predetermined, although it does not appear beyond itself, and which contains all the essential conditions for its actualization."[2] It is therefore already an actuality, but suspended; it is a guarantee of accomplishment, not an indefinite inflation of the possible.

In the case of scientific potentialization, one has to deal with a sense of the possible that is already actualized, as well as with that which is in the process of actualizing. Whence the current meaning of the adverb "virtually" as "almost." "This decision is virtually made" means exactly the opposite of a purely hypothetical exploration of eventualities; in fact, the decision *is going to be made*. The sense of potentialization that is our principle concern here—that of modern science—is completely centered on actualization.

It is not without interest to note that the expression "Science for power" would hardly have made any sense to an Aristotelian or a Thomist. St. Thomas writes, "Everything that receives something from another is empowered by relation to that other."[3] Thus, pure intelligence, holding its existence from God, is potential in relation to the pure Act. Science is first of all science of the Act or of being; it is the science of power only because it must recognize composite beings. But in no case could it be science *for* power. It would be, for its part, contradictory to subordinate itself to a lesser being (such as would be the case if it subordinated itself to a physical power, limited by definition).

Modern science inverts the meaning of the scholastic phrase: *operari sequitur esse*. The *operation* does not subordinate itself to an anterior being, superior in itself. It releases, for example, a universally valid law whose applications do not cease to confirm its necessity and efficacy. The appearance of the law releases a *region of potentialization*, that is say of new powers: virtualities turned not toward an anterior being absolute and in itself, but toward new results that will be obtained by the enactment (and in the application) of scientific explication.

Through this, one grasps the limits of terminological clarifications. It is necessary to distinguish the first meaning of power (possible) from the second meaning (effectuation), while fully acknowledging their link in linguistic practice. While it is necessary to identify the virtual at the junction of the first and second meaning, this does not suffice. These distinctions remain formal as long as one does not restore them to the ontological horizon that they presuppose. When Aristotle gives ontological primacy to *energeia*, it is at the heart of a hierarchical ontological order run by the Prime Mover. When we moderns think of the power of the rational according to its effectuation, this unfolds and increases in a horizon which is indefinitely open, without God or rule. The meaning of the possible changes with that of being.

In turn, the terminology follows the ontology when one coins an expression according to a new reality. This is the case with *potentialization*, a concept which renders particularly well that which René Char calls "science-action," but whose usage has been enlarged. When nuclear physics makes the fusion of the atom possible, it potentializes. When Boltzmann introduces statistical calculus in thermodynamics, he potentializes. When René Thom conceives catastrophic theory, he potentializes.

We see cited here, intentionally, quickly, and without chronological order or classification according to intrinsic scientific importance or technical impact, *discoveries* of a scientific character that contribute, in infinitely diverse degrees, to the enrichment of the already formulated possibility (the possibility that we possess, the scientific *corpus* at large) and to the transformation of the world (even in the case of the theory of catastrophes, which, according to Thom, has already permitted the perfecting of the equilibrium of American bombers). Here we rediscover the first and second meanings of power, but focused on the creation of innovation in the first meaning and the effective (technical) realization of the virtual in the second meaning. The onto-theological hierarchy having disappeared, the possible is no longer thought of for itself according to an eternal in-itself, but according to its productivity and its *exploitable* creativity (such is the "logic" of Research). In the techno-scientific complex, potentialization takes place twice: by a systematic exploitation of the possible and by an accelerated realization of the virtual. The first and second senses of power *fuse* in *potentialization.*

Certainly, a lover of ideas, a surviving Platonist, a thinker clinging to "disinterest," could still practice, even in techno-scientific operations, a purely speculative *epoche.* One could indeed isolate a theoretical object, for example, a "truth" of the fusion process in order to examine its theoretical contribution, to record it in the world of mathematical concepts or theoretical physics, and so on. One could also return to Aristotle, finding there a meaning of the possible that would allow one to think both the possible contingency of the future and the continuity of capacities such as one knows them (and, by consequence, maintain that they already had potentialization—*in a sense*—according to Aristotle). It is no less true that, even in meditating on future contingencies, Aristotle was far from detaching science from its principle orientation toward the constant and the supreme. Yet anyone returning to pure theory is not able to suspend the forward march of modern science with its potentializing drive.

Are the characteristics of scientific potentialization thus standing up and being adequately defined? This *dynamic possible* ought to determine itself further by being exhibited. In particular, precise analysis ought to make clear the role or the function of this placing in reserve (and of this preparation) of the effects of power in and through the capitalization of scientific knowledge.

Potentialization in its principle and in its phenomenality

The reader is now entitled not only to further knowledge about potentiality, but also to understanding what it is and to see it in some form of action. If one were to demand point-blank, "Show us potentialization," we would respond, "What is presentable, at least directly describable and analyzable, is such and such effect of power, or, more generally, the effectuation of power. Potentialization being the origin of the aforementioned effects of power, it is not *more* phenomenological than this." Potentialization, therefore, does not occupy an undetermined, unassignable theoretical place. It is not thinkable either in itself or for itself, outside of all total effectuation of power—otherwise, one falls again into a pure possibility, removed from any realization. However, the fact that it is not amenable to a pure eidetic analysis does not forbid a genetic analysis that reconstitutes it and permits an understanding of its eventual engendering, starting from episteme.

Grasped in its most general sense, potentialization designates the process of the engendering of power. What is of concern to us is the principle of scientific potentialization. We want to know if there is a connection between epistemological growth and the "explosions" of power that testify to the history, especially recent history, of science and technology. While this connection has been postulated since Greek science, this does not, however, minimize the significant threshold crossed at the dawn of Modern times by the "revolution in reason." I do not deny, as we will soon see, that technology could have a moment of autonomous development or even that essentially technical inventions could have had scientific consequences, but these multiple comings and goings, these "exchanges" between science and technique, do not appear to compromise this preliminary outline of potentialization in its general sense (and in its effectivity): it is science itself (starting from the moment in which it achieves the coherence, the apodicticity, and the universality of episteme) that crosses an absolutely decisive threshold in human potentialization, rendering *operations* that had been unimaginable, inconceivable, and unknowable before this epistemological penetration, *effectively possible* and virtually realizable. Of course, science does not do anything but potentialize purely technical effects, following a unique thread and according to one sense alone. It potentializes initially and above all in the first sense: it secures its own storing of information [*mise en mémoire*], its own reinforcement (here I am thinking in particular of the constitution of the Euclidean geometrical *corpus*). This constitution, which one ordinarily interprets as purely theoretical, is *already*, according to my hypotheses, the *intensification of a field of power* of Western rationality. This is, however, admissible only under two conditions: First, one must remember that this hypothesis is retrospective; one must not presume to attribute to the Greeks a dynamic sense of the possible which would have been foreign to them.[4] Second, one must understand that the holding in reserve [*mise en réserve*] of its own potentiality by Greek rationality is an excellent example of that which could be called the "principle of distance" between each scientific discovery and its effects of power. This example is obviously exceptional due to its radicality, insofar as Greek rationality was not allowed to be "contaminated" by effects that it was not incapable of conceiving. This principle distance was therefore set a priori, not contingently.

Is the principle of potentialization episteme?[5] Yes, but not entirely, since we have just seen that Greek rationality essentially potentializes in the first sense. That is to say that it still "lacks" something such that the second sense of potentialization is able to take it and make the power of the rational dynamic for its own benefit.

This "supplement" that reinforces potentialization reflects it upon itself by carrying it to a power (without limits) that up to this point was unknown. Is this supplement a particular scientific discovery: differential calculus, for example? Even if the most specifically modern potentializations (the example cited and, more generally, the principles first of Galilean and then Newtonian physics) appear to have had a transcendent influence in relation to isolated discoveries; none among them can be substituted for the method itself, nor can they be understood without it.

"It is the scientific method, insofar as its specification is highly reflective and self-controlled by the rational method, which is at the base of the common dynamic that spans the properly scientific field as well as the technological field."[6] Jean Ladrière hardly wishes to explain by these words a point of view that is particular or partisan. Rather, he wishes to formulate a universally acceptable position, as he must in a study commissioned by UNESCO. Rather than simply focusing on the "common dynamic" as that which unites science and technology (a point which appears to be agreed upon),[7] he rightly insists on the "highly considered specification and self-control of the rational method." There is *specification* because the method, far from invalidating ancient rationality, bases itself on its rules and takes inspiration from "the clarity and evidence" of mathematical reasoning: it is this which allows for the introduction of the concept of a surplus of rationality in regard to modern science. There is, furthermore, "self control" to the extent that the rules of the method or, more generally, of the scientific spirit are optimized through a constant readjustment between the results which have been obtained and the ideal of objectivity (itself based metaphysically and implicitly on the *cogito*).

The Method is effectuation, but it does not effect anything *in particular*. It makes possible a new regime of power. Manifested here is a decisive characteristic of scientific "progress": from the moment where epistemic potentialization transforms itself into operative potentialization, science integrates the increase of power into its own project as a strategic objective, and the related effects, inside and outside of science, are without common measure with the previously obtained "results."

Is this to say that operative potentialization can be conceived and organized without being based on the first type of potentialization? In historical terms, did Bacon, Descartes, Galileo, Newton, and the artisans of modern science "need" Greek rationality? Have they not returned to ground zero? If it was at all worthwhile to pose the question, then the response, at least in principle, should carry very little doubt. Is it possible to measure the loss if these great minds had found themselves without the framework that is solidly constituted by Aristotelian logic, without the mathematical models of Euclid and Archimedes? The modern beginning is a recommencement: the exploitation of the possible, its "virtualization." Mystery of the inefficacy of the "Greek miracle"? Miracle of the efficacy of some principles of Method that are almost too clear? Without losing any of their prodigious and admirable character, these two great moments of the West appear difficult to dissociate if they are thought in terms of an intricate (but differently emphasized) unity of the potentialization of the rational.

The four phases of potentialization

With the principle of potentiality having been posed, as well as its phenomenological reality imposing itself historically—along the great threads of the scientifico-technological effectuation of modern history with its irreversible institutionalization (Research as concept, but also and especially as social practice)[8]—it becomes urgent to give an overall view of the enormous process of potentialization (within and beyond History), by definitively putting into place the necessary distinctions that have been enumerated. Indeed, to better understand in what respect modern science *potentializes*, the preliminary semantic distinctions and methodological precautions no longer suffice. A more rigorous determination of the concept of potentialization is necessary, which makes more clearly apparent the different significations and their modes of historical (or transhistorical) deployment.

Potentialization is not strictly identified with any of the already isolated meanings of the possible. It leads to the effects of power, but is not reduced to them (except in the extreme condition when a nuclear apocalypse totally annihilates all the potentiality that it produced). All the "mystery" of potentialization resides in this: it proceeds from the possible and returns to it. But if it is only an enrichment of the possible, what role shall we assign to actualization and how shall we explain the "intensification" of the possible? For example, is the self-referential growth of potentialization similar to the process of capitalization, which increases the sum of money through the mediation of production and the exchange of merchandise according to the scheme A-M-Á? The reference to capitalism is even more enlightening in that capitalism is referred to as the process of potentialization, not the contrary. Potentialization is ontologically and economically much more originary than capitalization, an essentially modern phenomenon. Inversely, however, there is an ultramodern meaning of potentialization, in relation to which the capitalistic law of value appears anachronistic. How can this aporia be reconciled?

First of all it is necessary to give up the retrospective application of *a single meaning* of potentialization. This retrospective application is not an achievement only of the great philosophies of History (of which Hegel is the *paragon*), but also of all exclusively ideological theories.[9]

It becomes evident that one single response cannot be given to the haunting question that remains: "How is it that a new possibility appears?" Consultation with history, as well as reflection on the diversity of potentialization, diverts us from a simple response. Our point of departure was from the immanence of experience and I do not intend to leave it; fire, theorems, steam engines, and atomic explosions are examples of the possible that *virtualizes*, accomplishes itself, and engenders other possibilities. Without being reductionists, how can we account for all these processes that are so heterogeneous in relation to one another? The principle verification of the *growth of power* is given in its immanence. Invoking neither the absolute, nor being, nor even humanity as subject, this analysis intends to take only a *minimal approach*, an approach that is, however, already diversified fourfold by a global process that seems to reveal itself in a unique word: *potentialization.*

The distinction of four meanings of potentialization is imposed by a pleiad of statements concerning the heterogeneity of the effects of power seen as changes in the very nature of "power." The only general "law" that I allow myself to give—starting from the statement concerning the growth of power—is that there is a *proper logic* to the virtualization of the possible, according to four "moments," which, for the following reasons, I choose to call *phases.* The word *phase* (which still resonates the Greek *phaïnein*) links *appearance* and *development.*[10] It is certainly the case here. Each mode of potentialization is phenomenologically knowable.[11] It has, however, nothing of an atemporal idea, since it presents itself not only in time, but dynamically as a holding in reserve and/or as an effectuation of power. The phase supposes a minimal unity of appearance and disappearance. As a state delimited at the heart of a general development, can the *phase* itself be equated with an epoch? Not at all. Phases do not necessarily unfold completely at the heart of History. Moreover, they do not proceed as encapsulated unities, as do Hegel's world-historical "principles." Without doubt, they appear successively; no necessity, however, "suppresses" the first when the last has appeared. They remain today simultaneously possible and, to differing degrees, "real." They do not completely cease to potentialize, even if their "dominance" is unequal and tends to press catastrophically toward the maximum releasement of power (and all which it implies). One must be willing to excuse the brevity of the presentation of these four phases. It is justified by the desire not to stray too far from the central concern: contemporary scientific potentialization.[12]

Phase I. There is no focus on power *as such.* Potentialization is essentially *plural*—potentializations *take place.* The effects of power inscribe themselves more or less directly on the natural environment and on human society. Potentialization can be cumulative, but it radiates in different directions and reflects like a "halo." A systematic storing [*mise en mémoire*] of potential is impossible; it is too difficult and always incomplete. There is no capitalization (the growth of power ignores the systematic profitability of time), but only reserves that are either temporary[13] or difficult to evaluate, for example, the earth itself[14] and precious metals (and on another level, the acquisition of the designs or techniques of artisans). The rationality that manifests itself here is principally that of an astute[15] and/or classificatory[16] intelligence. The relation with History is either nonexistent or completely lateral (when this type of potentialization manifests itself at the interior of a world marked by the other phases.)

Phase II. There is not always a focus on power *as such*. Potentialization finds its unity in episteme which sees itself as disinterested, whereby the power of effectuation, almost totally *reserved*, remains unperceived even to the eyes of its agent. The effects of power that are directly visible in the natural environment and even in society prove negligible. This phase is essentially *potentializing* [*potentialisatrice*] (in the first sense mentioned above)[17]; it accumulates absolutely new possibilities without effecting them outside of the site at which they appear (the *theoria*), but places them in a narrative and arranges them in a *corpus*. This growth (potential) of an extraordinarily reserved power is made without a calculable relation to time. There is here no self-valorization of "capital," but a constitution and enrichment of the self-referential space of a new and apodictic scientificity: a deployment of "pure rationality"—mathematical, logical, philosophical. The relation to History is (potentially) not nonexistent, but proves itself to be incalculable.[18]

Phase III. Power as such has a thematic importance: the ideal is that of a systematic mastery of natural reality and society. Potentialization does not realize itself in a simple holding itself in reserve; it becomes directly *effectuating*. At the highest rank, thanks to the mathematization of the laws of physics, the Method is the universal operator of power. The effects of power are precipitated and systematized in this effectuating potentialization, which redoubles itself by storing in computerized memory scientific acquisitions, as well as in the transformation/exploitation of the natural environment and society. The mathematical mastery of time accompanies on the economic level a research which is being pushed more and more toward the profitability of work and, finally, the "rationalization" of all practice. Capitalism develops and functions as the *economic accelerator* of effectuating potentialization. The rationality of this phase is characterized by the methodological growth of rigorous scientific knowledge and by its positive contribution to technical progress. It is the height of rationalism. This phase cannot be found outside of the history of our modernity.

Phase IV. There is an exclusive focus on power as such. The effects of massive and extreme power become the exclusive goals of rational activity. Potentialization inscribes on its horizon the limits of all power: self-destruction. As the "hyper-telos" of the power-effect, this phase tends to exploit to its own ends all the "former" potentializations, primarily and principally science itself, but at the same time it places them in peril (erasing the inventiveness of Phase I, paralyzing the epistemological disinterest of Phase II, and even transgressing all profitability and all laws of value in Phase III). Through the self-appointing of new techno-scientific complexes, the rationality of this phase seeks to be always *operative* in a constant movement of homogenization between science and technique. It is also catastrophic, manifesting, in spite of itself, the limits of the rational through its complete reversal into the irrational.

This outline of the four phases of potentialization is only schematic. It announces or names justifications and illustrations at the crossroads of multiple facts and associations. I have not wanted to lay down a set of dogmas, but rather, to find in these phases so many *"grids of intelligibility"* in the appearances of potentialization.

Before setting forth the "illustrations" of the different phases, let me quickly show the play of associations and differences that regulate and attest to the coherence of this schema. Starting with the last factor, the relation with History, it is necessary to remember that a phase is not an epoch. Phase I ignores History, but comes to challenge it even in our citadels of "civilization." In asserting that we have not passed definitely beyond this stage, I am thinking here not only about the rare and fragile survivors (from South America to New Guinea), but about the fact that this type of potentialization has never really stopped in our lives or in our cities, even if there remain at times only microscopic residues. There is always that which Husserl has called an infrahistorical (or ahistorical) "stratum" within human history itself. It is not a bad idea to remember that these "wild" herbs continue to exist and that it is perhaps these which allow for survival.

On the other hand, if one wishes to identify Phase II historically, it is evidently necessary to respond that it corresponds first and foremost to the "Greek" miracle. It is, however, not restricted to it. The survival of epistemic disinterestedness in any activity is always Phase II: despised, threatened, stoic, but all the more unique. If Phase III is the historical phase par excellence, is it not necessary to associate it with chronological limits? It is evident that the modern epoch sees a surge not only in methodological science, but also in the *historical* consciousness of itself, drawn, of course, from the great philosophies of History. This is the phase of triumphant and ethnocentric rationalism. Having recognized that, where are we *now*? Still in Phase III and already in Phase IV (Phase IV sharing the same border as Phase III). The methodological instruction engenders an eminently conflictual situation which can be stated as the following: We leave History, in the sense that the complexity of our situation can no longer be judged by an epochal principle.[19] Whence arises the strange complicity between Phase I and Phase IV, such that it is able to flaunt History.

Rationality? Here we already find ourselves far from its triumphant presentation. But in the face of the "impotence" of Phase IV (if indeed the power of absolute destruction can be equated with impotence), it does not suffice to declare a tragic or disillusioned return to the power of Phase III. We must not underestimate the irreducible plurality of the language of the rational. The modern fascination with apodicticity has too often forgotten this. Likewise, too often it has been believed that the triumph of a certain Reason was definite and absolute. The denotation of Phase I signifies not only that "primitive" thought survives (in the "good" as well as the "bad" sense, my purpose hardly being to judge that), but also that it perhaps always *potentializes*. Is it necessary to exclude that rationality could be, in its own way, a reserve of powers (*dynamis* in a sense more archaic than modern) and at the same time a reserve of the possibles?[20]

If potentialization constitutes Ariadne's thread that unites the different phases, it nevertheless does not pursue this without ruptures so radical that power itself is transformed. In Phase I, physical force, moral energy and the technical capacity of an individual or a group dominate: nothing is more pregnant than this power for the contemporary world. On the other hand, Phase II proves to be of a completely different nature. We make an extraordinarily qualitative leap whose effects—constitutionally different—obey the "principle of distance" that I could formulate as follows: "The effects of an epistemological "invention" are not directly visible in the formulation (and comprehension) of this discovery. They reveal themselves in threads which are extremely different from one another, and which are chronologically distant from the initial event." This is a distance that, without disappearing, is much less clear in Phase III, since Phase III is characterized by a systematic project of effectuation that—without controlling or previewing all the effects of power— nevertheless thinks possibility a priori. Phase IV causes this distance to disappear: effects of power seize all that was in reserve.

Given these clarifications, the question of the nature of potentialization can—it seems—be specified starting from illustrations of the modes of potentializations.

Priority of technological development (Phase I)?

Paradoxically, the first detailed explanations concerning Phase I will be given principally from relatively recent examples, taken from eighteenth- and nineteenth-century histories of science and technology. One would expect, obviously, references to prehistory or, at least, to the potentializations of a *homo faber* close to the "life-world."

A fundamental difficulty presents itself: if we attribute to Phase I the dominant character of a priority of technology, is this not a surrender to a classic *retrospective projection* of technicalization from the "era of technology"? The classification effected by human paleontology of materials and types of tool-making appears to be the most ordinary and less easily avoided by this retrospective projection. But recently this projection was condemned in a phenomenological enterprise in which it is even less excusable: the priority Heidegger gives to implements (*Zuhandenheit* thought as more original than *Vorhandenheit*) would be, according to Hubert Dreyfus, the sign that the author of *Being and Time* was still under the sway of metaphysics, which is to say under the sway of technological thought (in the sense of *Gestell*), whose presuppositions he would later critique.[21]

How may we escape this danger? We must understand that this danger can solidify itself in two diametrically opposed fashions: either by a hypertechnicalization of the "primitive" or by its "infratechnicalization." To the objection raised from the point of view of the first hypothesis, I respond that Phase I is not an epoch, either positively or by default. The mode of potentialization that we isolate phenomenologically as originary does not concern only prehistoric man, but is much more recently perceptible and is undoubtedly still with us today. The technicalization of the mode of potentialization in Phase I is, therefore, nothing mysterious nor phantom-like; it can be known from simple and accessible examples. But this does not disarm the objection; to the contrary. The response must admit of a second, more radical aspect: the expression "era (or epoch) of technology" is confusing. It is tempting to argue that this expression is taken up by Heidegger, even though he admits that the essence of modern technology "has nothing to do with technology." This lesson deserves to be retained on two counts: from the viewpoint of the contemporary world, to avoid assimilating the specific mode of potentialization of Phase IV to the use of instruments or even to the hypertrophy of "bodies grown large beyond measure" by machines; and, from the viewpoint of Phase I, to rediscover the specific sense of an instrumentality that does not respond to the injunction of a methodical will to power and that does not insert itself at all (or only by accident) into the network of techno-scientific operations.

As to the second aspect of the same danger, it consists of a type of inversion of the retrospective projection, an idealized projection of Phase I, which would be assimilated (as with Rousseau) to an anterior state of nature of violent encounters and established domination. Such is not our purpose; again, it is not a question of making an ontological judgment on an original state, its dominating traits, or its "*laisser-être*." Of singular importance in the present work is isolating the *types of potentialization*, starting from a minimal formalization that does not at all prejudge the anachronistical assimilation of the modes of existence or of consciousness. Hence, the recognition of the limited potentialization permitted by "skill" in ethnological or historical contexts, which are very different from one another, allows for a formal ordering of elementary facts (effectuated solely from the point of view of potentialization).

How can this minimal formalization articulate itself? This is accomplished by recognizing, through a juxtaposition of operative discriminations from other phases, the *plural* character of the potentialization in Phase I: works (or "effects") are obtained by ingenuity, which seeks areas of domination (and not in view of a universal and definitive power). This recognition can appear to be too general; in fact, this is the case, but it has the advantage of disarming an important objection and, by the same token, allowing one to think the type of priority that one must learn to recognize in technicalization.

The objection is the following: the theory of potentialization, through the fact of its generality and its philosophical character, privileges the influence of abstraction and science (especially mathematics) at the expense of truly technological facts; it is, therefore, a theory that errors in its idealism, even if it is no longer a question of absolute idealism, but of an "operative idealism" that reconstitutes the ascent of progress around the principle elements of science itself, ballasted by a (metaphysical) will to power which is more and more explicit (in Phase IV).

This objection can be supported by precise and apparently conclusive facts, to only mention the scientific and technological history of the "industrial revolution" and the expansion of mechanization. Whether or not we systematize this history in the manner of Simondon (discovering an evolution with "ups and downs," which modified first of all the technological elements at the heart of the former ensemble and then the technological individual, leading to the formation of a new ensemble and so on),[22] we can agree with his independence and even his caution, in order to relativize the importance of scientific progress. Thus, in perfecting Newcomen's machine, Watt, behaving in a purely practical manner, concentrated on the problem of condensation and steam and benefited in this task from a purely technological perfection (Wilkinson's cylinder) permitting a watertight seal between the piston and the motor.[23] Thus, one century later, Gramme, a pure technician without a theoretical base, developed the first industrially usable dynamo.[24] Over the course of the same period, we could also cite the empirical development of carbon coal and of cellulose, which preceded the emergence of organic chemistry,[25] the various and intensive utilization of ferrous metals, upon which the theory of the resistance of matter is based,[26] the invention of alternating current, etc.[27]

I have purposefully used examples that are relatively recent, so as to allow no easy terms. It is evident that the more one goes back in time, the more this priority of technological progress clearly affirms itself, to the point of being the single factor of potentialization; one can still discuss the respective role of science and technology in the birth of the great discoveries and inventions at the end of the Middle Ages and the Renaissance. But in contemplating an Egyptian bas-relief dating from 2400 B.C., where the processes for making bread and beer are described in detail, how can we deny an empirical inventiveness—two thousand years before the conception of *theoria* as such—revealing the heritage of the ingenuity of the great Neolithic inventions?

Not only is this "non-theoretical" inventiveness not denied here, it is explicitly recognized and thematized in that which I call Phase I, constituting immemorially, since the origins of humanity, the sum of potentializations, which perhaps is not improperly called *culture* in its broadest sense. Can the most recent technological autonomies and priorities be ranked under the same rubric? From the point of view of content, certainly not. Our rapprochement is operative only from the *mode* of potentialization. Let us consider in this regard the following remark by Maurice Daumas concerning the inception of the great chemical industry, "The innovations are the result of the skillful activity of *ingenious moments*."[28] The emergence of type II and III potentializations has not suppressed an apparently anachronistic potentialization. There are always particular, unforeseeable, and precise factors, especially in the experimental sciences. Only the "survival" of Phase I allows us to understand the effective evolution of scientific-technological progress and to disarm the above-mentioned objection.

It goes without saying that type I interventions, in a complex process that is already strongly launched by the potentializations of II and III, no longer have the empirical purity the rhythm, or the "ontological" characteristics of the first break-throughs that opened up the possible. Their inclusion in an infinitely vaster process could make them appear minor. It is because, inasmuch as they are "technological events," they are radically different from the constitution of a universally epistemological space (Phase II) and from the constitution of a scientific method that systematically organizes the mastery of nature (Phase III).

To cite some examples: When Louis Slotin, an atomic scientist working at the heart of the "Manhattan Project," maneuvers the screwdriver to make the two hemispheres of the experimental bomb approach one another, permitting thereby the exact determination of the "critical mass" of uranium, he tragically illustrates (he died of this in May 1946) the unbelievably skilled character of work that is indispensable even at the heart of a huge, extremely well planned, and "sophisticated" project.[29] Yet, it would be absurd to argue from this for the renewed "priority" of technology in the sense of Phase I. This *techne*, if it manifests itself in diverse forms, is more and more subsumed by the dynamics of Phase IV. From habitual and marvelous gestures to skills well learned and repeated, they still exist today and are not at all "taken up" like those of the unfortunate Slotin (or, less tragically, like the small tasks that astronauts must undertake). But one must remember that the very character of Phase IV is the formalization of all that is retained from the anterior phases, using it to its own advantage. The Complex of Power makes use of everything.[30] That which is produced cannot be called a triumph of technology without some confusion (this is why we no longer prefer to circumscribe that which occurs today in such a manner, contrary to Jacques Ellul and even Heidegger). It is the triumph of an absolutely new process that has its mandate, not in the immemorial technical skills of human beings, but in the subordination of them to other factors of potentialization. I call this process Phase IV, for want of something better.

This focus, as incomplete as it is, has at least the advantage of allowing us to understand the complexity of the process of Power whose megamachines are only the most potent manifestation. Phase IV is absolutely not monolithic; for the moment, it lets the other phases remain in play. It must not therefore be reduced to its most visible mobilizations nor to its most provocative phantasms—from *Metropolis* to *1984*. But what becomes of Phase I? Does it remain in play?[31] I cannot give a definitive answer to this question, which is posed under conditions that are, according to Lévi-Strauss, completely comparable to those of contemporary revivals of primitive thought. Without going as far as Lévi-Strauss (when he writes that, thanks to the current world of communication, "the entire process of human communication assumes . . . the character of a closed system"),[32] without presupposing that much (because the differentiation of the modes of potentialization are less inclusive than the structural theory of language), we must learn to recognize this possible which surges from multiple masteries, technological and/or artistic, through which twentieth-century human beings reestablish and recast their amorous and conflictual bonds with life—at the margins or in retreat from the current dominating discussion. This is the ambiguity of this *techne* that defies Phase IV, but which may just as well be its hostage.

The "priority" recognized in technology is nothing other than this originary phenomenological characteristic of human being-in-the-world: the *Zuhandenheit*, which is to say, not the setting up of the conditions of the profitability of the Machine from a systematic and efficient rationality, but the proximity of the implements of everyday praxis. An implement, or anything that lends itself to use, inscribes itself on a network of returns: it is, more or less, "at hand" *for* such and such a task. And "the less that we just stare at the hammer, and the more we seize hold of it and use it, the more primordial does our relationship to it become," Heidegger remarks.[33] More originary than all knowledge, the "technological" preoccupation (which we name instead "manipulability") offers itself to humans as a network of more or less direct possibilities inscribed in their surroundings. It is clearly the very exercise of this first praxis that discovers the "mode of being of the tool," the means and the know-how that permit the elaboration of "technological knowledge" in that which is specific. It is on this phenomenological "ground" that Phase I manifests itself and involves itself in all potentialization.

The discussion of the eventual priority of technology over science as a factor of "progress" now appears to be distant. This detour was necessary so that we could clarify through specific cases what we understand by "technology." When the technological historian insists on emphasizing the important and at times precursory role of truly empirical and "professional" research, I can only grant him this. Yet, quite simply, Phase I has nothing to do with a closed system where everything would be defined in advance: it only gives that impression after the fact when the apprenticeship is viewed as a well-delimited task within a structure of fixed implementation. The original inventiveness is displaced in the course of history according to the transformation of the environment and the elaboration of implementation: from carved stone to bronze and iron, from the medieval artisan to the "makeshift job" of an apparatus in the twentieth-century laboratory. But we have also seen that the unmooring of another type of potentialization (systematic and universal) is unable not to reflect on the activity of Phase I, if only to seize it almost totally for its own advantage. This is what is essentially produced in Phase IV through an intense focus on the effects of the prioritizing of power. Already in the course of Phase III, where the respective roles of science and technology appear well distinguished and more wisely distributed, we are not able to abstract from the "spirit of the times" and from the constant comings and goings between theory and the empiricism of the modern epoch (we can ascertain this among the great scientists of the Renaissance, from da Vinci to Galileo). It seems that Maurice Daumas is correct in discerning the first beginnings of a "technology" as early as the sixteenth century.[34] The argument can work, however, only in one sense, and it becomes at the very least paradoxical—the more one advances in time towards the intense focus of Phase IV—to reintroduce a concept such as Gramme's "pure practitioner," a remark that is psychologically exact, but epistemologically incomplete (insofar as the "pure practice" of 1850 already capitalized to a considerable degree from scientific advancements which were integrated into technology itself). And from this, one is perpetually unable to effect an ahistorical comparison between science and technology, as if it were a matter of two factors defined once and for all, weights that one could add or subtract from the scales of history.

This field of study is immense, and the present inquiry does not pretend to be exhaustive; nevertheless, we must take up again the evidence cited earlier, in particular, Watt's example, around which a large part of the argument favorable to a priority of pure practice revolves. First, if we consider again the period mentioned above (the nineteenth century), we understand that Daumas himself honestly tries to follow Lucien Febvre's advice, in showing "the role of science in technological invention."[35] Continuing his line of thought, I point to Faraday's laws of electromagnetic induction, allowing for the technological development of electromagnetism (since 1832, the perfecter of Pixii's generator),[36] and the curiosity of physicists who will hasten the exploration of the "effects of the greatest possible number of combinations of metals for electrodes."[37] Would there have been the construction of alternators at the end of the century without the determining role of engineers who "possessed a strong mathematical foundation?"[38] Would refrigerators and combustion engines have been possible without the new science of thermodynamics?[39] In chemistry itself, a discipline removed from pure theory, Daumas does well to note the determining influence of the scientific work of Gay-Lussac (on sulfuric acid), of Chevreul (on fatty acids), of Murdock and Lebon initiating organic chemistry by the distillation of carbon coal, opening the way to the discovery of coal gas.[40] The movement will only be accentuated in the twentieth century; this can be seen in the multiple applications of organic chemistry in the areas of synthetic products and plastic materials. One can see, therefore, that the purely (or supposedly pure) "technological" branch of thought is largely counterbalanced by the influence of science with its different levels of effectuation.

I now return to Watt's example, eminently symbolic ("classic" Daumas suggests) and completely crucial, which takes account of the role of the steam engine in the industrial development of the eighteenth and nineteenth centuries. Is Watt truly admitting a "pure practice," and what does that mean? I again take up Maurice Daumas's study of the very foundation of Watt's work, a study which draws attention to Watt's analysis of the Newcomen machine, perfected by Jonathon Sisson. I cite Daumas's study at length:

> He (Sisson) will measure the volume of water vapor to the atmospheric pressure and will find that it would represent 1800 times the volume of condensed water corresponding to the same quantity of steam; he will establish that for every thrust of the piston the necessary volume of steam would represent four times the volume of the cylinder. This enormous expenditure of steam would entail a proportional expenditure of combustion and would not permit one to assure the

functioning of the machine. Watt learned from Joseph Black that the transformation of water into steam would absorb a large quantity of heat conforming to the principle of the latent heat of vaporization that Black had come to discover several years earlier. Then he sought to use the heat yielded by the condensation of steam to heat the water destined for the heater. . . . [41]

Even a non-scientist can understand that Watt poses the problem of understanding how to avoid a considerable loss of heat. The very posing of this problem presupposes that the question of heat had been quantified. This objective measure, which for us seems to go without saying (or goes without saying for Watt), was realized only at the beginning of the seventeenth century by thinkers who constructed the first thermometers (from Galileo to Torricelli and O. von Guericke). [42] This is the first acquired scientific knowledge that Watt definitely presupposed. Second, in a precise, quantitative manner, Watt compares volumes under a constant pressure. Doing this, he cannot avoid referring to the Boyle-Mariotte law, which is indispensable to him both for a general conception of his problem and for the fabrication of his cylinders; the inversely proportional relationship between pressure and volume obviously ought to be taken into consideration both outside and inside the cylinder. Third, the very statement that we just read takes into account Joseph Black's later (1760) work on the difference between temperature and the quantity of heat. [43] The solution of the problem, therefore, will be a *recuperation* of the quantity of heat that has already been utilized. Here the truly technological moment intervenes, as closely as one can isolate it, which consists of enclosing the cylinder in an envelope of steam; [44] this will be the condenser that will permit a truly rational and profitable use, within the economic conditions of the epoch, of the aforementioned steam engine.

It is necessary, therefore, to yield to the evidence: the effect of "steam engine" power does not come from "pure practice." Certainly, Watt closely studied the machines immediately at hand, certainly he confined himself to an extremely limited problem, and moreover, he resolved the problem perfectly, with an incontestable technological genius. But we have just verified that all of this would have been reduced to trial and error if it were not for the precise *conception* of the problem, which was possible through the achievement of scientific work on heat. When Watt poses the technological problem, his solution is already scientifically *potentialized*, which takes nothing away from the inventor.

What do I wish to demonstrate with this? Certainly not an absolute, idealistic privileging of "pure science." That would be the case if we had claimed to "deduce" the effect of the power under investigation from theory according to a necessary law. This is not at all my purpose, particularly owing to the distinction between Phases I and II. In the one, the potentializing factor is episteme, without the effects of direct technological power; in the other, the potentializing factor is a science that is no longer a priori purely contemplative, but is in contact with—thanks to the method—the quantified control of natural phenomena (whether by the intervention of the law of falling bodies, the conception of differential calculus, Mariott's law, etc.), and in this manner, consequently, solicits technological inventiveness. At any rate, whatever the degree of the "purity" of the science that potentializes in Phase III, *the fundamental determinant of development is no longer technology in the sense of the first phase.*

The recognition of the coexistence of these phases allows us to comprehend the nonlinear character of technological evolution: we see this in the fact that the potentializations of type I subsist in a world where the determining factor has been displaced. Watt has undoubtedly offered proof of a great *Zuhandenheit*, but the "technological event" to which he attaches his name does not occur principally at this level. It is not a question of denying the importance of factors by which true technicalization is evident (for example, the precision of watchmaking that is so important for the navy);[45] it is a question of placing this relative autonomy of truly technological development (these interplays of implements extended to complex flows among elements, individuals, technological ensembles) within the overall process of the movement of the potentialization. Moreover, what I am very clearly affirming is that there is no common measure between the factors of potentialization of type I and the others. If one suppresses the totality of the mathematical apparatus of Phase II, the moments of technological genius would never attain their apodictic and universal support, their "Archimedes' principle." If one suppresses the methodological implementation of Phase III, one would have only a juxtaposition of theoretical certainties and "sleights of hand," without any unifying project. But from the moment that these phases come together, efficacy is accelerated; it is the very character of Phase IV that comes into plain view alongside the intensification of this acceleration.

In what sense is Phase I first? In what sense is it characterized by a priority of technology? I can explain it only by a constant play of differentiations arising in and occurring through the other three phases. This will allow us to speculate that the inverse is equally true, and will allow us to hope that the deficiencies in this presentation of Phase I will be compensated for by the numerous "retro-references" in the discussions of Phases II, III, and IV.

Euclid or the first power of *mathesis* (Phase II)

Nothing is more elusive than the delimitation of the *theoretical* in order to understand the significance of Greek mathematics. I would like to show this in the case, entirely exemplary, of Euclid.[46] Certainly, one finds in Euclid neither numerical evaluation nor concern about the application of theorems.[47] Is this to say that it is a question of a pure theory, implying no ontological actualization? I will try to respond to this question by outlining a phenomenology of the basic Euclidean operations.

What happened in the positing of a geometric "perspective"? Nothing more than the constitution of a dominant referential (and universal) field of *potentializa-tion*. Thales and Euclid, in order to incline toward power? No: in order to *already* exercise power ontologically.[48] *Mathesis* is therefore not one preamble to the scientific revolution among others—it is its dominant potentialization (an effective operation *and* held in reserve).

It is precisely a phenomenology of the geometric perspective that Husserl wanted to write in *The Origins of Geometry*: to reactivate the donation of sense perception by which idealities *once and for all* acquire the objectivity and the universality that are constitutive of them. As Husserl himself remarks, "The Pythagorean theorem, [indeed,] all of geometry exists only once . . . "[49] This expression could be misunderstood if one were to take it to mean the contrary of what Husserl observed—singularity in place of universality. This "only once" is the definitive acquisition of a supratemporal "existence," that of the *apodictically invariant*.[50] Husserl continues down this path in the sense that he restricts himself to establishing the *forgetfulness* of the donation of sense perception in geometric rationality, calling for a reactivation of the "idealizing primal establishment of the meaning-structure 'geometry'."[51] Is it possible to account for the origin of geometry in presupposing the universality of human language? Husserl did not really succeed in accounting for this.[52] The threshold crossed by geometrical idealization remains no less gaping, enigmatic. This gap is not filled in by declarations of intentions and by programmatic enthusiasm.

The *suggestions* that follow reveal a different orientation: not aiming at a return to the ground of the donation of sense to Euclidean geometry, but to its constitutive and irreducible characteristics. Husserl undertakes the task of *joining* the life-world and the sphere of the apodictic; he aims at that which "humanity realizes in the totality of its concrete being with an apodictic freedom in an apodictic science."[53] I am not asking for so much; my enterprise is not as metaphysical; it is indeed the *inverse* of Husserl's enterprise, all the while still endeavoring to be—to some degree—phenomenological. What this inquiry seeks is less the origins of geometry (supposing that the origins are able to be found and that one could, one ought to "domicile" humanity there) than its *results*, which until now have been too infrequently glimpsed, its "prodigious practical utility" in regard to which Husserl recognizes in passing that it is "misunderstood."[54]

We open, therefore, Book I of the *Elements*. We read the definitions, postulates, axioms, and first propositions. Even someone who has no head for mathematics seems able to comprehend the initial condition for the intelligence of these "idealities." Ordinarily one calls it the faculty of abstraction. Philosophically, it presupposes a definitive *indifference* to ontological content: "A point is that which has no part."[55] A point, which point? Is it necessary to have read Plato to know that it concerns the point in itself? Euclid was a Platonist; Plato was fascinated by "beautiful mathematics": this intertwining has nothing to do with "interdisciplinary" difficulty. What is ideally present in this first definition—ontologically undifferentiated, logically differentiated—is the abstraction of mathematical being, an already im-manent property of *number*, yet hidden by its empirical usage. That this "being" is understood once and for all, whatever its empirical translations, is the incommen-surable innovation which we cannot—literally—*cast aside*. Potentialization is launched: irreversible (it is no longer possible to return to a *perfect* empirical na-iveté) and neutral (the point, defined, has "nondifferentiated" all the empirical at-tempts at the division of a grain of sand, a crumb, a speck, etc.)

To conceive of a point for all points, to conceive of "that which has no parts" (the logical atom) is already a highly speculative act. In terms that Hegel will use: to be capable of denying space.[56] Effectively, in holding to the letter of Euclid's first definition and without revealing an excessive complaisance with regard to the spec-ulative dialectic, we acknowledge the negation lodged in this first delimitation of the geometric horizon: the conditional negation of divisibility to infinity. If we wish to begin to "geometricize," we must admit the *ideal* existence of this being "which has no parts": such is Euclid's proposal.

This first mathematical being, punctuality defined as indivisible, is therefore far from being as "simple" as it first appeared, and, moreover, it always tends to mani-fest itself in the cycle of a mathematical and scientific culture in which infinitely more complex concepts ought to occur. The first phase of a rational "exposition" that will remain—up to Spinoza and beyond—the model of synthetic order, it pre-supposes an ontological beginning that is completely different from daily speech or poetic utterance. This *nondifferentiation* of an ontic field removed (abstracted) from the total life of being and from the beauty of the *cosmos* in order to be *differentiated* once and for all logically: "That which has no part" a priori mobilizes (mentally) a sector—still infinitely minimal—of being. Euclid (and with him, all of geometry) immediately *gives himself* an ontic space, which will be called geometric space (which I bring up here only in order to isolate the condition of ontological possibil-ity—which in no way reduces it to the definition of a plane)[57] and where one knows in advance that nothing will appear other than what has been announced and de-manded by the definition. This apparently negligible "act," which develops never-theless very quickly in a monumental succession of *Elements*, opens the unprece-dented possibility of a priori potentializing being, of de-fining a "reality" without equivalent in the empirical. With the ὅρος of the point breaking through a new ho-rizon, which is not going to stop extending itself until it invests the totality of being. Euclidean punctuality already potentializes the "Archimedes' principle," which Descartes will make the foundation of the indubitable method.

Certainly the abstraction of the first definitions is compensated for by the pos-sibility of *figuring* the defined. Even if the first figure is that of proposition I, the definitions are able to be immediately figured: they are not presented because they are to some degree evident, very easily representable. Imagine that "this which has no part" is not understood by the interlocutor: one point drawn on a board will make it immediately intelligible. Kant will grant this privilege to geometry: the *con-

struction of its concepts. Will this rest which the figure offers to the mind come to diminish the dynamism of potentialization? Only in part. Descartes will see that the help that it first offers will transform into an obstacle, and that it "is unable to exercise the understanding without greatly fatiguing the imagination."[58] The *operative* will no longer endure any "compromise."

Nevertheless, establishing definitions is far from tracing their visualization. Definition 4 indicates that "a straight line is a line which extends itself equally in respect to its points."[59] Arguing that one no longer finds here a trace of sensible representations (a taut string, a ray of light), Charles Mugler notes that the care that Euclid took to liberate the concept of straight lines from its representations "led him to a formula which was slightly obscure," and which even "defied the wisdom of its commentators."[60] It is true that our "economic" definition of straight lines as the "shortest distance between two points" dates from Archimedes, not Euclid. In this case, the economic principle allowed for the rediscovery of the geodesic representation of taut strings, and in some way turned abstraction back toward the sensible.

Whatever the operative significance of the different definitions, a supplementary step is cleared with the entry into the *apodictic* dimension, which is accomplished through the demonstration of proposition I: "On a given finite straight line to construct an equilateral triangle." The demonstration works thanks to the construction of two circles each having the same ray AB (the given segment), the first having as its center the extremity A of the segment; the second, the extremity B. The conclusion (the demonstration) is obtained by breaking down the difficulty into its elements, using axiom I: "Things that are equal to the same thing are also equal to one another." If Γ is the point of intersection for two circles, it is first necessary that ΓA = AB and ΓB = AB to demonstrate ΓA = AB = ΓB, which is to say that the triangle AB constructs on the given segment AB is truly equilateral. (See figure.)

What the new *apodeixis* offers in relation to the definitions can be summarized in three points. On the one hand, this demonstration is at the same time a construction, allowing for the repetition of the abstraction in the figure, but this time in the service of a demonstration, not simply of a delimitation; neither in the protasis nor in the conclusion do we find mention of the circle. This is the mediating figure of the demonstration. Nevertheless—second point—the fact that it concerns a construction does not taint the certitude that is attached to this type of demonstration. Suppose that one were to ask us to empirically construct a figure consisting of an equilateral triangle; even if we have good instruments, we can never be sure that they are absolutely (apodictically) equal. This certitude is obtained thanks to the mediation/construction of the circle, by virtue of definition 15: "A circle is a plane figure contained by one line such that all the straight lines falling upon it from one point among those lying within the figure are equal to one another." The *apodeixis* allows us to attain this truth that Socrates calls "irrefutable."[61] Third, the reasoning progresses by a series of operations that are perfectly able to be split into parts and

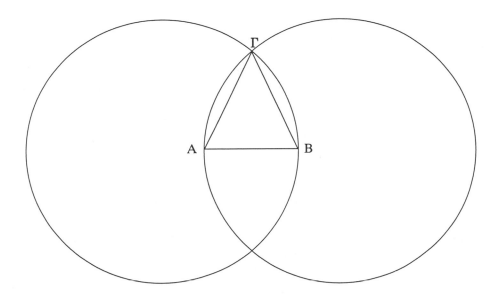

effectively ex-posed (it is the literal sense of *apodeixis*) in the previous positions (definitions, postulates, axioms) as well as in propositions (their demonstrations as the links that they maintain between them). Propositions are so many *elements*. Given that an element ought to be indispensable to the whole of the theory, one will have thus a systematic sequence of propositions that imply one another, following a synthetic order which is more and more complex. Thus, as soon as proposition II is demonstrated, the conclusion of proposition I intervenes and avoids a regression. Already the first proposition would imply recourse to postulate 3, to definition 7 and to axiom 1. The first propositions are the elements of a number becoming, by degrees, significant from other propositions; these are as much operations that extend the power of an indubitable rational, from the point to regular polyhedrals (just as, twenty centuries later, Descartes will prescribe going from truths that are "the most simple to the more complex"), as they are *potentializations* of the onto-logical field held hostage by necessity. Aristotle defined necessity as "that in virtue of which it is impossible that a thing be otherwise."[62] The *apodeixis*, discovered from this rational power, is the ex-position of this that is only able to be such that it cannot be otherwise.

One is therefore wrong to confine oneself to the theoretical character of the *Elements* without seeing that this theory is already *operative* at a degree that is rarely suspected. The most important fact is, indeed, that the self-limitation of the rational extends itself to the point of deploying itself in a systematic *corpus*, more and more comprehensible, insofar as Books XI to XIII begin to develop geometry in space, whereas the arithmetic books present the theory of rational numbers and then—in Book X—the incommensurate and irrational magnitudes (up to the irrational numbers of the second degree). Even if Euclid's work is, by nature, finite, nothing definitively fixes its limits: this mathematical space is ideally open, available. This is why innumerable renewals arise over the course of the centuries, resulting in, on the one hand, the repeated confirmation of Euclidean knowledge; on the other, its extension and "correction," from the attempts at rewriting the *Porismes* to non-Euclidean geometries.

What a dilatory tradition did not want to see, confining itself to the cliché of a "theory" without application, was, nevertheless, before its very eyes. Phenomenologically, since Euclid, geometry has been practiced, the *opératoire* is there under the gaze of the spirit; this possible which effectuates itself, this is the extension of the necessary, the possible inverted into its "contrary"—beyond contingency. The ruse of this "neutrality" is very particular: this potentialization effectuates itself through an "internal usage," from element to element, apparently without contact with the exterior, gratuitous, and innocent under the disinterested contemplation of the mathematician. However, already the constitution of this apodictic field—ontologically separated from the *cosmos* by its very "nondifferentiation"—in *expansion* perfectible (an already very modern characteristic of episteme, which Archimedes will exploit, leading to the creation of integral calculus). Here there is a more decisive character than that which Husserl insists upon: the fact that the geometrical operation is indefinitely *repeatable*. To be sure, in *The Crisis* Husserl clearly saw that universality is not merely always available in its identity, but open to an indefinite perfectibility. The "incubation period" of which Husserl speaks in regard to the principle of reason is not less for that which concerns the "application" of Euclidean geometry to the physical sciences. It is as if History had been anticipated to the same degree as the power held in reserve.

Now, what modern science reveals is not the discovery of a certainty in itself that is more absolute, more "scientific" than that of Greek geometries. *In his domain*, Euclid remains essentially irrefuted: Hilbert presents only an objection concerning a postulate.[63] And this proclamation that looks so modern—"Those who have science in its absolute sense ought to be unshakable"—is not from Descartes, but from Aristotle.[64] What does History reveal to those who do not view it too myopically? It will reveal a lesson that comes to confirm from an ontic and "positive" viewpoint that which will be manifested from the ontological and philosophical viewpoint, namely, all the great modern thinkers from Galileo to Newton and beyond begin with Euclid and Greek geometers. It does more than this: they take these geometers as models. Their goal was not to surpass the *Elements* in apodicticity, even if Descartes intends to complete "the analysis of the Ancients by the algebra of the Moderns." It was to extend this apodicticity to those domains that the Greeks and medievalists believed in principle to be out of reach of mathematical methods. This extension presupposed a metaphysical force that represents the establishment of a *mathesis* which is henceforth *universalis*, of a certitude applicable to *all of being*, of an operativity which is henceforth universalizable to infinity. But, to return to the facts, it is undoubtedly upon the apodicticity of mathematics—essentially Greek—that all these pioneers of modernity rely; Bacon and Descartes above all rely on it methodologically; Galileo and Newton, in the effective transformation of physics. The foundation of modern physics operates through the Galilean geometrization of nature, the substitution of the model of Euclidean geometric space for the ancient theory of place. The principle of inertia presupposes, as Koyré has shown, this identification of space with Euclid's "infinite homogeneous space."[65]

If the mathematical acquisition of the Greeks has incontestably served to effectively attack (whatever the developments or perfections) in the siege made on the ancient system of the world, as well as in the constitution of the new physics, it is necessary to draw out of it the consequences as regards our theory of potentialization. If every epistemological discovery potentializes, which is to say creates a possible starting from extremely selective effectuation, it is *starting from mathematics that an axis of maximum potentialization is effectuated.* As paradoxical as this seems, considering the distance of mathematics from the empirical, and considering also the extremely unequal "applicable" character of its theorems and calculations, it is the necessary and the calculable that increasingly dictate the possible, to the degree that modern science constitutes, assures, and accelerates its march towards the control of the totality of the real. This "privilege" of mathematics, already perceived by Plato, owes nothing to a type of arrangement between disciplines or to a prejudice; it is in virtue of the originary presupposition of the mathematic-geometric "view" that it is recognized. Mathematics are not the concern of this discussion, neither because of the Greek origin in its arrangements nor because of the modern caesura in its brutality: they are of concern due to *their* philosophy, and without this their clarity could be separated from the brilliance of Western knowledge-power. Mathematics are essentially potentializers: they constitute the most formidable conceptual "instrument" that man has ever had at his disposal; and yet, they would rest eternally in their divine repose of *kala mathemata* if there had not been the intervention—in order to orient and apply mathematics—of a "supplement" that changes everything and without which modernity would lose its profound dynamism: the method, in the (metaphysical) sense that I am going to elucidate.

The importance of this qualitative rupture calls for an examination of its principles and its mode of deployment. The entry into Phase III of potentialization can be explained only by a veritable rational revolution. I do not intend to undertake a truly historical study of that revolution here, nor do I have pretensions of saying everything about it. Of essential importance: to clarify the access to a new type of potentialization.

Rationality of mastery (Phase III)

The rationality of mastery cannot be understood without its principle and its substance: the method. How does this play—in Phase III—the essential and efficient role of the *operator of power*? This is what one must understand in recalling the role of the great initiators: Bacon, Galileo, Descartes, without losing sight—in other respects—of the limits of the effectuation of power in a phase whose ideal remains the sovereign mastery of rational man.

It would be a misunderstanding to believe that under the pretext that we are attempting to isolate the "driving force" of power, we take ourselves to be the advocates of an unconditionally invented antecedent of the method of scientific discoveries. We must concede to Koyré that no science has ever begun *stricto sensu* by a *tractatus de methodo* and that "the place of methodology is not at the beginning of scientific development but, so to speak, in the middle of it."[66] Nevertheless, whatever the delimitation of the methodology and its role in the march of science would be, it remains, as Koyré himself recognized, that the great modern scientific revolutions are "theoretical revolutions" and that their result was "to acquire a new conception of profound reality" that underlies the givens of experience.[67] This is essential for the very understanding of our present: whether the project of History has been tortuous or illuminated, whether science has advanced or not with a total methodological lucidity, we are in possession of principles—or the conditions of theoretical possibilities—without which modernity would not be thinkable.

What Bacon did more than announce, what he *enacted* in defining the new method, is the end of all the theatricality that had been given as scientific. The future of methodological science? It is the coincidence of science and human power,[68] thanks to the *operativity* of reason. No one has better formulated this program proposed to a reason that is sober and useful, progressive and certain than Bacon himself in his preface to *Instauratio Magna*:

> The end that is proposed to our science is no longer the discovery of arguments, but of techniques (*arts*), no longer the concordance with principles, but the principles themselves, not of probable arguments but of dispositions and operative indications (*designationes et indicationes operum*). This is why a different effect will follow from a different intention. To conquer and to restrain: over there an adversary through discussion; here, nature through work (*hic natura opere*).[69]

The precision of Baconian Latin indicates unequivocally the true mediation that will ensure the victory of man over nature: *opus* (not the great work, but work clearly delimited by the observation of nature). *Natura enim non nisi parendo vincitur*: this often-cited expression should no longer be understood anthropomorphically, despite the metaphor suggested by Publilius Syrus.[70] This would be again to form an idol of theater that believes that it is sufficient to see in order to know, to listen to nature in order to hear its true melody. Bacon clearly demands a complete inversion of method, but the necessary submissiveness is precisely no longer theatrical or contemplative; it is operative: "That which is an object in contemplation becomes a rule in operations."[71] The ruse of nature exceeds the theater, as it exceeds all dialectic and all directly causal logic. To measure the *subtlety* of nature, it is no longer sufficient to substitute modesty for pride; it is necessary to *enact* "a better and more certain operation of the intellect,"[72] which, in interpreting nature with efficacious instruments (reasoning not separated from observation), will delimit the *effective* knowledge of man. Nature has only to prostrate itself. It is an exact obedience to its rules that makes man its "minister."[73]

Is Bacon the inventor of the experimental method? He is much more radical: the scientific revolution itself is announced and enunciated through his pen. A scholastic epistemology would have us believe that it was first necessary to discover the experimental method in order to elevate it to method in general: Bacon in order to reach Descartes. It is true that the Baconian foundation, this dismissal given to idols of "fantastic" reason, constitutes a general repetition and even perhaps a preamble to the Cartesian foundation. But there are not two scientific methods in the modern sense. From Copernicus to Descartes there is clearly one and the same revolution, which attacks ancient rationality and reverses it, a reversal in which Bacon plays a central and essential role. The form of the reversal is, moreover, expressly presented by Bacon: far from restricting himself to opposing experimental induction to syllogistic deduction, Bacon called for an inversion[74] of the hasty induction that was practiced until then, to the benefit of an abstraction and of a progressive confirmation of *axioms*. It is on this condition that one will penetrate at the "deepest depths" of nature.

Galileo succeeds in this by setting in place the mathematization of nature through the intervention of a new thinking of movement. Descartes himself was not fooled here: despite his severity, he recognized that Galileo approached the truth in setting the task of "examining physical matter through mathematical reason."[75] This is possible only thanks to the breaking down of ancient cosmology and Aristotelian physics, a process that had already largely begun with Galileo. What is accomplished and realized by his effective practice (with its complexity and its "shift" that accounts for its singularity) is clearly an *ontological mutation*.

"To envisage Being in a new way": this evocation of the task of the founders of modern science is not from Heidegger, but Koyré.[76] Why should positive knowledge necessarily be in contradiction to ontological intelligence? And likewise, why, under the pretext of extreme acuteness and positivism, do we fail to recognize that which fundamentally and historically reconciles Galileo with Bacon and Descartes?

Husserl does not succumb to any of these detours when he identifies the scientific march inaugurated by Galileo: "By means of pure mathematics and the practical art of measuring, one can produce, for everything in the world of bodies which is extended in this way, a completely new kind of inductive prediction; namely, one can "calculate" with compelling necessity, on the basis of given and measured events involving shapes, events which are unknown and were never accessible to direct measurement."[77]

The generating of the parabola at the beginning of the Fourth Day of Galileo's *Discourses* offers an example of what has been advanced. An example among many others, it illustrates both the *operative enactment* of the Euclidean geometric potential (without excluding other contributions, in particular those of Apollonius and Archimedes) and the new technological potentialization that results from it.

Galileo's Euclidianism allows us to see Euclid's contribution from various points of view. Already Galileo's biography testifies, in a still distant but significant manner, to the constant study of Euclid.[78] But especially, in terms of the doctrine, it is enough to cite the famous passage of *Saggiatore* in order to establish that, if the great book of nature is written in mathematical language, "its characters are triangles, circles, and other geometric figures." It is certainly therefore the Greek, Euclidean geometric instrument that made a direct contribution to founding the method of the new applied natural philosophy (which is not to say, of course, that geometrization is made through a simple mechanical and schematic projection of geometric forms upon the perceptible world). This Euclidianism can perhaps be confirmed in other ways: in the form of Galileo's discourse, as well as by the biographical facts. From the first point of view, what is more rhetorical than a dialogue? Despite the introduction of abstract and figurative reasoning, the *Discourses* are not exempt from digressions (we know that Descartes deplored this); and yet we see that from the second *Discourse*, concerning the resistance of materials, an extraordinary caesura occurs, reinforced again in the Third and even the Fourth Day; the "ordinary" conversation suddenly surrenders its place to a series of propositions and theorems which obey the Euclidean model. The Third Day especially is only, so to speak, a treatise exposing the new laws of motion: universal truths necessarily linked as so many *elements*, truths which remain geometrically constructed.

At this point, the more specific problem of the parabola concerning the movement of projectiles is posed when one envisions an animated mobile with a "double movement, namely, a uniform movement and a naturally accelerated movement."[79] The mobile is supposed to move first via a uniform movement on a horizontal plane, then from a point b, it is to lack support and fall according to the law of naturally accelerated movement (but with its acquired speed). Salviati exposes the two properties of a parabola taken from Apollonius, and Simplicio challenges him: he does not understand, because he does not completely understand Euclid. To which Salviati, not without contempt replies: "In truth, all mathematicians who are even slightly advanced presuppose that their reader is at the very least in full possession of Euclid's *Elements*."[80] And he refers to the proposition from Book II of the *Elements*.

The demonstration which follows is not too difficult, neither for Simplicio, nor for us, but is, nevertheless, extremely interesting because it proves that the trajectory described by a weighted mobile, descending according to the complex movement just described, is nothing other than a demi-parabola. It does not seem necessary to give a detailed explanation of it here, the drawing will suffice to show the reader that the parabolic track of the curve *bh* is verified with an application of the second theorem of naturally accelerated motion ("the distances traversed in whatever time

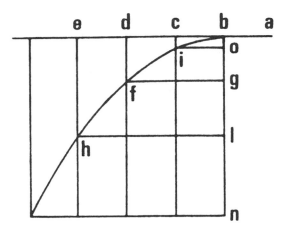

... are between them accounting for double the time, which is to say, as the squares of the same time"). Thus *df* is the quadruple of *ci* and so forth.

Analyzing the general structure of this proposal of the Galilean explanation, Maurice Clavelin shows that, as in the case of the apparent movement of sunspots, "it is clearly a matter of seeking an expression of the phenomena at the interior of a previously determined rationality." But in this specific case of compound movement, we have, as Clavelin further notes, a "taking charge by mathematical reason" of the selected domain of rationality. In other words, not only did Galileo not *experiment* anachronistically without having rigorously defined a domain of rationality, he constructed his physics of movement in "mathematical language," which is to say that he did so geometrically, recovering through mathematical and apodictic rationality the selected physical domain whose "phenomenality" will, after this, rigorously correspond to its mathematical expression (because the physical observation has been *mathematically problematized*). Geometrical potentialities (here Euclid is supported by Apollonius) are, therefore, rendered *physically operative* due to this *reading* of the phenomena that knew, in some way, to preselect the real so that it could then extricate the typology of compound movements.

A new, altogether significant potentialization results from this recovery of the physical by the mathematical, insofar as, in the case that concerns us, the movement of these projectiles, whose specific analysis follows in the Fourth Journal of the *Discourses*, is henceforth calculable a priori: the determination of the *impeto* (speed) in each of the points of a given parabola (proposition IV), the determination of the amplitude in relation to the height (proposition VI), the fixation of the optimal rapport between *impeto* and amplitude (corollary of proposition VII). More precisely regarding the last point: the greatest amplitude is reached at a 45° elevation. One sees immediately the application of this in ballistics: the curve of fire which projects the bullet the greatest distance is determined a priori and no longer by trial and error. As Sagredo said, "In learning this, reason infinitely prevails over simple knowledge arising from other relationships or even from habitual experience."[81] Salviati's response is equally worthy of citation: "The knowledge of a single effect through its causes permits the mind to understand and *to be assured of other effects without need of having the experience again*." Salviati refers here to precisely that which we call the "power-effect": not a brute flux of energy, but the *direct result* of scientific knowledge (whether this result be open directly to technicalization or not) by which power over things is definitively assured. Even more to the point, Salviati locates the cumulative character of the process of potentialization: a single effect which is mathematically known allows the securing of other effects (potentially indefinite in number). The example given is completely convincing; knowledge of the optimal shot fired has for a corollary the determination of the equality of shots fired whose elevation is superior or inferior to the same quantity at an angle of 45° (hence the equality between a shot of 7 points and one of 5 points).[82]

One grasps here in a privileged fashion how, since Galileo, modern science directly eclipses the conventional position on the problem of the technological application of a "theory." Mathematical theory *potentializes* insofar as it invests a priori a nature viewed in an operative manner. Technology will follow more or less quickly, and in a more or less sophisticated manner. This is of little importance. The *event* is this: mathematical knowledge nevertheless completely *conceals* a series of a certain type of phenomena whose mastery is definitely capitalized. The engendering of power by science is direct, although the power effects can be differed or unequally distributed: "It is therefore spontaneous and *without renouncing any of its theoretical purity* that the Galilean science of motion will establish the basis for the renewal of the technology of ballistics."[83] I can only join Maurice Clavelin in his remark, which also notes that the relationship between science and technology is completely changed by this enactment of the mathematization of nature: "a technology without anything in common with that of simple practitioners"[84] is not only the *result* of this science, but from beginning to end accompanies it in some way through a process whose general characteristic is operativity.

It is evident that this operativity allows a mathematical potential (Euclidean knowledge, in particular) to flourish practically or to actualize, which turned then toward idealities finds at the same time a new consequence (physics) and a new content (*time*, which was absent from the entire Euclidean schematism, constantly intervenes through its correlation with space in the formulation and construction of Galilean laws of movement, and a fortiori in the calculation of this "element" so essential to modern physics: speed). We read thus of the *effective translation* of this new mathematical reading of nature, which would never have been possible without a philosophical revolution in the very conception of that nature. The example of the Galilean mastery of the parabola demonstrates that modern science *potentializes* directly from a mathematical theory of motion. Power did not come to the Moderns by means of trial and error. On the other hand, Euclid would have been eternally passive in his *Elements*, if he had not attempted to render it operative through an a priori project of the philosophical and scientific mastery of nature.

A revolution such as the founding of scientific modernity cannot be reduced to the application of some particular rules concerning clarity and distinction, the order of thought, the verifications to be done, etc. As important as these methodological prescriptions of the *Regulae* and the *Discourse* are, they make sense only in relation to a fundamentally new project that Rule II formulates with the greatest clarity: *Circa illa tantum objecta oportet versari, ad quorum certam et indubitatam cognitionem nostra ingenia videntur sufficere.*[85]

This brief sentence deserves a close analysis: we find here the very spirit of the new method. Or in even more striking, perhaps too picturesque, terms: the prime mover of the excessive rationality of Power. Indeed, why would we seek the origin of this excessive rationality *elsewhere* than at the very place where it formulates itself in the most clear terms as the prior condition to all scientific truth?

First and foremost, my concern is with a deontology of scientific understanding, not with a description. Is this a simple approach to the *Regulae*? Without doubt. But it is not unimportant that the new theory of truth formulates imperatives or duties whose very "existence" is not justified as such. Consider the conception of truth exposed by Aristotle at the beginning of Book II of the *Metaphysics*: the attitude is the inverse. Instead of prescribing and excluding (because the deontic carries with it the restriction to *certain objects*), Aristotle proposes an *open* conception of truth, "No one is able to attain the truth adequately, while, on the other hand, we do not collectively fail"; and he cites the proverb: *who would fail to hit a door?*[86] Certainly this passage does not contain the entirety of the Aristotelian theory of truth; moreover, one will object, it does not concern the same truth. Of course. But the awareness of the limits of the rapprochement can only serve to underscore the extent of the contrast: the two conceptions of truth cannot be *compared*, properly speaking, for the very reason that they define two different universes. The Aristotelian view is receptive because it conceives a truth that has its basis in experience which is consequently cumulative: the science that it advances is still a περί φυσεώς ἱστορία; the grasp of the truth has nothing of the "certain and indubitable," it is a τύχειν that changes according to circumstances and luck.

In contrast, the Cartesian method is demanding and restrictive. Demanding because it is regulatory *for the will*; restrictive because it rejects objects that are merely probable. The last point is clear in the commentary of Descartes himself: "By the present proposition, we reject all knowledge which is merely probable." Excluded are all beings (included, the ideal ones) which are not worthy of being elevated to the rank of *objecta* (this remains implicit); but also excluded, this time explicitly, are *objecta* that are not susceptible to being known in a "certain and indubitable" manner. This selective vision is *disciplinary*: rejected are objects that are the predilection of metaphysics, theology, and dialectics. On the other hand, mathematical objects— more precisely, arithmetic and geometry—are not only found acceptable by Descartes, but become the models for truly scientific knowledge.

This double Cartesian condition appears evident today. But the contrast with the secular tradition, for which Aristotle is here the symbolic representative, ought to remind us that nothing is less evident—according to common sense—than the *evidence* of "certain and indubitable" knowledge. Aristotle appeals to experience. Descartes challenges it: if arithmetic and geometry are the most certain, "it is because they alone . . . deal with an object so pure and simple that they need make no assumptions at all which experience renders uncertain, but wholly consist in the rational deduction of consequences."[87]

The debate would be closed if science had stopped at the limits of mathematics, but Descartes turned to *other objects*, capable of having "a certitude equal to that of the demonstrations of arithmetic and geometry." It is at this moment that all the weight of the second part of Rule II intervenes: . . . *ad quorum certam et indubitatam cognitionem, nostra ingenia videntur sufficere*. Indeed, this defines a correlation of *objecta* with the *capacities* of our minds. Not that the objective becomes subjective, but its limits are brought back to its foundation: the human mind. *Sufficere* does not have an indeterminate or abstract sense here, but intervenes in its personal usage: in the sense where a person has the ability to complete a given task. But the first sense—to place beneath—has not completely disappeared.[88] According to Descartes, our minds are the *sub-jecta* to which objects correspond. Once again I mean that no license is given to what we would today call individual subjectivities. The sub-jectivity that involves the *sufficere* is both the metaphysical (and not the epistemological) foundation of scientific knowledge and the *delimitation* of its horizon: excluded are objects of thought which are *too difficult*, that is to say, constitutively exceed the capacities of our minds and, consequently, render the "focus" of certitude impossible.

The brevity of Rule II yields, in very few words, the essence of the new deontology insofar as it implies an economy of certitude. This rule does not appeal merely to the understanding; the group of exclusions that it demands mobilizes the will. In substance, Descartes argues that the will must reject all kinds of rich, subtle, and suggestive thoughts on behalf of an "object which is very pure and simple," comparable in its purity and simplicity to mathematical idealities. For lack of a more secure certitude (which would assume that the problem of the constitution of science is resolved), I would acquire by this methodological decision-exclusion exposed by Rule II, the *certitude of principle* of retaining as scientific only those "objects" that are constituted paramathematically, which is to say with an indubitability comparable to $2 + 2 = 4$. This will be illustrated, primarily, by Rule III, in which indubitability increases and founds itself in clarity and evidence, as well as by the very idea of *method* whose necessity is exposed by Rule IV.

If it is necessary to put in place such a strategic apparatus, it is because *rationality absolutely does not furnish it itself*. In his commentary on Rule IV, Descartes remarks that "there is nothing more futile than to busy oneself with bare numbers and imaginary figures."[89] Reflecting on the contribution of Greek mathematics, he asks himself why "certain traces" of the true *mathesis* contained in Pappus and Diophantus do not lead to the construction of a solid science. The answer is clear in its theory, however subtle in its formulation: everything happens as if the Greeks had lacked the *method*. Descartes says he is tempted to believe that this "art" had already been discovered, but was concealed—or rejected—because of its lack of subtlety.

This "certain general science"—the method—does not, therefore, have a rational importance and dignity superior to that of Greek mathematics, logic, and philosophy. This new discipline of thought proves to be very simple; it is perhaps even its extreme simplicity that heretofore has rendered it so unattractive. The sacrifice of the prestige of the illusory, of the unusual, or of the transcendent to the benefit of the utility of a methodological simplicity confirms the Cartesian "decision."

Descartes himself gives his very clear and significant opinion of the enigma that is always referred to by historians of thought: namely, why the Greeks, provided with a mathematics already remarkable and capable enough to construct certain admirable "machines," did not "do better" in applying their brilliant mathematical discoveries. The response is as simple as the remedy: everything resides with the method, this universal discipline of which algebra will be, for Descartes, the principal applied form. The Greeks did not have the method, would not or could not enact it. This is, fundamentally, of little importance. The lesson is nonetheless clear and taught with a puzzling assurance (whereas modern scientific practice is still so garbled): the decisive progress which modernity accomplishes has nothing to do with intellectual brilliance or technological ability. It required a new *discipline* whose philosophical name is method, and whose mathematical name is algebra.

The revolution in reason is not a revolution *of* reason. It is made possible by a "decision" that is, in itself, undoubtedly not irrational, but whose degree of rationality is extremely difficult to determine. Is this new method which is going to found modern science and which efficiently links science to technology more "logical" than the "spirit" of medieval or Greek science? Is it not more logical and wiser to allow each reality its specificity, to respect the hierarchical order of being, than to make mathematics the model of a universal science destined to extend itself without limit? Could one even maintain that the project of indefinite rational mastery has an insane side, as some proclaim at the end of the twentieth century? The revolution in reason cannot be explained logically, formally, or mathematically. The method is not *deduced* syllogistically or otherwise. It is posed as a *new exigency*, which the will ought to take up and carry through to a successful conclusion.

Among its great inventors, as among its "executors," engineers, and technicians, the idea of mastery essentially articulates Phase III. One will not understand the latter by juxtaposing method and technological innovations. What is a phase? We recall that it is a mode of potentialization: the appearance, the reserve, and/or the effectuation of power. The unity of involution of Phase III can be grasped above all from that which stands in radical novelty to Phase II: the enactment of a methodological a priori aiming towards the effectuation of mastery.

What is mastery? A power which *at will* disposes of its potential. We will be "as masters and possessors of nature", insofar as the "mechanical arts" will be at the service of our sovereign freedom. In the deployment of mastery, humans as voluntary subjects place themselves in the center of a network of procedures for which they are at once author, supervisor, and recipient. This project finds its privileged chosen field in classical science—whose determinism satisfies its aim of definitive coherence—and in the thermodynamic model, whose technological center is the motor (and not the regulation); a person desires therefore to be the absolute master of machines, not their partner.[90] As will be the case with Jules Verne, the human imagination will seek to reconcile the contradictory demands of a perfect *self-control* and of an unlimited dynamism in the expansion of the fields of domination.

Phase III still maintains an air of "wisdom," if not modesty, in comparison to the acceleration of Phase IV toward the limit of all rátionality. But the supposedly sovereign rational subject can discern only that the very dynamism of its method is distorted with a displacement that will dispossess it of its mastery. Not being an epoch, Phase IV has no "date of birth," properly speaking. If, according to Marx, it announces itself as early as the completed expropriation of the worker to the benefit of capital whose circulation transcends all accumulation, then it confirms itself in the "automatization" of capitalism itself, whose centralization and bureaucratization strips the entrepreneur of his initiative, and even the bourgeoisie of its own privileges.[91] More generally, rational autonomy refers less and less to the person who commands machines and enterprises, and is more and more dependent on the auto-regulation of technological networks themselves: Gabor sees this mutation inscribed in the structures of innovation (less and less based on the satisfying of human needs and desires).[92] The passage to the "systemic" and to an informational rationality (no longer uniquely energetic) appears to be an indisputable criterion for crossing the threshold of Phase IV. But we are going to see that the perfections of auto-regulation do not reabsorb the tension imposed by the dynamism of increased power.

Phase IV and the future of power: Research

> ... all that is effectively possible is effectively realized, even if it entails
> looking to a later time for the use that one can make of it.
> —Robin Clarke[93]

> The essence of that which we call science today is research.
> —Martin Heidegger[94]

Described and analyzed since the inception of this inquiry, techno-science appears as a massive, organized, vast Complex in expansion. However, in becoming a

System, Phase IV perhaps makes us largely forget the death-dealing explosion that accompanies its detonation and which could well stimulate its inescapable end.

After the explosion at Hiroshima, what might have been the responsible attitude of a concerned humanity towards the mastery of its power? At the very least it might have been to take some critical distance and impose a delay on this beginning of a new era. If, as Michel Serres argues, the destruction of Hiroshima is to our "new history" what the fall of Troy was to ancient history, then does not the event and what followed merit immediate circumscription and meditation? But this time actual and *potential* reality completely prevail over myth. What is different this time is not humanity, which has always been unconcerned with the lessons of History. What is different is that the danger has swung into the future and is complete. Troy flamboyantly foretold an epic, inaugurated a history; Hiroshima closed history, announced a permanent and absolute danger. The event eclipses the myth, defies the narrative, even and especially in the unforgettable *Hiroshima, mon amour*.

Anxious prudence could have been replaced by a measure of reason: the decision for a general moratorium and putting into place extremely strict international controls. Instead of trembling before one hundred thousand suns, one became a Kantian. What did we witness? The sacred shudders of the first witnesses became an object of curiosity in the chain of newsflashes. The horror itself, detailed and commercialized, became boring. Consider Bikini: a bathing suit was made of that. The race had already started anew, one justification following on the heels of another: Hitler, the Japanese, Stalin.

What is this "rationality of the irrational," this ever-increasing destructive stockpile of micro-engines, MXs preparing to outdo the SS-20s and their successors, which today defies the imagination? Philosophical rationalism, for once allied with common sense, can only ask a bankrupt and impotent question: where are you, sovereign *Nous* that Greece admired in its Sages? The minimal rationality of the equilibrium of terror and game theory remains. The extreme rationalization of the plans for the execution of different programs, brutally contrasted against the extreme irrationality of fundamental choices, continues. The Vietnam war and the Apollo Project are typical examples of this disjunction within rationality itself: a fantastic programmatic on the one hand, and on the other, a demand for instantaneous decisions in the face of every idea, however contingent or subjective, such that it defines national prestige at any given moment.

These considerations, as disabused as they are (why would they be enthusiastic?), do not seek to nobly moralize; they draw our attention to the insufficiency of the conceptual means that contemporary rationality applies to itself. To speak of the "instrumentalization" of reason, for example, satisfies at first glance, but quickly reveals its insufficiency. This easy formula presupposes that technology triumphs only by virtue of machines and equipment, a conception that we have already shown to be doubly faulty; to reduce the efficacy of modern technology to its most visible aspects (effects more than causes) is to misunderstand the specificity of this technology as a Complex of techno-scientific power. On the other hand, the massive appeal to the human sciences in order to correct the damage or harmful effects of development is a facile solution whereby one reinforces the very process that one wishes to delimit. Instead of taking some critical distance in relation to the process of potentialization, one adds another level. Alibis or avatars of the deployment of Phase IV. The growth of Potential ensues, at a level with catastrophe in defying the understanding, in disturbing the compass of intelligence—but not without every possible pledge on the side of "rationalization."

It is possible to grasp three stages in the "regulating" of scientific rationality by Power. The most evident, the most easily denounced, is the utilization of techno-science toward military ends; examples of this are numerous: considerable budgetary allocations, sophisticated weapons, intense informational efforts devoted to capturing the military-industrial secrets of the enemy. This exploitation could not operate at the amazing level that it actually does without a "techno-logic" which is much more direct and rapid in its exploration. Only five years elapsed between the invention of transistors and their industrialization; the effects of power, both civil and military, have accelerated almost as fast. Tomorrow, it seems that the laser ought to be able to increase destructive possibilities in a prodigious leap, both on the planet and in space. Science and technology reinforce their interaction, functioning increasingly together within the "moments of prediction" of planners and specialists in Research and Development. The third stage, the true core of this process is the "operationalization" of rationality: the theory of systems, modalization, global networking install themselves inside intelligence and quickly destroy (or cynically exploit) all curiosity that is too rebellious. A single concept functions with surprising efficacy in carrying out, both theoretically and practically, this manifestation: Research.

A transparent concept, Research has not changed in definition: "the effort to find something," as *Le Dictionnaire Robert* indicates. But, in less than forty years, its reality has been transformed to such a point that it has rendered suspect, if one does not exercise caution, all discourse that intends to heal our evils through the reinforcement and intensification of Research. Certainly there existed in France before World War II a "national office for scientific, industrial, agricultural and inventive research," a precursor to the CNRS (created in 1939), and undoubtedly Louis de Broglie (in *Physics and Microphysics*) and Georges Duhamel (*The Time of Research*) announced the moment when the state would take charge of scientific research, but at that time there was neither the unification, nor the systematization, nor especially the prioritized intensification of scientific research. During the sixties, there were ten times more people involved in research than before the war;[95] the level of funding is without comparison to figures before the war. A specific trait of Phase IV, Research becomes an absolute *system of the optimatization of science*, of the programmatic of its work, of its results and of its technological impact. As J.-J. Salomon states so eloquently, "Pure science is no more than an element among others in a *system* which constitutes the activities of research."[96] Naive minds justify this evolution through reflections of common sense, of the sort, "It is clearly necessary to organize," or, "For real development, it is necessary to have materials and credit." As always, "nothing completely fails to hit the door," and there is some truth in these reflections. But systematization does not have any common measure with truly scientific exigencies: by definition, it is solicited, stimulated, organized; this accomplished, Research is caught in a double paradox. Its essence *inverts itself*; it becomes the complete opposite of a free quest, seeking to fix in advance that which it ought to discover. Moreover, truly scientific knowledge finds itself shaped by a new science, Research and Development, which assumes to know better than itself, even perhaps everything, what it ought to look for, or at least how it ought to look for it.

Of course, nothing is so simple: there are irreducible elements of irrationality in creativity, and some great discoveries—such as penicillin—have been made with ridiculously meagre financial means, outside of any massive worldwide planning. In fact, René Thom notes the "decline of generalized productivity" in scientific research, which he situates at the beginning of the fifties, precisely at the time when we began to inject considerable sums into laboratories.[97] For his part, Nicholas Rescher also establishes the quasi-exponential increase in the cost of scientific production, drawing from it a principle that he calls "Planck's principle," which is largely inspired by a statement from Planck: "Each advance of science increases the difficulty of the task,"[98] for which he gives a mathematical formula (the productivity of scientific research solely corresponds proportionately to the quantity of allocated resources). He concludes from this an inevitable slowing down of the progress of "hard" science, and sooner or later a zero growth, impeding the great investments of power.

Be that as it may, the misleading results of R-D for pure science are undoubtedly not to the same degree for the State, technocrats, the military, and business, who expect from Research a direct stimulation of the development of the economic, military, etc., and who program it in the hope of being able to generalize the performances obtained for *science-based industries* in the United States in the course of the 1960s. In France, Research has even become a kind of panacea for social, economic, and even cultural problems, a type of "post-crisis" laboratory.

Important debates take place between specialists, but also with politicians, concerning the profitability of research expenditures and even concerning the respective importance of "basic research" and "applied research." My project is not to participate in these technological or ideological jousts, but to place them at a necessary critical distance in order to discern their philosophical lessons.

First, concerning the limits of an excessively technological debate between "basic research" and "applied research," it appears that J.-J. Salomon has reason to denounce the extraordinarily absurd character of measures of profitability as "Project Hindsight": a systematic investigation into the profitability of scientific development from the point of view of new armaments invented in the last twenty years. Only 0.3% of scientific undertakings (themselves accounting for 8% in relation to technological "developments") would have been "profitable".[99] Not only has this inquiry been contradicted by others (such as the study "Traces"),[100] but it is characterized by a kind of incurable shortsightedness: it proceeds as if all the "hardware" of armament technologies had been able to be developed without the fundamental foundation (and dynamic) of modern science. I have already noted that the effects of the power of science are marked by two essential traits: first, their incommensurability; hence the, at times, considerable "distance" between a discovery and its direct or indirect technological effects. Second, as Salomon again notes, the setting of a deadline for the end of a study is completely arbitrary.[101] Therefore, it could be that "utilizers"—industrial or military—are technically correct in assuming that basic research is very costly to them (it is true that it is very expensive, particularly in physics), but this does not prove that the global increase in power does not principally result from this basic research. Otherwise, it is evident that governments would be only too happy to confine themselves to those projects which are technologically and financially well delineated and which have an assured profitability. If they do not do this, at least not yet excessively (in spite of the relative decrease of funding in Research and Development since 1967), it is because they have their reasons. It seems, indeed, that the general stimulation of scientific research allows a nation to remain a member of the "club" of ultradeveloped countries, ready to utilize as well as to make propositions, at the different levels of techno-scientific activity.[102] The direct application of nuclear fission has, therefore, imposed a new model of development, wherein scientific potential plays a driving role; this does not seem able to be invalidated anytime in the near future.

Coincidence? The systematic rise of Research began with Hiroshima, as if this explosion had been the rallying signal for every will to power. Governments believed that all power would surrender itself like nuclear fission, that scientific knowledge was henceforth a fantastic expertise *based on this model.* This is not true of all *science* (which uncontestedly flourishes *as such* in Phases II and III) or of *technology* (which above all flourishes in Phase III): whence arises the nostalgia of scholars for "pure science" and even that of engineers attached to a mode of production dislodged by this competition? But *it is still true enough that through force of will one verifies it* (this is what Research and Development provokes): the investigative thrust *globally* oriented toward Power. This is what happens in Phase IV, in which the governmental Megamachine ultimately directs [*hyperfinalise*] science toward power.

The complexity of problems tied to Research (including the technological problems of programming, coordination, and interaction between events that are quantified, in the future, etc.) ought not to conceal the fundamental "philosophical fact": the concept of Research effects a rupture that seems irreversible, because it is linked to the particular "rationality" of Phase IV, namely, the optimization of potential. Or even, as Salomon states, the maximizing of the possible is compensated by a "minimization" of costs (at least in principle).[103] The systematic exploration of everything possible is a decisive characteristic of the potentialization of Phase IV: in this regard no region of the real is taboo. From one day to the next, anything can become the object of research. Nevertheless, the possible is so vast that the aforementioned exploration is constantly compensated (and oriented) by an exploitation—equally systematic—of everything *effectuable*.

It would be too metaphysical or idealistic to represent Research as *truly* exploring entirely the possible, an enterprise which would be indefinite and "Borgesian." Research protects itself against this. The goal of Research and Development is precisely to select the *exploitable possibles*. The maximalization of a possible, previously enunciated, operates inside the realization of *well-defined objectives*. Research and Development is, in actual fact, research *for development*. And in this perspective, Research itself is a potentialization of science—and not for the love of science itself as certain soothing discourses would have one believe.

Through the prejudice toward this systematic domestication of episteme such as it was able to function in Phase II or III, the power of the rational becomes, in Phase IV, the rationality of power. Research is henceforth the execution of science, because science ought to proclaim itself as Potential (from power). The effectuation of this process consists in *holding in reserve* [*mise en réserve*] actual power according to a similar logic, if not identical to the logic of deterrence. In the global strategy that this implies, the effect of nuclear power is indefinitely delayed and stockpiled, not due to wisdom, but because it is the only way to *further* increase power without blowing up the entire System. Clearly, this potentialization is at once speculation about the possible (in the sense of game theory) and, in this very capacity, the effective increase of dangerous potential—without which it would cease to be credible. With some allowances, the same thing occurs at the global level of Research and Development as the systematic increase of the potential of power (scientific, economic, etc.); one systematically explores the possible, one lays wagers on it, all the time effectively increasing the potential for development. Research becomes the essential instrument of technological "advance," the decisive weapon of worldwide economic "competition." One recognizes this language, our ears ring with it; indeed, to dispense of the interpreter and demonstrate its logic would be easy.

Why Research? One hardly needs to avoid the response: because it produces tomorrow's power. But will this again be research of a superior power ad infinitum? Or is Rescher correct to predict the end of hard and "synthetic" research in favor of "analytic" discoveries, a kind of new triumph of scientific intelligence wherein Power will be blocked at its maximal point of expansion? In the two hypotheses, it is perilous to sacrifice *the entire scientific thought with all its possibles* to the type of potentialization of Phase IV (either definitively or temporarily death-dealing). Likewise, it is deplorable to wish with the best intentions to professionalize the University at any cost: Pierre Aigrain yesterday,[104] the socialist government today have ceded to the dominant current, which consists of a worldwide planning of science at all levels. Undoubtedly, it is as if it were destined that operativity will be gener-alized into techno-scientific truth, and that each will have its place in the anthill.[105] Within this "fatal strategy," science (piloted by Research and Development) be-comes the "chief investigator" of the techno-scientific complex.

Can it be otherwise? One will notice my phenomenological caution (which does not proceed from a will to cheap consolation, because the current worldwide evolution is not the least bit reassuring); everything happens *as if there were a des-tiny* in this race for power. We remain phenomenological in the sense that we guard ourselves against crossing the ontological threshold, from which an eschatology would propose itself. This reserve will not prohibit *all* ontology; one can better see this in interrogating the possible.[106] It is certainly the case that ninety percent of to-day's science is at work for power. It is certainly the case that the process is launched on such a vast scale that one cannot see who is able—or how—to stop it. It is also now certainly the case (the latter more philosophical) that this evolution represents the last phase of a multiform and complex potentialization that leads rationality to a point where it risks no longer functioning except in the name of power.

But does the type of potentialization of Phase IV completely exhaust the pos-sible? It risks sterilizing it, as it risks gathering for its exclusive benefit all remnants and potentialities still alive from Phase III and above all Phases II and I, thus knowing how to convene pure science and technology in the exclusive sense of the increase of Potential. The possible is held hostage by the power-effect.

"Science does not think."[107] Does not Heidegger's famous reproach, recently quoted again by René Thom,[108] apply to science as such? It would be useless to jus-tify this from the race for power, which has been inscribed *in ovo* in the very essence of modern science and, from Plato, in episteme. Everything that our genealogy of the power of the rational has established can be found in Heidegger's meditation. Yes, science has been constitutively enlarged through power since its beginnings, but incommensurably *at a distance* from its possible effects—and ambivalent in re-gards to the technological domination of nature. Yes, "the evil genius of the indus-trial system has found its cradle in the very essence of modern science." It is not Heidegger who notes this, but J.-J. Salomon,[109] among others. Yes, we are today faced with a new configuration of Power—which one calls *Gestell* or Phase IV—and no one knows exactly where it will lead humanity.

Yet, this configuration is so powerful in itself that when surrendering to it all the keys of the possible, one does not attribute more to it than it already has. Sci-ence-for-power: a new phase that must be understood as such, and not by a simple combination of its terms. It could be, indeed, that science itself, having contributed to this event, is threatened by that which until now had assured its control: *theoria*. We have established that Research in its actual dominant sense is the complete op-posite of a revival of this. Science-for-power: this formula could also lead us to be-lieve that science has finally achieved power, an anachronistic formulation of a new distribution of power (which overwhelms its possessors at the same time that it effaces its origins; the Oppenheimer and Sakharov affairs, among many other examples, each testify to this).

Science, in Phase IV, will no longer speak the language of Good and Evil, as episteme would also not do. If its most widespread effects of power can appear demonic, we must not forget that the possibility of evil lurks everywhere, including in the turning of the Good against itself, for example, in the dialectic of good intentions. Moreover, and above all, it does not seem certain that the possibles of Phases I and II are invalidated by the appearance of Phases III and IV—despite the specifically dominating, invasive, almost exclusive character of the latter.

To think what science does not think—before thinking that which it is unable to think—is to think that which it is able to think, *all* that it is *able* to think—in its *different* phases. To think the faces of potentialization is also to repotentialize thought, to redistribute its cards. Is the possible able to be analyzed once and for all, even in its infinity—as Leibniz argues? Potentialization creates new possibles, unimaginable yesterday even if they were possible. But this proves that the possible is infinitely richer than the potentialization *of Phase IV* has revealed: evidently rich in the power of these phases as in other reserved possibles.

We have established that science *potentializes* principally from the origin, that this potentialization is actualized in Modernity by the decisive catalyst of the Method, that it is unceasingly reactualized and displaced by a more and more contemporary form of this method: Research. What remains is to consider an incisive objection: does not contemporary science exhibit traits that are so novel that they are able to escape the dominating destiny that seems to seal the development of classical science? We rediscover the question of the power of science, the guiding thread of our genealogy of rationality. Like its unparalleled effects, hope clutches at the smallest signs: what if contemporary science were, *in spite of everything*, liberating? What if the "new alliance" that it prepares could prevail over its evil geniuses?

6

Science between Power and Alliance

Here is the context in relation to which we want to situate the metamorphosis of science. The context is that of classical science whose successes can be seen as tragic, and which we today no longer call our science.

—Ilya Prigogine and Isabelle Stengers[1]

Dust to dust, the last word of philosophy. We will see the light or we will trespass among the thousand suns of our infernal reason. Passing this threshold, we begin to speak of immortality. Of the new science.

—Michel Serres[2]

Certainly we lack the method at the outset; but at least we are able to dispose of the anti-method, wherein ignorance, uncertainty and confusion become virtues.

We are able to have all the more confidence in those excluded from classical science because they became the pioneers of the new science.

—Edgar Morin[3]

LIKEWISE, THE SAME rumor in Epistemopolis: "Science is no longer what it was. The New Science exists; we have encountered it. The world has already fundamentally changed; the future is going to sing again in laboratories and lumberyards!" Who would not be delighted to finally learn this good news in these times where the sordid dispute has infected everything and where—at best—banalization and "normalization" are law?

Good news or the Good News? The cantors of the New Science have an inspired and quasi-religious tone: *it was daring to venture* this title, *The New Alliance*. The authors had their reasons for doing this: one wants to presume this before testing its strength. What is certain is that the *idea* of the New Science offers itself with a great simplicity and fascinates as much by its clarity, its unity, its apparent evidence, all the more because it is buttressed by complex realities, protected on all sides, but whose total-power creates an infinite reverence. It is so easy and consoling to believe that scientific work never ceases to discover "the new and the reasonable," and that there is in it, in reserve, a recourse for our uncertain and threatened future! We want to hope that the solution to the perils, which have for a large part come from science (the rupture of much of the stability provoked by scientific-technological development), have already been prepared by science itself.

Yet, the idea of a New Science is perhaps not exactly a "scientific truth"; it does not substantially add to apodictic knowledge and established facts. Can it even claim the stature of a scientific concept, next to the others, more or less hypothetical, verified or not? Neither its content nor its mode of dissemination exhibit the ordinary exercise of science.

Its content: is it a collection of facts or a set of truths that science hands over to us as part of itself? What is, truly, the status of this "new science" in relation to science as such? In what does its novelty consist? Is it new *only* through its results? Hypothesis or theory? Fact or myth?

Its mode of dissemination *inverts* the customary procedure of a scientific discovery, first established in a narrow circle, communicated to the community of scientists, then possibly to a larger public. In the case of this unified idea of "new science," the interlocutor is immediately the vast audience of reasonable people interested in science; one presupposes (or pretends) that the scientific underpinnings have been acquired in specialized and authorized circumstances; the pieces cited are supported by proofs that are numerous, impressive, and seemingly above all suspicion. What more could one ask?

This *je ne sais quoi* that is lacking in the almost overly sensational presentation of the New Science, is this not a necessary philosophical analysis? If this idea is not *strictly* scientific in the sense that there is a theorem, a verified law, or even a theory, why not recognize its philosophical character? What harm would this do? Undoubtedly, none; but the obligation for a critical reflection on the very idea of "new science" would make sense. For now I assert that in not presenting this idea *as such*, one economizes on this philosophical work: the announcement of the (new) Good News, by short-circuiting the doubts and uncertainties of the critical moment, permits a direct linkage of science with ethics.

If the New Science has somewhat mythical connotations, which relay and relaunch the great scientistic hope, the beguilement no longer operates at the outset; without doubt it is revived, but in a more subtle manner. Lyricism makes itself more furtive, religiosity allows itself to be sensed more than it asserts itself, the pavilion can no longer hold the merchandise. In order to be haunting, the myth must be accompanied by an impressive mass of documents, hope must give the impression that it is allied with objectivity.

My hypothesis, therefore, is this: a new ideological complex is constituted. It is neither pure myth nor one scientific fact among others. A hybrid too recent to be fully understood, it deserves examination with the weapons of critique—not of polemic. Is it a matter of the myth of the end of our millennium? Or, on the contrary, is it a matter of a new epistemological configuration that must assert itself? Certainly, between these two extremes lies a complex truth. Every myth sustains conflictual relations with reality: it is necessary to establish how these roots grow.

The epistemological "truth" *speaks* a science which already exists and which loses strength in systematizing itself. This is an occasion to reflect on the ambiguous status of this difficult "discipline," at the most delicate junctions of proven knowledge.

I do not bestow any superior right on the authors whose theses I am going to analyze or discuss, and who would authorize me to call into question the spiritual and ethical hopes nourished by certain scientific breakthroughs. The only right claimed here: to test the philosophical coherence of this idea of new science, delimiting its place, and isolating, as much as possible, its constituents. I take up, therefore, the question that I have up to now resisted: starting from the incontestable "points of rupture" that can be isolated in modern science, is it permissible to sketch a *unified outline*, to exhibit a single configuration of the novelty of science, a novelty *as such*, inaugurating a specific epoch? It is first necessary to give shape to these comments by getting one's hands dirty, noting the new scientific realities.

The new scientific realities

It is an unbelievable venture to want to give a summary of the new scientific realities in just a few pages! The methodological aporias prove themselves to be considerable. Where do we begin to isolate the significant novelties in the long chain of scientific progress? Which order to adopt in order to classify and understand the aforementioned novelties? How to treat an ever more considerable mass of information?

Despite these difficulties in theory and in fact, the enterprise continues to be attempted, and by the best minds. Without doubt, it corresponds to a somewhat indispensable demand: to understand contemporary science, even partially, is to throw some light on the agent of many transformations; it is to draw closer to the source of technological power; it is to gain a point of view on our collective future. The quick panorama that I am going to attempt assumes only to *prepare* the discussion on the interpretation of the New Science as qualitative mutation. For the sake of convenience and in order to confine itself to the essential, the exposition will limit itself to two grand disciplines (mathematics and physics) which today share, always in a privileged manner, the incontestable status of apodictic or positive sciences.

Prigogine and Stengers provide a guiding thread that seems quite useful in finding one's way in the different labyrinths: that which they call "classical science" (which is to say, modern science, from the seventeenth to the nineteenth century) was itself constituted from "a central conviction": the conviction that the *microscopic is simple*, ruled by simple mathematical laws."[4] This science, for example, Newtonian physics, was deterministic and mechanist. Its laws sought to describe the objective reality of the world "in terms of deterministic and reversible trajectories." The fiction, imagined by Laplace, of an omniscient genius capable of knowing, at instant t, the position and speed of each element constitutive of physical nature, would symbolize the project of this universal objective science, perfectly determined and ultimately realized. Certainly, one could object to Prigogine and Stengers, arguing that infinity and chance play an essential role in the scientific intelligence of the seventeenth century, as Pascal and Leibniz have demonstrated. Are the infinitesimal, integral calculus, or the calculus of probabilities "simple"? They already represent an admirable quantitative mastery of the phenomena of movement, acceleration, series, chance. We nevertheless grant Prigogine and Stengers the insight that the already considerable degree of complexity in the reasoning and procedures of classical science does not taint the general project of a definitive mastery of the understanding of the universe from a small number of simple laws, as is evinced in the myth of the Great Watchmaker, still very compelling up to the end of the eighteenth century.

Retaining this idea of classical science, we will inspect the decisive fractures of this knowledge in its principle domain, the mathematical field. Jean Dieudonné notes that, from Euclid to the nineteenth century, deductive reasoning continues to be marked by a double proximity: in relation to intuitive evidence, and comparatively in relation to experimental givens.[5] One could also object here that Galileo, for example, established the law of falling bodies only by breaking with the evidence of common sense, which was in large part supported by Aristotelian physics. It remains no less true that contemporary axiomatization decidedly and definitively severs every connection with intuition. Dieudonné recalls that, in the second half of the nineteenth century, the progress of Analysis apparently engendered "these mathematical entities, stupefying contemporaries, which are curves without tangents, curves filling a square, surfaces applicable to a plane without rules, the first specimens of a gallery of monsters which continue to grow until our time."[6]

Not only is the effected formalization, beginning with Hilbert, going to permit at once the reformulation of Euclidean geometry and the axiomatization of non-Euclidean geometry, but the deepening of the "crisis of foundations" (Dieudonné argues that mathematics had known nothing of such scope since the discovery of irrationals) is going to lead, with Gödel's theorem and afferent theorems, to the demonstration of the internal limitations of formalism.[7] Modern mathematics can no longer ignore the paradox; it must do even more because it includes this paradox in its field of research: Mandelbrot demonstrates that Brownian movement was the first of the 'fractals' to be studied.[8] The concept of the fractal (or fractary) dimension permits the mathematical description of "paradoxical objects," such as the coast of an island of Brittany, a crater on the moon, the distribution of stellar matter, etc. In a comparable but certainly much more mathematically complex manner, the theory of catastrophes, attributed to René Thom, establishes the formal condition of the appearance of discontinuities in a given system; or even, as René Thom explained in a conference in Nice, it allows us to predict (mathematically) the "leap that the system must make in order to escape destruction."[9]

This far too rapid inspection of contemporary mathematical revolution already allows us to imagine the numerous possibilities for application in the physical domain, but suggests above all that the image of nature and the universe has been profoundly modified since the end of the nineteenth century. Indeed, as it has often been noted,[10] the physical world appeared to have been dominated and definitively synthesized about one century ago; the main branches of classical science—including chemistry—had attained a point of equilibrium and apparent perfection: "All matter is reduced to 92 elements, all facts of physics are translated into several differential equations, is there a more splendid construction for human reason?"[11] In a few decades, this beautiful edifice (already weakened from the inside by Gibbs and Bolzmann's mechanical statistics, for example) is cracked, and then completely shattered: we cite the successive discoveries of the electron, X rays, then radioactivity. At the conceptual level, the theory of relativity dispels the notions of absolute time and space: mechanistic actions can no longer be conceived as reproducing themselves instantaneously, but are relative to the speed of light, a universal constant that is the limit of all movement. The principle of determinism is really questioned only in Max Planck's theory of *quanta*, which, according to Heisenberg's expression, "forces to give a statistical formula to laws."[12] For his part, Heisenberg demonstrates that Planck's constant constitutes the "lower limit of approximation of the product of these two indeterminates" (position and speed).[13] Prigogine and Stengers note, however, that "the physician does not discover a given truth that silences the system, he must choose a language."[14] But of utmost importance to the authors of *The New Alliance* are the extensions and consequences of this new physics in the understanding of time and life. Newtonian physics ignored life and attempted to demonstrate the laws escaping time; on the contrary, statistical thermodynamics takes into account the irreversibilities resulting from entropy, the fundamental limitation of life. Thus are Prigogine and Stengers able to affirm that contemporary physics succeeded in an exploit even more incredible than pushing back the outer limits of the world, knowable up to the difference of a scale of 10: the conception of a universe in constant evolution, where "strange objects, quasars, pulsars, galaxies explode and disintegrate," and where stars "blaze into *black holes*",[15] a conception that recognizes even more clearly that "each complex being is constituted by a plurality of time, extending one upon the other according to subtle and multiple articulations."[16]

Thus we have not had to remove ourselves to some distant Museum of Discovery in order to gather these new scientific truths. Occurring as a succession for the last two or three generations, these truths now constitute the enlightened and quite impressive *corpus* of contemporary scientific knowledge: an unprecedented intellectual landscape, which looms above both technological successes and philosophical perplexities.

For the most part, these novelties do not merely defy good sense: they are disconcerting to every definitive view of the truth, blurring the line between reality and fiction, confronting humans with the undecidable. Is it therefore not justifiable to give a unified—probably provisional—concept, in the spirit of a new Science? This is undeniable. What remains is nothing less than to test the internal coherence of this idea, and to critique the subsequent mythical connotations

Idea or myth?

How do we test the coherence of the idea of the New Science? Jean-François Lyotard responds to this inquiry in his book *The Postmodern Condition*, which is presented as a report *on* knowledge. Treating the "research of instabilities" in contemporary science, Lyotard erects a topology which is more complete than that which I have been able to do here. But he introduces in this book a striking systematization in connection with the new modes of societal organization and with his "pragmatics"; the result is that these instabilities define a new culture, a new epoch, a new condition, which the epithet "postmodern"—coined on the other side of the Atlantic [the United States–Trans.]—unifies and qualifies, for lack of something better.

Lyotard judiciously situates science as "conflict with narratives."[17] The basic opposition between science and fables is surmounted by that which Lyotard calls the modern "discourse of legitimation"; it is the discourse that science holds "to legitimate its rules of play,"[18] the moment when science becomes, *nolens volens*, philosophy. The great discourses of self-referential legitimation have been Hegelian or Marxist dialectics, Saint-Simon's rationalism, and hermeneutics; we add here positivism, which Lyotard strangely omits. How do we define the novel configurations of knowledge in relation to these "discourses of legitimation"? Lyotard responds, " . . . I define *postmodern* as incredulity towards metanarratives."[19] Therefore, the subjective disposition underlying this is, through its negativity, privileged. Is this subjective support the noetic correlate of the New Science? The new knowledge, defining itself by its heterogeneity, has no unity other than the nominal and historicist.

Such a presentation is not satisfying for the following reasons. On the one hand, clearly the "incredulity towards metanarratives" does not correspond to the *specificity* of the new knowledge." Lucretius and Voltaire provide the proof—already faced with metanarrative. Even if we admit that the totalizations of the nineteenth century were the only metanarratives worthy of the name, does it suffice to react with incredulity towards them in order to discover the theory of relativity or the logical theory of limitation? A second objection, certainly more fundamental, is added to this first objection: What is Lyotard himself presenting in *The Postmodern Condition* if not a metadiscourse? A report is a discourse and it is a discourse *on* knowledge. This "metanarrative" recounts the *saga* of incredulity, dispersion, and instability. It certainly does not *reproduce* the metadiscourses of the nineteenth century. It reverses them, denies them, disperses them; we can readily concede this. But is the denial sufficient? The book ends: "This sketches the outline of a politics that would respect both the desire for justice and the desire for the unknown."[20] Here again, more invigorating than ever, the wind of History carries the mythic charge of the future. Where has incredulity gone?

Lyotard himself, therefore, also constitutes a metanarrative, but he is very careful to ask himself about its rules, divided as he is between the commensurable (the pragmatics) and the incommensurable (the "paralogic of inventors"). Furthermore, under his pen, Gödel's theorem becomes "a veritable paradigm of how this *change of nature* takes place,"[21] which is to say, it includes within the scientific discourse those paradoxes which constitute it. If there is a change of nature, certainly the new scientific spirit has a positive, delimited content in relation to which "incredulity" appears as a very frivolous accessory. Is this a change of nature relative to classical science, or to a fundamental and originary project of rationality? Lyotard does not pose this question, any more than he questions the status of language that he uses in his report on knowledge. The unity of contemporary scientific knowledge is at once denied (as metanarrative) and affirmed (as a new historical constellation and new "nature" of rationality). On the one hand, he demands incredulity; on the other, he reintroduces a messianic scientist who does not say his name.

The classical science/new science distinction in Prigogine and Stengers supports the distinction we have noted between modern and postmodern knowledge in Lyotard. Here we find the same dichotomy, the same affirmation of a qualitative mutation characterizing a new epoch. To be sure, Lyotard places more emphasis on the heterogeneity of languages in the postmodern culture. Nevertheless, the New Science is not any less greeted in an analogous manner and invested with hopes which are very closely related. Yet, the authors of *The New Alliance* carry it one step further, properly ontological—if not religious: they maintain the thesis, indebted to contemporary science, of a reconciliation of man and nature. One does not have to look elsewhere in order to find the content and form of the reconciliation that is prepared: the content is the "fragmented universe, rich in qualitative diversity and potential surprises,"[22] which contemporary physics and biology discovered; the formal vehicle is none other than the totality of scientific languages becoming the "poetic ear" of nature.[23] The "disenchantment of the world" comes to an end; the cold, inert, determined universe of classical science yields to fresh visions of nature and the cosmos. "*Life, destiny, freedom, spontaneity*"[24] become the themes of this symphony of the new world that plays henceforth the metamorphosis of science.

I have suggested that the border thus crossed is no longer that which separates one epistemology from another—as Michel Serres claims when he evokes the upsurging of a (new) Novel scientific spirit—but rather the boundary between epistemology on the one hand, and ontology—if not religion—on the other. This boundary is at once crossed and confused by the authors of *The New Alliance*, who, beyond exposing and directly interpreting scientific novelties, undeniably postulate a thesis *on* science and its intrinsic relationship to the world (and, in so doing, postulate a thesis on the world itself). At the same time, it is perhaps legitimate to discern here the reintroduction of a myth, insofar as they use the narration of an inaugural gesture, the announcement of an era which restores the time, life, spontaneity, and treasures of the primitive life that modern humanity seems definitively to have renounced. Despite this nonnegligible dose of lyricism, the inquiry is carried out with a seriousness that is lacking in the neo-mysticism of Raymond Ruyer's essays in *The Gnosis of Princeton*.[25] The inquiry is also conducted with more rigor than Edgar Morin's *The Method*, which characterizes the crisis of twentieth-century science as "the upsurge of the non-simplifiable, the uncertain and the confused."[26] The non-simplifiable, the complex? Without any doubt. The uncertain? In relational or statistical limits, assuredly. The confusing? This is another matter altogether. One can find other lapses of this kind in Morin, for example, when he interprets Benard's vortex, convection cells appearing in a liquid heated to a certain temperature, as signs of a universe that "constitutes its order and organization in turbulence, instability, deviance and improbability."[27] According to Morin, it is almost Disorder *in itself* which gives birth to order. There is nothing of this in Prigogine and Stengers. What they do with skill is to prepare their epistemology from ontology in order to better enhance the very image of the scientific enterprise.

Thus they reverse (word-for-word, concept-for-concept) Jacques Monod's conclusion in *Chance and Necessity:* "The old alliance is broken; man knows at last that he is alone in the immense indifference of the universe from which he has emerged by chance. As with his destiny, his work is written nowhere. It is up to him to choose between the Kingdom and darkness."[28] Why does Prigogine, when taking apart Monod's rationalism, judge the ascetic and stoic disillusion of the great biologist to be dangerous? The answer is given from the beginning of *The New Alliance*: "Modern science inspired the awe of its adversaries who saw in it an unacceptable and menacing enterprise, and of its partisans who engaged in research so heroic that it was necessary to make a tragic decision in order to take it up."[29] The last is none other than Monod's "ethic of knowledge," alone against all consolations. Does he remain prisoner of a "classical"—and mechanistic—"conception" of the world?[30] Is Monod Prigogine's bad conscience, Mephistophelian, a "spirit that always denies" that which the New Faust wishes to celebrate?

How do we think this confrontation? We already have our first lesson: two great scientists, equally competent, have visions that are metaphysically and diametrically opposed. Science, for Monod, was always the pursuit and success of the same demand for explanation held from Thales to Einstein; it was, finally, a triumph over the entire planet of an "austere and cold idea," imposing itself not by reason of its intrinsic truth, but due solely to "its prodigious power of performance," a substitute for the old, fleeting magical powers.[31]

Who is correct? Metaphysically, Prigogine does not assert himself over Monod any more than Hegel has "reason" over Kant; nor—in the two cases—is the inverse true. In the last analysis, who can decide if nature "speaks" or not? Each decides in his own conscience, the last temple of "philosophical faith."

What is, however, possible and desirable in the limits of this work is a critical recovery of the *intermediary series* and connecting reasonings that allow us to pass from epistemological statements to philosophical options. More precisely, the difficulty can be formulated in the following manner: Does the dominant movement of contemporary science conduct itself so as to liberate classical models? Or further, does the incontestable metamorphosis of science change the fundamental project?

Critical reappraisal

It appears that we are able to carry out our critique at three levels: the first, relatively superficial, where it is simply a matter of correcting errors or, at least, of correcting hermeneutical slips; a second level where more fundamental epistemological options will be set out; and, finally, a third plane where the effective functioning of contemporary science will be faced with the mythical connotations of the New Science.

At the most directly accessible level, our critique benefits in several ways from the superficialities that our interlocutor allows himself. Edgar Morin's partial and biased presentation of Bénard's turbulences has been discussed. Other extreme glosses or slips permit Morin to interpret the relation of incertitude as a success of the "anti-method," if not of confusion.[32] Under the pretext of describing instabilities, must one have a destabilized vision? One could yield to vertigo in the face of the prodigious expansion of the borders and the unsuspected windfalls of contemporary science. But is this vertigo to be extolled? Above all, is it the constitutive armature of this science, and is it through this that science pursues its formidable course? Russell in the *Principia Mathematica* and Gödel in his 1931 article remain more or less rigorous logicians. Although the former develops the paradox of the totality, it is at the interior of a formal system; the latter does not remain satisfied with postulating the impossibility of formalizing the non-contradiction of a system: he demonstrates it. The integration of paradox and the undecidable into logical discourse does not mean the abandonment of the fundamental principles of rational thought or the rules of inference; on the contrary, it can appear as their ultimate and subtle ratification. In the same way, it would be arbitrary to transform fractal objects in flux as adrift: it is the very nature of fractals to be calculable and calculated. Undoubtedly, Morin and occasionally several others allow themselves too much superficiality in order to sustain the idea and reinforce the myth of the New Science. If the evidence is so clear, why these manipulations? In any case, our critique would be too limited and itself too superficial if it were satisfied with reprimanding these theoretical pilferings.

I indicated above that the second level of critique attains a truly epistemological dimension. It concerns itself with analyzing, essentially in Prigogine and Stengers, the theoretical organization that permits disengaging the New Science as an autonomous unity. First of all, the rupture between contemporary science and classical science is not really *demonstrated*. Certainly, it has been established that the operative schemes of classical thought (its "archetypes" as Kuhn argues) have become dated (its cosmological model, its conception of time and space, its ideal of knowledge). But does this signify the invalidation of many of its procedures, its essentially positive results (confining ourselves to contemporary physics: the law of falling bodies, the utilization of differential and integral calculus in the problem of limits, whether it is applicable to the speed of mobiles and of particles, the pressure of fluids, of work accomplished by constant or variable forces, etc.)? Likewise, Euclidean geometry is not purely and simply dismissed by non-Euclidean geometries, but, on the contrary, is extended (and formalized by Hilbert); hence, the fundamental rules of logic and of mathematics and the laws of classical physics continue to reign over large areas of scientific activity, conditioning our objectification of daily life. Can it be argued that the rules of elementary arithmetic are invalidated because one is no longer in the habit of computing mentally, but has automated these operations through electronic calculators? All allowances being made, it is this type of paralogism that is adhered to by certain sectors of the New Science. From an important change in the organization of experience (and perhaps the vision of the world), they infer that a radical ontological mutation is taking place *in science itself*. A more prudent approach would permit perhaps taking the side of the evidence, which is to say not transforming hastily the "complexification" of science into a radical metamorphosis of its essence and its fundamental project.

Has science itself changed *in nature* (as Lyotard, for example, suggests)? I am not alone in doubting that it has. Monod would also contest it. And René Thom would strongly agree that the "scientist is a priori deterministic: to recognize a fundamental indeterminacy in phenomena (as one at times does in quantum mechanics and as it is repeated in emulation of the neodarwinist theory of evolution) is an anti-scientific perception par excellence."[33] Has Einstein not already affirmed that "God does not play with dice"? He would consider the general laws of nature as relatively invariant according to the transformation of Lorentz.[34] The theory of *quanta*, often invoked, certainly teaches that one cannot predict the "precise moment" of an energy emission, for example; but it does indicate a degree of probability, the statistical delimitation of the phenomena whose degree of certainty is therefore going to depend on the scale.[35] Heisenberg indicates that one again finds the same problem in the determination of the critical mass necessary for an atomic explosion. Similarly, the problem resolved by the relation of incertitude can only be posed for a science which, having attained an extraordinary degree of precision,[36] seeks total certainty, and—failing to attain this—succeeds in evaluating the constitutive limit it comes up against, taking into account its own intervention in the observed system. This is still an ultimate victory of precision. The discussion surrounding the alleged indeterminacy of the New Science revolves to a large degree around the interpretation of probability: in utilizing statistics, does contemporary science diminish or augment its hold on things? The facts leave no doubt and are settled in favor of the second hypothesis, which Kant already chose when he affirmed that the "*calculus probabilium* . . . does not contain probable but perfectly certain judgments concerning the degree of the possibility of certain cases under given uniform conditions."[37]

Returning to Prigogine and Stengers, their epistemological presuppositions are not without difficulties or even contradictions. Pretending to follow Koyré, they insist on the fact that "it is the *experimental dialogue* which constitutes the original practice of what is called modern science";[38] but Koyré himself shows, particularly in regard to the experience of Pisa, that Galileo would never have formulated his law of falling bodies if he had confined himself to observation and experience.[39] Moreover, the authors of *The New Alliance* acknowledge some difficulties when they themselves adjust the priority given to the experimental dialogue. Modern science, they argue, "engaged in the experimental dialogue, but starting from a series of presuppositions and dogmatic affirmations."[40] Indeed, without the presuppositions of the possibility of a mathematization of nature, starting from a projection of Euclidean space on nature, one has difficulty seeing how modern physics could have constituted itself. Remove the mathematization, and this physics collapses. One has the impression that Prigogine and Stengers are reluctant to frankly acknowledge this truth, in the same way that they situate the origin of science in an immemorial dialogue with nature, where—here again—the epistemology soothingly allows them to turn their eyes from this decisive threshold, the discovery of the apodictic dimension by Greek mathematicians.

There remains the aforementioned third level: taking into account the *effective functioning* of contemporary science as "techno-science." Even if one could admit, with Prigogine, that today's science opens a possibility that is able "to create the circuits of a culture,"[41] this immense fact would remain at the planetary level; scientific knowledge is canonized, capitalized, and systematically stimulated to the almost exclusive ends of the increase of military, economic, and other forms of power. It is easy to point out, unfortunately, that the non-dominating seeds and the poetic possibilities of the New Science still carry too little weight, if they shall ever matter, in the face of the stupefying march toward the objectification of all orders of phenomena, in the face of the unbelievable increase in the control of technological operations at the planetary and spatial level (high-precision electronic materials, the use of perfected radar and lasers, surveillance and telecommunications through satellites, unprecedented archival information systems, etc.). In the face of this almost exponential and perhaps irrational increase of a *certain* rationality (and this, despite what one rather lightly names the "crisis"), is Prigogine's silence really reassuring or even justifiable? It appears to him "dangerous" and even "more than" dangerous—agreeing with Heidegger that "the scientific project accomplished what it announced since the dawn of the Greeks: the will to power hidden in every rationality."[42] But, on the one hand, Heidegger does not exactly say this, because, according to him, the dawn of the Greeks escaped precisely the will to power; on the other hand, one would expect of defenders, convinced of rational and positive knowledge, that they methodologically and precisely discuss the thesis of a continuity between scientific mastery of phenomena in their constant laws and the technological acquisition that permits a transformation or control over the aforementioned phenomena, rather than peremptorily rejecting it in the outer shadows of obscurity. Is the less dangerous attitude one which discreetly turns its eyes from the effects of the power of science or considerably minimizes them, therefore, renewing at the end of this century where one believed it had become impossible, scientific idealization, which had declared in the nineteenth century a premature triumph, certainly condemned forever?

Why this impulse to come to the aid of a science which so many contemporary realizations claim to "supplant", and why this touching, pleading sermon in these matters where one could wait in the coldest and most definitive serenity? In order to respond to this, perhaps it is necessary to ask the question: what does the myth serve?

A new scientism?

It would be naive and in bad taste to accuse supporters of the New Science of executing, start-to-finish, a totally deliberated strategy. Surely, we have established that Prigogine and Stengers do not hide the stakes of their book, namely, to end the "terror" in the face of the new aspects of contemporary science, to reconcile contemporary humanity with this science, to prepare a new "cultural dialogue" in which nature would poetically "speak again" through, and thanks to, science itself. At the same time, *The New Alliance* names its adversaries: the militants of the "anti-science" movement (a confused troop where none other than Pauwels, Bergier, and . . . Heidegger clandestinely appear!).

However, this work does not concern itself solely with *The New Alliance*. Above all, the need to hope is vital; it slips its way into each of us, even when we deny it; science has become such a source of power, such a formidable key to the future, that it would be consoling to believe that it is not reduced to this "abominable motor of a new history" that Michel Serres tries to reflect upon in "Thanatocracy."[43] In *fact* and at its origin, if a myth articulates itself around the New Science, it is the collective unconscious which carries it beyond the masked intentions or assumptions of some authors.

Be that as it may, this myth is scientistic insofar as it attributes to science, and science alone, the capacity of engendering a creative future. It is necessary to note that, in affirming that science "makes known things as they are, resolves all real problems and suffices to satisfy all the needs of human intelligence,"[44] scientism does not say everything about *contemporary* science; on the contrary, it passes over in silence (or presents only in a systematically favorable light) this destiny of power whose ability to destroy through nuclear armament is a perilous and perhaps suicidal issue. One could excuse Renan and undoubtedly even Le Dantec in 1911 for ignoring what journalists prudently call the "relapse" of scientific knowledge. After the two world wars, after Hiroshima, this is no longer possible. The merit of Michel Serres is his contribution to the shattering of the "consensus of silence" that gagged scientism. To cite him: "Humanity is collectively suicidal. . . . The monster of the collective unconscious has reached consciousness; this would be but a small matter if it had not come to reason, which does everything to dissimulate it. You will always find an ideology, a system named scientific, or a consensus of silence to hide this fact from you."[45]

Why is it so critical to shatter this law of silence, even if it is necessary to make better laws in thinking the motives, in thwarting the mobiles? Because it is here that we slide into our future and it is here that we locate this question that we address to the New Science, if we know to listen to it in spite of the magnificence of its exorbitant powers. Is the destiny of power inscribed in the essence of rationality and in all rationality? Or better, is this destiny only established in the specifically modern conjunction of the mathematization of nature and the will to certainty, a conjunction sealed with the word *method*?

Depending upon the response (that the theory of the phases of potentialization prepared), one will assign varying recourses to thought.[46] If one responds affirmatively to the first question, one will only seek "salvation" in a disengagement from the rational as such; it could take the form of a brutal rejection (one thinks of hippies and of the most radical ecologists), or of a progressive apprenticeship in a thoughtful and more original poetic language. If one leans, on the other hand, towards the other hypothesis (the destiny of power truly actualizes itself only in the metaphysico-scientific project of Galileo, Bacon, Descartes, Newton), one will be less inclined to reject the rational as such, and more disposed to watch for indications, in rationality itself, of a movement that permits escape from the fatality of an exclusive and obsessional growth of power. Without expecting everything from the New Science, one will not necessarily refuse its teachings.

To the end of orienting ourselves toward this second hypothesis, I have tried to incriminate a total and hasty confidence in a science, which, under the pretext that it discovers new horizons and appears still rich in potentialities, sees itself suddenly embellished with all the prestige that one confers without control or prudence upon the culture to come (as if it could directly produce this culture). *Exploited in this sense*, the New Science becomes a scientistic myth, useful for remobilizing intellectuals posing to desert the scientific community, or, more globally, in order that an ill-informed public that scares easily regains its confidence in science, scientists, their institutions, and their projects.

The authors with whom I have taken up a dialogue on this subject are too intelligent to have sacrificed everything to this kind of scientific militarism. All the same, I have already pointed out the nonnegligible nuances that subsist between them. While Serres despairs and all his arguments push him to see the New Science as intervening only in extremis to save the situation or permit it,[47] Prigogine and Stengers, Morin, and even Lyotard locate themselves in the new scientificity with the satisfaction of the collector who takes a tour of the property and promises to create a beautiful interior from such magnificent walls.

What makes for difficulty in the conclusion of *The New Alliance* is not the poetic perspective that is beneficially opened (through which a vulgar scientism is illuminated); it is rather the pronominal which can allow for the belief in a kind of poetic self-engenderment of scientific knowledge. Let us recall the text: "Scientific knowledge, drawn by the dreams of an inspired revelation, that is to say, a supernatural revelation, can today simultaneously expose the "poetic ear" of nature and the natural processes in nature . . . "[48] Wanting to rush ahead of schedule, does one not short-circuit thought, its perplexities, its questions, its long patience? And does one not fall into a poetization of science that is too direct and too easy, which risks—like science fiction—drowning some irrefutable illuminations in the waves of a purely imaginary "consumer"?

For its part, the fall of "The Thanatocracy" is as ambiguous as it is brutal. To conclude, I must return to this text that constitutes one of the epigraphs that has captivated me from the beginning.[49] It is to the complete credit of this "new science" that it itself recognizes the supreme power to say "ashes to ashes." It is, literally, to make it the trustee of the last word, and of the final forces, of philosophy. Certainly, *to speak* of it is not truly to repeat it, which Michel Serres protects himself against as well. But salvation is there, on the side of the New Science, as if henceforth there will no longer be, for the human spirit, another side [coté] (in the Proustian sense) where freshness and vital inspiration are found. Is the messianic lyricism that accompanies a spectacular rallying sufficient to summon our "infernal reason" and the formidable functioning of its triangular motor (the industrial complex articulated by scientific research in its quasi totality, both finalized in "military applications")?[50] The originary innocence of reason has been proclaimed, treason denounced, the "rabble" accused. We would like to know more of this violent history, our history, to understand a bit better this genealogical uncertainty of a reason gone mad, ultimately to be certain that we do not wager the rest of history on the New Science because of its singular novelty (an alibi of its compromising triumphs), as if to find a conclusion at any price. If we did know more, we would be better able to take up the incisive questions raised in this passage: "Whence does our course of calculated suicide arise, what has made our reason a reasoning of death, how is it that this theory borders on terror? . . . "[51]

Does the rumor in Epistemopolis calm things? Nothing is less certain. If we listen to it attentively, and if we look at it closely, it is clear that science has declared itself new for more than three centuries. The *Novum Organum* remains emblematic of a scientificity that set itself against the principle of authority, Aristotle's cosmology and logic; but unquestionably, it ought to be thought more profoundly as constitutively oriented toward the New: "To be new: this which belongs to the World comes to be conceived as a Picture," Heidegger notes.[52] The conquest of the world, its objectification by man, who has become explicitly a *subjectum*, is an enterprise constantly undertaken by a science which wants to do research, a technology which adapts itself to be more efficient, an economy which aims towards expansion, a politics in quest of superpower under the guise of progress. In this regard, the postmodern condition does not appear to break with modernity, but to prolong it, in continually deploying more before its dynamism. Is its most urgent task to further accelerate the movement, in narcissistically celebrating its own myth? Let us rather slow down the pace, if there is still time, so that the New can engender this incalculable miracle: a human thought capable of entering into a dialogue with science— old and new—and perhaps deriving from this dialogue a counter-power [*contre-pouvoir*) to its excessive power.

Thus, in reference to the phases of potentialization, there is no longer any doubt that science is not situated in a single place. Although scientists have difficulty admitting it, science *potentializes* in every case, however unequally. Placed against mathematical potentialization and its epistemological deployment, it *reserves* the possible in the sense of Phase II; aiming at a determined and arranged mastery of the real through the intervention of technological mediations, it *understands* and *transforms* the world, on behalf of a humanity thought as rational and sovereign, following the trajectory launched by Phase III. Such is still often the ideal of scientists, very close to those of the nineteenth century; but here *functioning* more and more at the heart of a technological Complex, and at the almost exclusive service of the increase of Power, ensnared by the logic of Phase IV. I have not only critiqued a "myth" in the New Science, insofar as it contributed to obfuscating this formidable force exerted on science as Institution, but also—from within—by its "modalization" as Research and the convergence (almost pushed to homology) between scientific operations and techno-industrial operations.[53] Whatever the degree of influence of Phase IV, the passage from Power to Alliance cannot be easily made, if it is the case that this does not reduce itself to a production nor to a techno-discourse. If it were to be profound and durable, how could a rediscovered Alliance with nature surrender itself completely, produced as one concept operative among others? It would be wise not to exclude the possibility of such an Alliance, to safeguard its chances, to think its access. This will be the task of "contiguity."

7

From Absolute Power to Contiguity

IF THE MOST recent aspects of science do not diverge essentially from the destiny of power sealed by Phase IV, are we simply left with the task of preparing an appraisal and analyzing prospectively a development that is overtaking us and that appears ineluctable? To be sure, the Complex of Power allows itself to be understood in its functioning (and not in the secret motives of its "finality"), but must we be satisfied with this? It is not certain that pure and simple acceptance of this fact is desirable or even *tenable* for a thought worthy of this name. Through the vitality of its hopes, scientism itself bears witness, despite itself, to the impossibility of a suppression (or definitive integration) of every interpretation (even if reduced to an ideology). What possible comes to contest the exclusive domination of one configuration of the power of the rational? Is it that of the rational itself, or that of a thought exceeding all rationality? Will the answer be even more complex if it does not allow this alternative?

As the conclusion of this work begins to take shape, it is clear that one question cannot be avoided: the question of the eventual *recourses* that thought can envisage in the face of the massive and apparently irremediable (despite the "crisis") reinforcement of the Apparatus of power. An even more delicate question is whether thought emerges as taking part in this process, engaging itself in this destiny of power, betting its resources on the operational efficacy of calculus, information, game theory and systems, winning nothing, consequently, by still pretending to be pure—except as the renovated abstraction of a philosophical dignity inefficacious and almost unreal.

When this question was brought up earlier[1] in regard to Ellul's interpretation, I argued that he did not minimize the following danger: to make an "adversary" of Technology (or of the Complex of Power) which is simultaneously omnipresent and all-powerful such that one gives up the possibility of adequately knowing it from within, and in the face of which one risks adopting a Manichean attitude. A metamorphosis of the "fight" between the Prince of this world and the Kingdom, the dialogue with the techno-scientific world, therefore, risks reducing itself to a kind of sterile standoff or to an exorcism of Evil by the forces of Salvation. Although the attempts of Ellul, as well as those of Heidegger, which are superior, do not appear to be reducible to this (we will still have the occasion to verify this as regards

Heidegger), I must ask again the question of recourse in such a way that we are not led to a kind of inverse technicism that credits technology with a "perfection" whose very automatization would block all chance at "liberation." Is there a future worthy of this name only outside the power of the rational?

On the other hand, we have seen scientistic hope, endlessly revived, which expects that science itself, through its emancipating creativity, will liberate us from the weight of determinist and functional rationality. This is the new Manicheanism, the new theory of salvation. We forget that science itself "does not think" (according to Heidegger's words), but also that it functions *as if* the science of Phase IV had the epistemic purity of Phase II, an abstraction that is always possible, but which, nevertheless, remains an abstraction in the state of the contemporary world.

Must we constantly research the most subtle choices from the perspective of the possibles reserved in the rational and beyond it: rational-irrationals, undecidables? Already it is apparent that the phases of potentialization have a relative coeffectuation and that a play subsists between them, at the same time that contingency is given free rein (or at least a certain course) at the interior of each of them. Is this sufficient? Is it not necessary to presuppose an originary potentialization that is now preparing Phase V, through which a power of the rational would be achieved that is free from its phantoms and the risks of power for the sake of power, a power of the rational that would be self-controlled and even open to the ontological enigmas which until now have been the most unrecognized? This wager, an accomplice to our hopes, still appears too metaphysical: it returns us to a philosophy of history whose prospective character does not diminish its fragility—or its need for security.

If the outcome of the game remains uncertain, if rationality is not excluded from it (without being the absolute trump card of the game), it is necessary to patiently reconsider how the cards are distributed and to reconsider to what degree they still depend on the will (to power).

The first fundamental possibility appears to us to be playable by rationality itself: the realization of a sovereignty by which the power of the rational assumes all its phases, thinks them in unity, and reconciles them with the world and with itself. Hegel was the precursor of this possibility, which is perhaps still held in reserve. We must, therefore, return to his conception of "absolute power" and justify an interpretation that breaks with most of the accepted presentations of his project today. The System does not prepare for planetary technology; it is not principally the vector of Phase IV, but constitutes a *speculative lock* on the general movement of the potentialization of the modern West (the relation between contemplation and potentializing vectors attaining, according to Hegel, almost the inverse of what it is according to Descartes).

Can this first possibility still be offered without being exclusive? We will not make one card the trump, deciding the whole game. The exploration of the possible will be pursued, and from other diverse recourses, less metaphysically sovereign, but which will reflect and think most closely the contemporary disquietudes that will be presented. They will be the stakes of a new critical effort imposed by the contradictions and impotencies of a rationalism pushed to the absolute. This is why I will place the different versions of this second possibility under the sign of "The Critical Resource."

It will be necessary, finally, to return to a thought that the dialogue touches on constantly, although most often implicitly: the Heideggerian interpretation of Technology. The intelligibility of the confrontation between the Apparatus of techno-scientific power and the ontological Secret will allow us to measure a danger (the Manicheanism of a theory of salvation), and to prepare for a chance called "contiguity."

Of the three routes thus presented, none can escape Power, none can be reduced to a techno-discourse or abandon itself to the illusions which are paralyzing for thought. Each deserves to be tested and explored. But, in their differences of ideology, none demands an unconditional celebration, none constitutes a dogmatic block. We will discover the opposite, that "contiguity" simultaneously welcomes the play of the phases of the potentializations of the rational and, at the limit of this, it listens to the ontological difference. Contiguity is not juxtaposition: we must apprehend here the proximity of differences, including elucidating the simultaneous nature of the possibles still held in reserve.

It would be better, therefore, not to dwell on announcing the dialogue that will follow, insofar as "contiguity" is not a situation that can be anticipated. It is necessary to live it in order to think it, to think it in order to live it.

Absolute power

We recall that the principle objection that we have consistently confronted is that there is not a direct engendering of modern technology by pure rationality, nor are megamachines engendered through techno-scientific rationality. At every level of analysis, we will be tempted to increase the methodological prerequisites as the traces of irreducible ruptures: Does not the history of the sciences and technologies make an inventory, impossible to synthesize, of inventions, procedures, extraordinarily unforeseeable research that bears witness to the irreducible and inexhaustibly rich heterogeneity of the "real"?[2]

The supreme merit that I recognize in this objection (to which I believe I have responded by the theory of the phases of potentialization)[3] is that it comes to contest, in its very principle, all neo-Hegelianism, that is to say, a new metaphysics of rational reconciliation integrating differences and contradictions. Not that this neo-Hegelianism is absurd or superficial: I do not exclude its speculative possibilities.[4] But I believe that it can *have value* only for a thinking under the sway of the most dense contradiction, the most bitter and corrosive irony of a truly adverse thesis. The speculative absolution will perhaps protect, therefore, all of its philosophical

chances, but in a place different from that which we assigned it at the outset—and
a place different from that which it has assigned itself.

To explain this point: Hegel gave so much probability to his speculative genealogy that it has become an almost inevitable model (or foil). It seems that Heidegger himself did not know to resist entirely this fascination in his interpretation of
Hegelian thinking as a phase of the accomplishment of metaphysics.[5] The hypothesis that I will examine in the following pages is, indeed, that of a continuity of engendering (of technology by metaphysics), such that the Hegelian achievement
ought to be thought as the *summum* of the representative position and of the "placing at the disposal" of being. Is this thesis of a Hegel—"to prepare for technology"
(or even of a homology between the speculative Absolute and the Spirit of Technology)—tenable?

There is no other "solution" than to discuss it. It can only result in more clarity
when carrying out our own genealogy: we will better see why and how this must not
be confused with the Hegelian "model," if it is true that the dynamic of potentialities is thinkable according to a different modality than *Aufhebung*.

"The principle is that spirit dominates." For Hegel, this actual domination of
spirit is, properly speaking, no longer a *Herrschaft*; in any case, it no longer dominates one thing over the other, but rather, it dominates itself. True domination is,
therefore, auto-mastery [*automaîtrise*] (we differentiate from the mastery of self
[*maîtrise de soi*] whose connotations, in French, are too psychological). Nevertheless,
even this vocabulary of mastery is used only at certain moments:[6] in this last passage, as in *The History of Philosophy*, Hegel contends that "power, insofar as it is
absolute . . . is not mastery over another but over oneself, a reflection on oneself, in-
dividuality." Mastery comes to dwell in a hierarchical ontological structure (logic,
nature, spirit) handed down from the entire metaphysical tradition. It is never liberated for itself. We must lend a particularly attentive ear to the Hegelian texts in order to extract this presence that *goes without saying* in a "philosophy of spirit," but
whose "evidence" becomes suspect when the power of the rational unleashes itself
in unconditional domination of the earth.

"The principle is that spirit dominates": we will see that in this fascinating
phrase, Hegel says both *that which he wants to say* and, *more*, that which he does not
want to say.

Paradoxically, we must begin with the *more*: up to this point, our investigation
has been led, more or less visibly, by this *not-said*. If the affirmation of spiritual
domination is an *avowal* of domination, where does it lead? It leads us to recognize,
in the Absolute, *an ontological structure of unconditioned mastery*. Auto-referential
and exclusive, it is not only the Absolute that prevails in ontological density and
quality over every other being or over every other position of being, but in ruling
the legitimate approach, it is the alpha and omega. The great difference that we

must make between exterior domination and spiritual domination is that the first can always be counterbalanced or "neutralized," whereas the second cannot. Therein lies the absoluteness of the Absolute: there is no "counter-power." From the ontological point of view, there *exists only* auto-mastery or, more precisely, nothing but the process by which the auto-mastery of the spirit "verifies" its superiority and efficacy. All being and all truth are hung—shielded from the chaos of nonsense or insanity—from this ray of sense: auto-mastery, another name for the Absolute.

Let us now take a certain critical distance. This is hardly difficult: the great majority of Hegel's readers, since his death, do not accept the identification between the divine and the rational. We find ourselves in this bizarre—in truth, impossible—situation: the entire frame of the work still holds up, due to its brilliant intelligibility of the real, but the keystone, which Heidegger has correctly defined as onto-theology, is no longer recognized. If, therefore, we do not lack references to critique Hegel, the difficulty becomes more subtle: within this quasi unanimity, where do we grasp a point of solid support?

The refusal to crown theologically the speculative can be reformulated in view of a major convergence that takes rationality as the principle axis: *the positivity of the rational abolishes itself in making itself absolute.* Positivists, Marxists, and analysts are going to recognize themselves here in more or less neo-Kantian positions: the kind of rationality that the Hegelian speculative enacts escapes the limits of determined, properly scientific knowledge; it is not sufficient to invoke rationality in order to surmount the antinomies of pure reason, such as the very opposition between understanding and reason. And, it is not sufficient to prepare a metaphysical claim in a rational method in order to bend metaphysics to scientific exigencies, in order to demand from this rationality that it present credentials of truly positive belief. For Hegel, the supreme skill consists in merging a more or less positive knowledge of the state of the world (from art, the sciences, etc.) with a rationalized theology, which would not receive any attention if it were taken up for itself alone.

Primarily, it is in the name of positive exigencies that speculative onto-theology finds itself challenged. What could be more normal in this age of the triumph of positive sciences and their technological application?

My position begins with this quasi unanimity of the scientific community, accepts the rejection of speculative theology at the philosophical level, but differentiates itself in this: in the Hegelian acknowledgment of the "divinity" of rationality, there is no regressive mysticism. There is, on the one hand, a "rational core" whose dialectic is without doubt the most incontestable form; on the other hand, there is the premonition of a state of the world marked by the total power of the rational.

At the first level, what is of interest is not the application of a general but uniform schema (that Engels will carry to caricature): it is the aspiration, authoritatively postulated, of responding rationally to the incessant defiance of the contradictions of the real; it is the aspiration to the rational knowledge of the Whole. In this regard, Hegel has the same respect for rationality that every thinker has for objectivity and the constancy of truth; he suffers a true repulsion with regard to subjectivism.

At the same time, we know that there is no will to prophecy in Hegel. But how shall we remain deaf to that which he calls the work of the Spirit? How can we regard with an analytical and detached eye the immense scientifico-technological transformation, without making the connection between all these phenomena? Ought scientific knowledge, under the pretext of positivism and rigor, close itself off to every synoptic view concerning its own evolution, its results, its massive power— and even perhaps its impotence in mastering itself?

If there is a danger of the loss of rigor from the Hegelian side, the inverse danger is perhaps no less: the crumbling of argumentation, of analyses and of statements (to which a mind as brilliant as Russell, despite himself, bears witness). A rationality that is too concerned with its rationality comes to lose all sense of totality. In Wittgenstein, the critique of language becomes so pointillist and fastidious that it disperses in its own fleeting image and leads (in recognizing it at least implicitly) to an almost total impotence in the face of the problems of the actual world. From the Marxist point of view, one will say that this is not the case; in fact, the Hegelian concern for totalization has instead been displaced to the area of human praxis and reduced to ideological formulas, where the immediate concern for transformation prevails over the complex knowledge of actual reality.

I do not intend here to deify the enormous technological revolution, but, in recognizing its power, to identify its *regions of realization*, without neglecting the tensions or the resistances that subsist or are created. In this regard, the Hegelian moment is decisive, but in a manner which is perhaps more subtle than a fatal progression toward an ever-accruing domination. The question becomes: What happens when Hegel, giving an absolute ontological privilege to rationality, accords this rationality a specific superiority that has the character of a divinity, understood here as all-powerful? To what superior truth does he raise this thought?

To those who would proclaim that all rationality is divine, and that all divinity is rational, it would be necessary to respond, from Hegel's point of view, that these far too general propositions are speculatively determined truths only in the sense that "the effective is rational" and vice versa. When the divine is ontologically supreme, there is no ground upon which it could refuse the privileges and enlightenment of rationality. Or better: the totality of Hegel's effort is to demonstrate the speculative identity between rationality and the Christian mystery, and this not only in *The Philosophy of Religion*, but in all of his works. When he writes at the end of *Lessons Concerning the Proof of the Existence of God*[7] that "all speculation is a mystery for understanding," this clearly implies that developed rationality no longer knows the mystery (in the sense of an unknown content that one must accept in its entirety), but assures its own profundity as the full revelation of that which still imposes itself as obscure mystery to the understanding (or to the believing consciousness).

The Philosophy of Religion determines the degree of the rationality of the divine, from the religions of nature to the truly speculative religion—Christianity. All divinity, therefore, is truly rational, but in view of its departure from natural abstraction and in view of its self-determination. The Hegelian doctrine essentially establishes itself so clearly and so consistently that it is stupefying to see it so constantly misunderstood or openly falsified (as with Kojève): in the first paragraph of the *Encyclopedia* we read, "It [philosophy] . . . has first and foremost objects which it holds in common with religion. Both have *truth* as their object, and in the most elevated sense—in the sense according to which God is the truth and He *alone* is the truth." The speculative identity of the *content* of philosophy and religion proceeds on an equal footing with a difference in their forms. This difference itself is not merely formal (this would be the case if philosophy were only *ancilla theologiae* and had only to prove the existence of God), because when the conceptual form determines itself, it first becomes autonomous to the point of posing its radical difference in relation to religion, then it poses its very opposition to it (in the enlightened consciousness of critical philosophy in the Age of Enlightenment), and finally it surmounts this opposition by a return to the identity of the contents, the ultimate conquest of speculative rationality.

Not only does Hegel not have any doubt concerning the rationality of God, it is in the true sense the flesh and blood of the System. Marx himself recognized this, and never claimed—as Kojève would later—that Hegel was an atheist, but saw that it was necessary to critique Hegel's still mystical rationalism, in order to extract the "rational core," and to apply it to the socioeconomic transformation of the world.

Hegel, after all, granted to the believing conscience the right to rest on a piety that affirms and assumes its final character before the divine mystery; philosophy does not have to impose itself as an obligatory goal and universal destiny. In this sense, in spite of its monumental character and the exigencies of the speculative method, the System does not attempt intellectual coercion (in very modern terms: nothing terroristic). In substance, he proposes only this: "If you are capable of following the movement of a concept up to its logical and spiritual completion, follow the thread of my reason and allow yourselves to be drawn into the circle of the circles of Knowledge. If not, you could content yourselves with contemplating the *same truth* revealed to your understanding and offered to your piety." The System remains open, in that it wishes to be conciliatory and welcoming under the sign of the community of Saints and the invisible Church. There is only one supreme truth: there is only Spirit, whether it be worshipped as holy or posed as philosophically sovereign. In proclaiming the rationality of the divine (Christian), Hegel did not want to annex religion to his own method; we must see at which point the inverse movement is significant: the enormous speculative effort in order to think the mystery, the submission of philosophy to the superiority and the unity of the Spirit.

Without doubt, one will object, the Kojèvian position does not stand up in the face of Hegelian texts: Hegel never assumed to be the Sage at the dawn that would measure the Absolute, and it is a very suspect move to want to attribute to his authorship some conclusions that only a new configuration of reading imposes. However, if one returns to the speculative identity between the divine and the rational, if one envisions it from the perspective of the *divinity of rationality*, does one not have some cause to ask whether the speculative play does not turn in favor of the rational (if it is no longer the case, this time, of interpreting it in humanist and atheistic terms)?

One can, it seems, legitimately speak of a deification of the rational in Hegel, on the condition of respecting the terminology and economy of the System. This would be questionable, if one wishes to attribute to Hegel some mystical transfiguration of the cult of the goddess Reason. But, those who take up the very terms of the System understand that the principle attributes of the divine qualify the rational: life,[8] eternity,[9] and especially, total power. Philosophical understanding, in the strong and nonpejorative sense, is "absolute power,"[10] and the method seeks "the supreme force" [*höchste Kraft*] or rather, not merely the unique and absolute force of reason, but also its supreme and unique instinct [*Trieb*] to rediscover itself ... in everything."[11]

Yes, the System protests in many unambiguous pages, rationality is living, eternal, all-powerful. But do not understand by this abstract attributes. The all-powerful, in particular, exercises only itself in proportion to the determination of the rational. Since *Phenomenology*, a type of *law of intensity* is formulated which "is only as great as its exteriorization."[12] If rationality poses, resolves and absolves all contradiction, if all actual reality is illuminated dialectico-speculatively by its light, why refuse to recognize its divine all-powerfulness?

This Hegelian recognition of the divine character has nothing of an artificial deification. It is postulated in the affirmation that the Absolute wants to be immediately "near to us" and that the true is a universal. The dialectic of the law of exterior-interior intensity is valid for God as for man, for religion as for knowledge. The deification is, therefore, inscribed in the total structure of the System. From the moment (and this moment cannot be dated, returned to the circularity of Knowledge) when the System is not the reflection of my final understanding, but the life of All, rationality has all the actual privileges of the divine.

If it seems, from the point of view of common sense, that the balance tilts, in speculative identity, rather toward the side of rationality, this is because, reasoning in the final mode and considering Hegel's work in its very human results, we lost a bit the view of the intrinsic realization of the power of the rational in the System. In giving an absolute ontological privilege to rationality, Hegel maintains a specific transcendence which has all the characteristics of the divine, and especially of total power. For the author of the System, the balance could only be equal between divinity and rationality, as between rationality and effectivity. For him, it is not exactly the *deification* of the rational, but rather, "mutual recognition" of the Supreme and of the speculative.

The preceding pages could appear to lead to the following contradiction: Is Hegel the thinker who leads metaphysics to its dominating fulfillment, preparing the total exploitation of the earth, or the thinker who believes that we are able to recognize and admire a complete, divine truth and absolute method? The paradox of this interpretation: it is in returning to the Hegelian *said* [*dit*] and to its literal meaning that our interpretation attains its end, when every Heideggerian hermeneutic effort was to flush out the not-said [*non-dit*] of the System and to force the absolute structure of auto-mastery to reveal from itself its limits.

Between the "explicit Hegel" and the "unthought Hegel" there is not a contradiction if we bear in mind that the text of a great thinker sets forth, once and for all, a certain number of dogmatic propositions. Now, in spite of its character, which is monumental and sure of itself, the System is as overdetermined as rationality, in whose favor it testifies. The power of the rational is certainly the advent of an unconditioned structure of auto-mastery, which *dominates* in the complex of the real, but it is also—resource of the possible—the serene vigil of a gaze turned toward the eternal.

While I agree that it is necessary to take speculative theology seriously, I do not wish to suggest by this that the privileges of ultimate truth ought to be granted at the expense of the other connotations of the Hegelian proposal. The divinity of the rational finds itself understood here in the same sense as the permanence of the task "to conceive that which is" *in contiguity with* "thinking that which holds itself in reserve."

We will perhaps better understand this in reflecting on the limited validity of the arguments or intuitions which lead to minimizing Hegel's rational theology. In the debate between right and left Hegelianism, the divinity of the rational, in its interest and in its proper dynamic, has not been taken into consideration. I confine myself to the question that does not yet go to the heart of the matter: Does the System agree with religion, or with its rationality, which leads it to the inverse result, to liberate man from the divine? What a strange destiny for Hegel! As if the excess of ambiguous riches, sustained and united by his genius, could only be shattered into adverse positions after his death. I did not remain with the position of Hegel himself, because I believed that his rational theology was only a simple *repetition* transposed into speculative-dialectical language (some approve of this, others reject this pseudo-retrieval).

I present here the inverse of this tradition of misreading: the fact that Hegelian philosophy was proclaimed "theology and divine service" is considered neither as a stylistic clause, nor as a simple conservative thought, nor a fortiori in an apologetic spirit. The status of this "proclamation" is intraphilosophical: the elevation of human consciousness to an intrinsically superior, auto-nomous truth. This truth, certainly, is supposed to be *the same* as that of the Christian mystery, but uncovered, exposed, and understood solely by the resources of philosophy. The spirit of love frees itself there in the serenity of the concept. Humans, raised to this superior dignity (having become philosophical), see the divine being inhabiting the order of the world, henceforth revealed to themselves. The play of this accomplishment is unceasingly reiterated: forever lying in wait, watching anew for the astonishment of absolute being, the contradiction that the spirit never fails to surmount—divinely.

I intend to acknowledge only a specific sense of the divine accorded by the total-power of the rational (comparable to Aristotle's Θεῖον, as the explicit references assembled by Hegel himself indicate). Or again: "the thought of a God insofar as it lives from the intellect," as Plato wrote,[13] opens a superior domain that must not be annexed or explained by anything exterior, but only recognized and respected.

This orientation was put aside by Heidegger: not only in what concerns Hegel, but in all of metaphysics: "How does God exist in philosophy?" This question in *Identity and Difference*,[14] in accordance with other works of Heidegger, clearly allows us to understand that which will be no less explicitly shown in this treatise: the divine, which rationality seizes and which it intends to integrate, is empty of its life and strength, of all that made it divine. Already in *What Is Called Thinking?*[15] it was said that *logos* does not repress *mythos*, but occurs because "the god withdraws." "Heidegger the obscure" could not be clearer here: between the divine and philosophy, there is only incompatibility. The alleged Christian philosophy (understood here as Hegel's speculative theology) pushes us into a sterile impasse (which does not lead into the inviolate as a "*Holzweg*"), a kind of dead *annex* in the philosophical empire. Since then, Heidegger has demonstrated, in a perfectly coherent manner, that rational metaphysics has a totally dominating image, especially since Descartes and scientific modernity, wherein the real is objectified from a new absolute foundation: the *cogito*. In Heidegger's view, the contemplative side of metaphysics was there only to "sidetrack" what really happened: the setting-to-work of the principle of reason and the control of the world.

I wish to acknowledge this minimum of coherence: from the moment when the rational appears to spare certain possibilities of recourse (but not these alone), when an impervious separation between its deployment and the ontological enigma is refused (whose overture to the divine is harmonic), it becomes important and perhaps fruitful not to underestimate these recourses (these relatively non-dominating potentialities of rationality) in the history of rationality, especially in its recent history. If the overture to divinity is possible, why neglect it? If it is a likelihood—perhaps among others—in the play of possibles, what do we gain by closing ourselves off to it?

In the final analysis, Hegel shows himself to be Janus, because rationality itself has two faces. Its dominant face is so imprinted upon the world that we forget its other side. In Hegel, the power of the rational absolutizes itself until it becomes the unconditional certitude of itself and of the Whole. But "to recognize its limit means to know its devotion":[16] the rational also shelters within itself a *sovereignty* made from self-limitation and, in this barely visible devotion, something like an enthusiasm. Thus, this self-styled philosopher turned toward the past allows us to benefit from a *double anticipation*: the preparation of an unprecedented domination, and the appeal to a new sovereignty. At this difficult juncture: the rational in its power is henceforth absolute.

The critical resource

Fidelity to the Hegelian inspiration is seductive. At the same time that it inspires the quest for a new rational sovereignty which appropriates the conditions of domination while surmounting them at the heart of a superior "self-domination," it demands taking into account a historical-world reality that has been in a profound upheaval since the achieving of the System: not merely the drive of industrialization, planetary technicalization imposing a reexamination of the speculative conception of the present, but the correlative rise of the irrational, violence, and unprecedented risks revelatory of apparently insurmountable contradictions.[17] More radically still: the spirit of these new times leads us to pose a type of question that was inconceivable in Hegel's time, namely, the interrogation of the power of the rational which is the substance of this research. Certainly, our title remains Hegelian and preserves the triumphant allure of the proclamation of the fundamental rationality of the real, but it is weighted by the bitter irony of the Reversal of the rational into the irrational, by that of imperialist grimaces of generalized rationalization, by that also of the menacing absurdity of a programmed apocalypse. Rationality must confront the inconceivable on its own terms, along a path that it has marked, but which it recognizes only with extraordinary difficulties, if it is indeed still possible. Nothing assures us that the rose can bloom again at the "crossroad of the present,"[18] if it is true that the present is in the process of leading us toward still unexplored paths of suffering, beyond the pain of any cross.

To admit that speculative rationality still reserves possibilities within actual thought is one thing, to accept that this thought has no other goal than encasing itself in a new System is another. This is an insurmountable contradiction in Hegelian terms: it is the spirit of the times that renders the System impossible.[19] The only rational route that allows us to find a mediation, for lack of a solution, is called critique. But critique itself is possible and justifiable on several fronts and in several ways; rational critique of irrational forms of domination can develop into a self-critique of rationality up to a "diacritics" wherein rationality allows itself to be circumscribed by a thought which exceeds it. Recent history has shown how fragile and, in the end, how hopeless these "recourses" prove themselves to be, having no supports other than mysticism, subjectivism, and gregariousness. The Complex of Power plays from these superficial "reactions," to which, under no circumstance, would a philosophical inquiry worthy of its name give carte blanche. We now encounter a critical moment that entails appreciating again the importance of the Frankfurt school, principally Habermas, but already important since *Dialektik der Aufklärung*. It appears that the strange reversal that Adorno and Horkheimer make in relation to triumphant rationalism is at once very symptomatic of the most legitimate contemporary restlessness, and full of theoretical weaknesses that are important to address (because they risk engendering misunderstandings that "contaminate" our present inquiry).

Adorno and Horkheimer's text *Dialektik der Aufklärung*[20] plays a rather particular role in the history of contemporary ideas. Often cited as if one must render it an obligatory homage, it appears to be little read, especially in France; one uses it, but one distances oneself rather quickly, as if from an almost cursed writing.[21] Is it because the work is unfinished, composite, with a difficult style that is more brilliant than analytical, more literary than philosophically methodical? Is it because of its antirationalist violence and its extravagant formulations ("the *Aufklärung* is totalitarian," for example)?[22] Is it due to its iconoclastic joining of Kant and de Sade, thirty years before Lacan?[23]

In any case, the following remarks only very partially repair these doubts, omissions, or silences. These remarks intend to confine themselves to what is for us essential: to circumscribe the theoretical importance of this book and explain, if not its weaknesses, then its fundamental ambiguity. The theoretical importance? One will always be able to discuss the accuracy of this expression. It remains no less important that the *Dialektik der Aufklärung* touches a nerve when, departing from the Baconian program, it unmasks the will to power ("Knowledge, which is power, knows no limits");[24] when it gives an account of technological efficiency that proceeds not according "to concepts and images, to intuitive happiness," but according to the *method*; when it opposes to the Enlightenment the rude counterpart of domination (technological and social), which assumes the satisfaction of the ideal of autonomy and, in the end, the "triumphant unhappiness" that extends across the planet to the same degree as the progress of rationalization. Not only is rationality not "beyond domination," it essentially potentializes it. Furthermore, it itself does not escape the dialectic that it has succeeded in formulating as the philosophical focus of its activity. For this double thematic, Marx and Hegel have been largely, but very freely, used as contributors: the first, because the dialectic of the Enlightenment is a dialectic of bourgeois society; the second, because of the famous analyses of *Phenomenology* (above all, domination-servitude and, of course, the critique of the *Aufklärung*) brought together in Horkheimer and Adorno.

The weakness of this book appears only when we place it somehow in question, submitting its occasional inextricable knots to the rude light of analysis. Thus, the very concept of *Aufklärung* is utilized with a total misunderstanding of History; we see this in the analysis of Homeric society, an analysis that is undertaken with no hesitation. That the *Aufklärung* is equivalent to rationality in general is already begging the question; but here Ulysses becomes the symbol of mastery. Ulysses the "enlightened"? Ulysses representing a patriarchal society? The two accounts of rationality are not only obscure, but converge in their own contradictions. Charles Mugler has skillfully unmasked *The Origins of Greek Science in Homer*:[25] from such a precise piece of work, the passage from primitive potentialization to truly epistemic potentialization can undoubtedly be illuminated; but everything is thrown into confusion if viewed in terms of the anachronistically retrospective concept of *Aufklärung*. The same contradiction appears from the sociological side; if Ulysses is "oppressive" because he exerts patriarchal power, does all power of this type provide an opening towards the *Aufklärung*? The suit tailored for this poor Ulysses is truly too large for him, and all the more so as the differences between bourgeois and patriarchal "oppressions" remain imprecise.

The other key concept, the dialectic, is equally confused: to affirm *in general* that rationality returns in domination would be sufficient only if we acknowledge a magical virtue to this idea, which would be scarcely rational and even less critical. The Hegelian dialectic perhaps desires to embrace too much in its superhuman effort to say the Totality, but at least it develops it, enriches it in function of the precise and specific contradictions that it must surmount. It is, therefore, an always determined *process*. In Adorno and Horkheimer, the *entire* truth of the dialectic of the Enlightenment is already contained *in ovo* in an allegory, that of Ulysses reserving the pleasure of listening to the sirens' song, all the while bound to the mast, all the while watching his men rowing the ship, obligated to subjugate themselves and to work without pleasure.[26] But if everything is already there, why History? And above all, why social history reduced to such a generality, such a far distance from the lessons of Marx as well as those of Hegel?

"Unity remains the solution, from Parmenides to Russell," proclaim our authors.[27] This dictum of unity is still very little that of a dialectic, if it remains the same throughout twenty-five centuries! If this dialectic does allow for an antithesis, we would like to know when and how it can begin. Perhaps with the *Dialektik der Aufklärung* itself? Why not? The truly revolutionary impulses appear to be dedicated to pluralism and imaginative creativity (themes that will make fortunes). But why then do our authors formulate an argument, contrary to the *Aufklärung*, on the basis of the spirit of the system they themselves denounce: is one any less narrow-minded than the Enlightenment itself, under the pretext that one has treated it as "dictator"?[28] Or, if the dialectic is purely "negative" (a concept Adorno will later elaborate), I do not quite see how his skepticism agrees with the "sublating" of the *Aufklärung*. "Critical theory" will come to fill the void which remains brutally open in the *Dialektik der Aufklärung*.

What is fascinating in this book is precisely what is sobering in the flaw that marks it: a dogmatic intellectualism severed from its own contradictions, it locates its own inspiration in the *Aufklärung* which makes it despair. It yields to the great temptation to polemicize against reason, to reduce critique to a pamphlet. Initially it is easy: we first gather evidence to overwhelm the adversary, but the prosecution is so crushing that the defense rests without a voice, the accusers withdraw, the dialectic becomes tautological. The result is that we no longer know how to judge the apparent value of the fundamental arguments. The *Dialektik der Aufklärung* turns in a circle, as does every systematic antirationalism trapped in Reversal. It demonstrates through this precisely that which is not necessary, that which the power of the rational renders impossible: an enterprise that is globally *reactive* vis-à-vis rationality. This is, therefore, our confession; we have attempted to extract the lesson from this miscarried "dialectic", to differentiate rather than attack (whence the terminological precision, the distinction of the phases of potentialization), and to specify the field of the rational (instead of endlessly mixing up the sociological with the epistemological, without knowing exactly to what degree social domination determines the *Aufklärung*, or, conversely, if this determines social reality).

"Public, unrestricted discussion, free from domination, of the suitability and desirability of action-orienting principles and norms in the light of the sociocultural repercussions of developing subsystems of purposive-rationalization—such communication at all levels of political and repoliticized decision-making processes is the only medium in which anything like 'rationalization' is possible."[29]

Thus Habermas sketches his aspirations in terms of his discussion of Marcuse's thesis in *Technology and Science as Ideology*. This text could almost be signed by Marcuse: it is a political will of emancipation that orients rationality in its "common" sense, *Wertrationalität*, surpassing *Zweckrationalität*. Weberian "rationalization" is thus taken up again from a theory of decision that reintroduces the primacy of the ethical. The truly rational is that which is judged worthy of being rational by an active democratic community. Is this restoration a Kantian categorical imperative? Yes, with two important reservations: values do not impose themselves from the essence of pure reason (although practical rational interest as such preserves its archetypal role); they submit to the exchange, which is to say to public critique (itself illuminated by a rational effort). Furthermore, Habermas argues that the discussion ought to be led "to the light of the sociocultural repercussions of the subsystems of rational activity": in function of an entire body of information on the exchanges and ruptures of equilibrium which operate in developed society—of whose complexity Habermas is ever more aware.

More generally, the convergence with Marcuse can be noted at two levels: critical and positive.[30] For Habermas as for Marcuse, the critique of domination in its contemporary form can no longer operate from the proletariat whose practice is no longer revolutionary: critical theory now functions as an oppositional moment that Marxism would postulate as immanent to industrial society, but which proves to be failing.[31] On the other hand, the critique would not be limited to a dispute concerning the abuses of political authority (in the manner of the philosophers of the eighteenth century); instead it consists of closely analyzing the erosion of traditional authorities and of institutional frameworks to the benefit of new technologies of control and the manipulation of an omnipotent bureaucracy and centers of "technostructural" power.[32] According to our two authors, the critical analysis of domination appeals to a higher and purer rationality than manipulative rationality, thus exploiting the reflections of the *Dialektik der Aufklärung* on the ambiguity of the concept of "rationalization."[33] This results in a convergence of philosophical intentions at two levels: immediate practice and essential ends. It is a matter of immediately undermining the "legitimations" of advanced capitalism,[34] of questioning, for example, functionalist ideology as the more or less behaviorist valorization of individual performance; Marcuse and Habermas find themselves in agreement in supporting and inspiring the "student uprisings" at the end of the 1960s. More fundamentally, the primacy of *practical* rationality (*Wertrationalität*)—obtained by the liberation of communication—simultaneously reactivates the heritage of the *Aufklärung* and the Kantian ethical ideal, but within the novel conditions of a society where technological interest prevails. These technological conditions explain Habermas's liberation of an *emancipatory interest*, whose auto-reflection plays a mediating role in the world of domination. Integrating the Hegelian critique of Kantianism and the lessons of *Phenomenology* on the formative role of *Bildung*, this implant of a new interest into the sociocultural process itself allows us to avoid the confrontation between the empirical and the transcendental or between the social and the subjective. Auto-reflection constitutes the immanent and yet dynamic connection by which rationality *guides* practical freedom in an effective critique of its conditions of realization: "the identity of reason with the will of reason" is thus accomplished fully and without exterior constraints.[35]

We see that, if the philosophical project has benefited equally from the heritage of the Frankfurt school according to Habermas and to Marcuse, the first liberates more systematically the conceptual apparatus and its philosophical implications than does the second. Habermas provided and followed a considerable body of information and conceptual reformulation in a manner which bears more and more the imprimatur of scientific seriousness and at the crossroads of philosophy, sociology, and the political: it is the strength and weakness of critical theory.

We must now consider how this project of an immanent critique of its objects has recently developed, a project which unfolds as a *critical auto-rationalization* on the double front of the historico-philosophical (the argument with Weber) and the socio-ideological (the confrontation with Parsons and the discussion with Luhmann).

Habermas clarifies his debt to Weber by pursuing the analysis of the key concept of *rationalization*, which he strives to show in all its richness and complexity. One will understand that this type of work holds my attention because it joins with my concerns.[36] We will consider it here primarily from the view that is currently held: that of perspectives offered to reason itself. In other words, what rational possibilities taken up by Habermas allow for the emergence of the Weberian retrospection?

To understand Habermas's work on Weber's thesis, we must again take up an essential trait of Weber's *rationalization*; this process is not comprehensible *from a single point of view* (as a pure and simple increase of knowledge and capacities), but must also be understood *from the point of view of the life world*: "a growing intellectualization and rationalization in no way therefore signifies a general growing knowledge of the conditions in which we live."[37] The vital counterpart to intellectualism is the *disenchantment* of the world. In reading this, our unconscious rationalist reaction consists of taking our part in this disenchantment and accepting the inevitable, nothing more. Weber's attitude shows itself to be infinitely more scientific, and at the same time more far-reaching. Disenchantment is not a contingent event, a kind of inevitable debris that we must leave along the route of History, without returning for it. It is a *constitutive* phenomenon of our developed societies. As will be seen equally in the Husserl of the *Crisis*, scientific-technological rationalization does not happen without a loss of the *sense of life*. Insofar as it is a matter of the origin of the modern scientifico-technological and capitalistic world, this process of rationalization is absolutely not reducible to an ideal engendering (ideological in the Marxist sense), but already operates at the double level of the intellect and the life-world. Karl Löwith explains it very well; rationality, according to Weber, is an "originary totality,"[38] an attitude in the face of life, translating itself into both science and religion as well as into the economy. We see in the outcome of the process, at least such as we are able to analyze it, rationalization taking over the entire society, instrumental rationality uprooting the rational consideration of values themselves (*Wertrationalität*) and threatening the most intimate moment of individual freedom, which the first phase of rationalization had powerfully contributed toward liberating. We have here the workings of the phenomenon of *bureaucratization* that the Frankfurt school will denounce under the name of "total Administration."

Habermas does not refute Weber's thesis in either its theory (it is necessary to understand modernization as rationalization) or in its reduplicated formulation (the vital reverse of rationalization is the disenchantment of the world). He intends to rethink and reformulate this theory thanks to "ameliorated conceptual instruments,"[39] which are made available to him in part by Mead and Durkheim (a theory of communication and a sociology of solidarity), and in part by Parsons (a theory of social systems). Without reviving here bit by bit the methodological analyses that form the substance of the enormous second volume of *Theory of Communicative Action*, let us at least sketch, in its three great areas, the actual work done by Habermas in order to *reappropriate* Weberian theory.

Habermas shows that Weber's essential methodological deficiency concerns the concept of rationality according to an end (*Zweckrationalität*) that is not adequate in accounting for rationalization either in its origins or in its "results."[40] In its origins, instrumentality is certainly not absent, but it is structured according to possibilities of communication and of levels of activity where it is already possible to establish distinctions; the life-world is already mediated, as Mead's theory shows.[41]

In modern society, properly speaking, including its most recent forms, instrumentality is again omnipresent, but it is not adequate to explain the growing autonomy of systems of organization that subordinates all individual ends:[42] this is why a theory of systems must intervene, one that allows us to understand at once the internal logic and the equilibrium of the ensemble of the components of modern society.

In this regard, Habermas wants to complete, from the perspective of society, the Weberian schema of rationalization, which, according to him, is excessively rooted in the structures of consciousness (in Parson's terms: in personality and in culture).[43] But although he sometimes gives the impression of accepting everything linked to Parsons, he adopts his theory of social functions only in order to integrate it into a theory of communicative action (he researched this juncture—which distinguishes it from pure functionalism—for a number of years.)[44]

We recall Parson's schema establishing the links between social subsystems and "primary functions"[45]:

Subsystems	Primary Functions
Social	Integration
Cultural	Maintenance of Models
Personality	Goals
Behavior of the Organism	Adaptation

Looking at this schema, we see that *Zweckrationalität* is explicitly dominant only in the subsystem of personality (with the inevitable interventions in the cultural subsystem, taking into account the role that personality as such plays in the latter). From this schema comes Habermas's idea: Weberian rationalization ought to be understood in a more complex manner than Weber himself understood it, taking into account above all the truly social factors of integration and of differentiation (which modernity never stops developing). This results in a double restructuring of Weber's schema: Parson's structural model allows us to distinguish the three subsystems at the heart of the process of rationalization—society, culture, and personality. The differentiation between the capitalist economy and the modern economy is produced in the heart of the first.[46] The second, acting from universal principles, seeks to autonomize science, art, and ethics.[47] It is only in the third subsystem that the Weberian analyses of the methodological conduct of life, the secularization of the Lutheran notion of *Beruf*,[48] truly take on meaning. On the other hand—the second tier of this enrichment of Weber's scheme—the "mediation of the lifeworld" (founded upon, and realized in, a theory of communicative action) allows us truly to extract part of the importance Weber gave to this vast domain of attitudes, affections, activities—private or public—not yet economically or administratively systematized. We thus obtain an "un-coupling" [*Entkoppelung*] of the life-world and the two principle subsystems of modern society (economy and administration) that follow a structured network of exchanges for which money and power (that is to say, power as coercion or control) are the two mediators.[49]

Modern society is, therefore, a complex whole that can be understood from its genesis (rationalization) on the condition of restoring to it its dual face, institutional and "motivational," as well as from all its constitutive differentiations. According to Habermas, actual social evolution is characterized by a "complexification" of the social system and a rationalization of the life-world (with a correlative differentiation of the two spheres in relation to one another).[50] The Weberian paradox of a rationalization that evolves into the irrationality of an unbridled bureaucracy does not explain this in any direct way as an unconditional triumph of a single form of rationality (instrumentalization) over ethical rationality, but only as the social pathology (or, in Marxist terms, internal colonization) of a vast "uncoupling" between System and the life-world: "A paradoxical relation no longer exists between different types of organization of action, but between different principles of socialization."[51] "There is certainly rationalization in the life-world, but it does not emerge merely through contempt for values. Social evolution is even more advanced, because it leads to the "instrumentation" of the means of control,[52] which is to say to a level of integration that is extremely far-reaching (because interjected into the models of action); it leads to the differentiation and autonomization of the "domains of formalized action," where the question of value (even of the end) is impossible, neutralized a priori. The Weberian opposition between the mechanism of bureaucratization and life is far surpassed, because the life-world is directly rationalized, undifferentiated according to formal norms (whose real legitimation is shielded from individuals). The indifference toward *ends* and *values* is explained by a general lack of differentiation of the rationality of action as such (and of its spheres: culture, society, personality), to the benefit of a *systematic rationality* that is interested only in functional relations, the efficiency of plans, the techniques of decision-making, and the solution of strategic problems.[53]

This rationality does not propose any end *in itself*, and is not subordinate to any value: its end and its value are its own efficacy. What is new is not this retroactive inveigling of its inheritance, which has always existed as an immanent tendency of bureaucracies (as Mumford shows with regard to the first megamachines: Babylonia, Egypt, Rome); it is the unprecedented degree of integration attained by this systematic functioning. These are the "relations of indifference" that install themselves between the Organization and the personality, with regard to culture (which loses all living connection with traditions and becomes a by-product of ideologies and of the *media*), and to society itself (because the social is in its turn endangered, divided into separate and neutralized domains of action).[54]

One could thus believe that Habermas has totally surrendered to Luhmann's functionalism, the latest version of Parsons's systematic conclusions. When integrating the structural models of this school, does he himself not systematize to the extreme his general vision of successful societies being those that are in the last stage of modernization, that is to say, rationalization? In fact, despite the revisions and the "re-translations" to which he subjects Weber's theories, he remains closer to Weber than to functionalism. The following reasons account for this proximity. For Habermas, as for Weber, rationalization is a process that is always open (and whose dynamic and indefinite character is constitutive); this openness signifies principally that the *empirical* conditions remain, despite everything, decisive, and do not allow themselves to be enclosed once and for all in a theory, however systematic it may be. Therefore, just as the *spirit* of capitalism does not play an absolutely exclusive (nor purely idealistic) role, having been relativized at the heart of a more complex systemization, likewise functionalism unilaterally discloses itself, since it believes itself to triumph without conditions. But there is the future, there is experience that allows humanity to escape the absolute closure of the phantasm of total Administration. In his very theory, Habermas has above all preserved the chances for communication (which he also names "interaction"); on the one hand, in demonstrating with Mead the originary character of communicative action (never suppressed or invalidated by an ulterior evolution), which, consequently, *remains* the veritable nerve of the life-world, without which the very tension—vital tension—of the society would collapse; on the other hand, in showing that the "uncoupling" between the System and the Life-world does not play to the exclusive benefit of the massiveness of the System: the Life-world remains symbolically structured, and Luhmann is wrong to "trivially" presuppose that this living symbolic is retrieved and integrally colonized by systematic functionalism.[55] The very fact that the means of control, money and power, remain necessary (with their effects and countereffects) demonstrates that society is not, and undoubtedly cannot be, completely integrated into a formal systematic. The domains of action protect a relative autonomy.

The title of the second volume of *Theory of Communicative Action* is *Contribution to the Critique of Functionalist Reason.*[56] Its status remains explicitly critical. Habermas explains that it conducts itself in a critical manner, as much in regard to contemporary social sciences as vis-à-vis the social reality of advanced societies. He is critical of the social sciences because they lack critical distance (reflective and historical) toward their objects, and he is critical of developed societies because they impede their spiritual potential (literally: "they do not exhaust their potential for study for which they claim to be culturally disposed) and they "surrender to an uncontrolled increase in complexity."[57] In this way, consequently, the critique comes to the aid of the "communicative infrastructure of a more and more rationalized life-world."[58]

We find, therefore, the rational perspectives that we sought in approaching Habermas; they are the perspectives of a critical rationalism linked to a political progressivism (Habermas now critiques neoconservatism), and they seek the realization of a dialogue "without constraints." The *Theory of Communicative Action* does not fundamentally change the philosophical project of critical theory; it gives it more of a scientific seriousness and even more of a practical import (especially in relation to Marcuse's "utopian" discourse); it reinforces the "weapons of critique," but it remains exposed to objections that are formulated from various sides against it. More fundamental than the Marxist objections that reproach Habermas principally for having gone too far in detaching the domination that he critiques from class interests (because these objections that can be partially justified at the empirical level do not directly touch upon a theory which challenges the new complex concretions of social reality, among which Marxism is more and more outmoded), or those of Luhmann, which claim to include the critical moment in its very scientificity (nothing more), it seems to us that the recovery of the philosophical status of critical theory ought to begin from either hermeneutics[59] (since the latter *comprehends* the critical moment without making it an end in itself) or the transcendental conditions of interaction (if it is true that, despite his acknowledgment of the debt he owes to Fichte, Habermas does not manage to "derive the distortion of communication, that is to say, ideology and institution, from the transcendental body of interaction, or from the single fact of domination)[60], or even from a critique of Habermasian concept of communication. It is in this last direction, rather close to Rüdiger Bubner's orientations, that we direct our own remarks. Despite his efforts to establish his theory of communicative action in the life-world, Habermas's thought appears to be very fragile, insofar as it presupposes an ideal of transparency as the ultimate horizon of communication. From the outset, as Bubner writes, "the presupposition of an unlimited rationality, apparently to be paired with an ideal dialogue, cannot be obtained from successful communication as such."[61] A double disjunction imposes itself: between dialogue and practice, between dialogue and rationality. But above all, this objection ought to be taken up at an even more fundamental level: that of the theory of truth. Habermas believes to have sufficiently enriched predication through a universal pragmatic rooted in the life-world; but truth continues to be thought as adequation to full presence (with all its sociological by-products, for example, the Parsonian concepts of adaptation, integration, etc.), and never as dis-closure. Despite some cursory and insignificant allusions to *Being and Time*, paragraph 44 of this book appears to be ignored. As with Husserl, the mediations of communication are inscribed between two poles, the opacity of the *Lebenswelt* and the pure transparency of the rational ideal. But the mode of the appearance of truth—prior to adequation—is absolutely not indicated, even in the way of a presupposition. This results in a deep-seated inability to think the very withdrawal that implicates all demonstration: the reification of a language relegated to the exterior of its "communicative life," necessarily to be thought from the very essence of truth. This incapacity is only partially compensated by self-reflection, a moment inherited from German Idealism, but itself not critiqued in its principle.

To be sure, this critique goes beyond Habermas, since every theory of communication (or even every universal pragmatic) remains metaphysical in the Heideggerian sense. Appraising language and action from universal models, it preserves no opening to the very reserve of *a-letheia*. Nevertheless, to the degree that Habermas does not limit himself to designing a scientific theory of existing society, but also deploys a "communicational" *telos*, he advances to the point where the essence of the truth is at stake. To praise communication without constraints remains a rather naive ideal (and takes over, in an uncritical manner, an intrusive policy), if we disregard a double handicap to which this communication is exposed: the limits of knowledge in its activity investigated through the theory of information, and revealing in its theory the fundamental constitution of truth—from its appearance exposed to this Law of Reversal, which we have already seen ruling at the heart of the rational.

Indeed, the strength and weakness of Habermas is completely located in the very sense of *critical theory*, the legacy of Horkheimer. His strength: a continual surpassing, a perpetual enrichment of information, the immanence of an auto-reflexive moment that introduces—a rare event in the sciences, even the social sciences—the philosophical outlook into the sociological inquiry. His weakness resides in this always concomitant ambiguity between the truly scientific enterprise (as well as the critical resources that are immanent to it) and the ethico-political project (which I do not see that he needs in order to deploy a "sophisticated" theory of society), without which the philosophical disquietude would be so keen as to shake the foundations of an extremely *activistic* reflexive rationality. A supplementary danger: critical theory risks becoming an end in itself, the critique becoming satisfying to itself and thus mimicking the reflexivity of a subjectivity which is always self-referential (whether it be transcendental or not). Despite his execution and his improvements of Weber's theory of modernity, Habermas is probably not at the level of Weber's best intuitions; certainly, he lacks the critical distance to make full use of them. Weber had grasped the point at which the "dependence of capitalism in relation to modern science" is essential,[62] thus subordinating social or material factors to a more decisive potentialization. Weber had also been able to discern in rationalization something more than an accumulation of contingent factors or even an ensemble of significant facts: he had known to see there a fundamental trait of History, an enigma on which all reflection has stumbled: our destiny.[63] It is in profiting from this beneficial critical distance that the immense body of Habermas's work can find its direction: to enrich rationality due to the critical contribution, and not obtaining it to the benefit of a rationality which is ever more self-referential. But is he the only one in the contemporary world to find himself in a similar place, that is, at the border between the scientific and ethico-political, at the naked backbone where rationality is exposed to every breeze, including that of hope? Among the recourses, is there not here yet again, quite close and nevertheless dis-quieting, Heidegger's thought? It is in benefiting from the Heideggerian distance (vis-à-vis rationality itself) that critique can insert and expand itself, without repudiation, into the situation of "contiguity."

The *Ge-stell* and the Secret

The peculiarity of the Heideggerian interpretation of contemporary "technology" has been perceptible from the first line of the text of *The Question concerning Technology*: " . . . we shall be *questioning* concerning technology."[64] This questioning opens a way of thinking, and not an enunciation of theses. To question: neither an attitude nor a rhetorical formula, but this "piety of thought"[65] which will be maintained at all costs by Heidegger, separate from all the "problematizations" cherished by the technological world. The second unique feature—insofar as a classification makes sense: technology is essentially placed in relation not to skill or creativity, but to truth. From the second page of this lecture,[66] it appears that the current and traditional conceptions have been very correct [*richtig*]; however, they are not profoundly and completely true [*wahr*]. It is only in understanding technology as disclosure [*Unverborgenheit*] that we will arrive at appropriating essential being.

—"Among these peculiarities, you have forgotten what is only interested in the *essence* of technology. An essence which, in order to be no longer Platonic, does not remain distant from the phenomena. Perhaps Ellul would mix description and explanation, principles, and consequences, but at least he could pass to the riddle of innumerable facts: Heidegger appears to ignore them superbly."

It is by design that the quest for essence has not been counted among Heidegger's peculiarities. The quest for essence is truly philosophical. In the example he gives at the beginning of "The Question of Technology," Heidegger accepts remaining Platonic in appearance, but this appearance is perhaps relative. Does he not, at least, immediately introduce the *distinguo* between essence, in the traditional sense, and *Wesen*, which is truly Heideggerian: "When we are seeking the essence of "tree," we have to become aware that that which pervades every tree, as tree, is not itself a tree that can be encountered among all the other trees."[67] The provisional dismissal of the phenomena is only the necessary detour of an enterprise that seeks to "preserve" them. If Heidegger distinguishes himself on this point from Ellul, it is by the decidedly philosophical character of his thought, which remains, at this initial moment, in the grain of the Platonic tradition. In regard to the phenomena, Heidegger effectively rediscovers them, for example, when he evokes and analyzes the situation of an electrical plant on the Rhine; it is another question to determine if he "saves" them totally and exclusively, a question which must be, for the moment, put aside.

Regarding technology, Heidegger becomes Heidegger through the following four *views*:[68]

1. "Technology is a mode of revealing." [*Technik ist eine Weise des Entbergens.*][69]

2. "The revealing which rules in modern technology is a challenging." [*Das in der modernen Technik waltende Entbergen ist ein Herausfordern.*][70]

3. "Enframing means the gathering together of that setting-upon which sets

upon man, i.e., challenges him forth, to reveal the real, in the mode of ordering, as standing-reserve." [*Ge-stell heisst das Versammelnde jenes Stellens, das den Menschen stellt, d. h. herausfordert, das Wirkliche in der Weise des Bestellens als Bestand zu entbergen.*][71]

4. "The essence of technology is in a lofty sense ambiguous. Such ambiguity points to the mystery of all revealing, i.e., of truth." [*Das Wesen der Technik ist in einem hohem Sinne zweideutig. Solche Zweideutigkeit deutet in das Geheimnis aller Entbergung, d. h. der Wahrheit.*)[72]

Certainly, it is a matter here of four essential *views*. Not just in the sense of simple opinions or eidetic inspections (whether Platonic or Cartesian): these are views in the sense in which Heidegger says in *The Turning*, "When insight comes disclosingly to pass, then men are the ones who are struck in their essence by the flashing of Being."[73] We now return to each view and submit it to a diacritical examination.

1. This first approach to technology appears far too distant. Is not everything which concerns humanity a kind of disclosure: the natural and the intelligible, as well as technology? To respond to this objection, it is first necessary to explain three key words in this short phrase. "Technology" designates here neither "technological objects" in Simondon's sense, nor even, more generally, the technology of technicians (technology as a means). In fact, Heidegger envisions technology not in its phenomenality, but in its essential unfolding. Still more precisely: this unfolding is the essence of modern technology, its precondition (which we have seen is already present in the physics of the seventeenth century).

As important as this is to grasp, it reveals only one "facet," not the sole and unique truth. Modern technology tends to become exclusive and totalitarian. Originally, it was only a guise of ontological disclosure, since Being, according to Aristotle, and, still more radically, "truth," originally gives itself in "multiple ways." This given is not a "modality" in the sense that André Préau would unfortunately translate it: we situate ourselves here prior to the possible, to the real, to the necessary.

This essential unfolding with which we are concerned is a "revealing": *a-letheia* in its originary sense and not principally as correctness of judgment. There can be no understanding of this short sentence (apparently soothing in its "generality") without a complicity of thought with the breakthrough realized by Heidegger in paragraph 44 of *Being and Time* and, more completely, in *Of the Essence of Truth*. "The essence of truth is the truth of essence,"[74] that is to say, the dis-closing of the very being of the thing in what reveals and conceals. Technology itself as a "kind of revealing" appears—as every unconcealing—"sheltered and sheltering."[75] It already gestures toward this secret [*Geheimnis*], which it reserves in its own *demonstration*.[76]

2. There is a brutal rupture of this co-alliance of demonstration and reserve, of appearance and secrets: the demand [*Herausforderung*]. This is not as originary as the "revealing" *in* itself. If technology is only a "kind of revealing," modern technology is itself a very particular "revealing": that which demands. In what sense? We must reread the complete phrase: "The revealing which rules in modern tech-

nology is a challenging that puts to nature the unreasonable demand that it supply energy that can be extracted and stored as such."[77] What is of importance here is precisely the relation that, since the seventeenth century, modern man shares with nature: an imperious relation that no longer has anything of a "letting-be," of an opening to the revealed-concealed, but instead, submits the real, through objectifying it, to methodological and precise investigations.

Is this a Heideggerian mythology that is doubly arbitrary? An unjustifiable personification of modern "truth": the method? An accusation of violence launched at the least violent knowledge: objectivity?

It is true that here Heidegger forcefully marks the rupture that truth in the modern sense inaugurated. A rupture that is perhaps too radical, since between truth as "letting-be" and truth as *certitudo*, there is truth as *homoïosis* and adequation, to which Heidegger gives justice in many other texts (in particular, "On the Essence of Truth"). But modern "truth" is not "personified"; if it becomes subject, it is because Heidegger wants to show that the new relation, in spite of the important and decisive part that humans take, is not (no more than truth in its originary sense) a fabrication of human beings.

As to the second objection, concerning "violence," what it enacts is so decisive that one can read this entire book as a response to it: it is indeed the question of the power of the rational. Does this accomplish with modern scientific methodology a new regime, a super-rationality whose operativity cannot be compared to anything that preceded it? I do not see how it would be scandalous to affirm this. What is surprising is that it was necessary to wait until Heidegger to measure the intensity that would assume the force of the method. One admits this radical rupture as a piece of evidence, provided that one labels it as the triumphant ascent of "the progress of consciousness." This clash of cymbals is shocking only under a certain light. The Heideggerian presentation is perhaps no more "mythological" than rationalism's great "books of culture," written from Hegel to Brunschwicg by way of Comte and Renan.

It is necessary, in any case, to defuse the hostile connotation that *Herausforderung* risks. This "command" is none other than the exigency of the method in the sense of Bacon, Descartes, and Newton. There is no confusion, in the spirit of Heidegger, with brutal and exterior violence. But when one shows even now how the communal sense—ours as well as others'—resists intuitively apprehending the Galilean law of falling bodies and the principles of inertia, how can we continue to assume that principles of modern science go without saying? They are "evident" only for a *clara et distincta perceptio*, which we know has been dearly and strenuously gained through all the habitual and easy practices of our reason. Nonetheless, Heidegger saluted Descartes's lucidity in the *Regulae*,[78] and he wrote a commentary on the first propositions of Newton's *Principia* with a clarity and precision that one rarely finds among historians of science.[79] Heidegger's very particular style must not make us lose sight of the coherence of the sustained thesis and its historical and textual rootedness.

3. For a reader rebelling at this style, the worst has not yet come: we see this with *Gestell*. Do we find here the apex of the jargon of the ponderous Heideggerian affectation? It is not, however, this against which we will conduct our criticisms.

Gestell attempts to state in one word the principle and destined unity of modernity. That is not so easy. *Gestell* in current German designates a frame, a scaffold, a structure. In using this word to say and to think the truly modern "sending," Heidegger knows that he does, in a sense, violence to the language, and he explains this.[80] The prefix *Ge* signifies "that which gathers" and the root *Stellen* (which is consonant with *Verstellung*) diverges simultaneously toward *Bestellen* (to set in order, to command) and *Herstellen* (to produce, to reproduce, to make).

Translating *Ge-stell* by *Enframing* gives the advantage of articulating, at least implicitly, the reflexive project in the ideal or technical arrangement. As imperfect as every translation is, it is undeniable that command (*Bestellen*) is privileged in *Gestell*. Certainly, it is not insignificant that the real be available and held in reserve (*Bestand* is not only resource in the material sense). But what is foremost in the modern "sending" is the exigency that requires reality, including humans, to enter into this general framework of command—resource, production-consumption. The "command" is—at the theoretical level—the objectification of the real as knowable and the elaboration of "problems" in order to establish rigorous parameters. At the practical level, the command places nature into a framework (although void of all intimate force or "occult cause") as the field of experimentation and collection of materials for techno-scientific operations.[81]

The famous example that Heidegger gives to illustrate *Gestell* is that of a hydro-electrical plant on the Rhine. Here again one could cry mythology: Have you ever seen a plant that "summons" a river to surrender to its pressure? And what naiveté to oppose it so antithetically to the old windmill, as if the latter were only an offering to puffs of wind and not an object made with will and patience, in order to master ever so skillfully the energy of the wind! It could be that Heidegger forced the antithesis; but there is also, on the part of some of his detractors, a kind of deliberate passionate refusal to listen to what is said. One pretends to believe that Heidegger presents a "critique" of central electricity; however, he does not give himself to such foolishness. His project is the inverse, or rather, it is completely different: it is to allow us to see the ontologico-epistemological presuppositions of the *project* of mastering natural energy. Heidegger is in a position of strength when he believes that the representation of nature as a source of energy is a (philosophical) precondition of its exploitation. Elsewhere, Marx never said the contrary. For example, in a footnote in *Das Kapital* he acknowledges the role Descartes played in founding a "practical philosophy"[82] and, consequently, (partially) admits the role of philosophical preconditions in the transformation of the world. Heidegger would be an idealist if he had been contented with representation and directly inferred the transformation: this is not the case. One is unfair to Heidegger when one reproaches him for his "anachronism," as if it were forbidden to take some distance vis-à-vis the techno-scientific project! There is an essential difference between traditional technology and modern enframing: for the latter, thanks to the principle and universal character of the project, it is the *totality of the real* which is, by rights, objectifiable and manipulable. Dieudonné, renowned mathematician, once argued against this by pointing to the wheat silos in the Roman Empire.[83] This is a very empirical objection (for a mathematician), which reduces the stakes of the question to a problem of storage, certainly very old. But one could reverse the question: Is it sufficient to store great quantities in order to obtain the conception of a general mathematization of nature? And if all these stockpiles were already techno-science *in nuce*, why was it necessary to wait so long before this would systematically and universally develop its operativity? For us, one of Heidegger's essential merits is having noted with extreme rigor, beyond all historical statements, the radicality of the modern rational revolution, without which contemporary Technology could not exist and would have remained storage or skillfulness, whether this be as admirable and impressive as the great Roman public works, the Greek or Hindu temples, or the Egyptian pyramids.

4. The Heideggerian meditation on the *essence* of technology leads toward the origin of this first "view" that we had assigned to him: the un-veiling. In principle, its ambiguity is not different from that of all presencing that reveals and conceals. Why then speak of a "heightened sense" of the ambiguity of technology? On the one hand, it is not a question of a confused ambiguity (for example, the ambivalence of *"Das Man"* in *Being and Time*); on the other hand, the "quality" of this ambiguity is perhaps due to the degree of its unprecedented character: already, according to Aristotle, art produces what nature cannot accomplish; a fortiori modern technology obtains results that were inconceivable several centuries ago; and it is undoubtedly full of the effects of power that we are not yet able to conceive. To the unparalleled unveiling, the secret profoundly accrues. Heidegger remains faithful to this "law," which one could certainly contest in its strict "proportionality," but which is not of interest for a thought which seeks to probe the "unthought" of metaphysics.

If the mythical thought of traditional societies preserves its secrets and inscribes them in a cycle of initiation, if the techniques of artisans have their own secrets, transmitted from father to son, this is not yet either science or modern technology. A truth is not considered to be scientific as long as it is not submitted to the scientific community (it must at least be deemed worthy to be presented to it): universality, communication, the submission to critique are indispensable and vital factors for research. Technological procedures are assuredly less directly accessible, indeed often obscured by industrial secrecy or reasons of State, but here it is a matter of temporary occultations, destined to obtain decisive strategic advantages, an advance on an adversary. Modern technological work is nothing less than teamwork, enacting far-ranging plans, demanding multiple verifications, inscribing themselves, therefore, on objective and objectifiable structures.

There is nothing Heideggerian about this reminder, but it was necessary in order to clarify the distance between every ontic secret and the secret that the essence of technology *as such* is capable of harboring. It is paradoxical, indeed strange, that the technological framing leads to the secret of Being. But it does not lead there directly. When Heidegger asks, "Do we see the light of Being in the essence of technology?"[84] he valorizes neither machine nor procedures; neither does he devalorize them. In order to lead to the secret, ought not the framing be "placed outside the circuit," in the fashion of Husserlian phenomenology? There is certainly a decisive passage for this way of proceeding, but Heidegger does not refer to it as an instance of the transcendental *cogito*: it is the high sea of "this which is given to think" that comes to crash against human perplexities. And it is in this risk of a thinking exposed to unparalleled disclosures of the techno-scientific complex that the encounter between *danger* and *salvation* is played out,[85] the "stellar course of the secret."[86]

We have reached the moment when the presentation of my own questioning has become possible. Up to this point, a certain number of objections that appear to me to have avoided touching on the heart of the Heideggerian thought concerning technology have been refuted. Is it too presumptuous to claim to have reached this sensitive point? To believe this would be to place Heideggerian thought on a kind of pedestal, where it would be shielded from all criticism. If this thought actually has an exceptional status and merits a very particular consideration, it is less because of an accumulation of uniquely intellectual difficulties—of a composite order—than by virtue of its demanding an opening to the unprecedented, an opening which could not be deployed without a constant critical acuity.

To begin from the strength of the adversary, to penetrate it in order to elevate it to a higher degree, such was the spirit of Hegelian refutation. I do not believe that it is desirable to "refute" Heidegger in this way, because that would presuppose the possibility of a dialectical and speculative integration into a superior unity. Or, at least, in order to initiate a truly critical dialogue, is it necessary, as Bourdieu has done, to caricature the Heideggerian position, allowing us to believe that Heidegger rails against technology and has nothing more to propose other than a kind of neo-conservative nostalgia for the primitive?[87] (Heidegger, to the contrary, states that "there is no *demonizing* of technology.")[88]

The axis of our "diacritic" will be completely different, almost the inverse: we must remember that Heidegger attributes an enormous amount, perhaps too much, to technology—as regards the essential. We have come to see that the grasp of the essence of technology as sending [*Geschick*] unveiling opens to the depth of its secret and to a "salvation": " . . . the rule of Enframing [*Ge-stell*] cannot limit itself solely to blocking all illumination of every revealing, every appearing of truth. The essence of technology *ought* rather to harbor in itself the fullness of that which saves."[89]

When Heidegger opens himself up at this point to the essence of technology, he goes so far in this direction only because he recognizes it as the place of destiny, the ultimate face of the destiny (forgotten) of Being. It is always the secret of Being which is in play here. The unity of destiny, the appropriation of the gaze of man in the gaze of Being.[90] *Hic Rhodus, hic saltus.* What appears to us doubtful, at the very least worthy of being called into question, is less—as Derrida argues—the reintroduction of the Proper than the epochal unification that all this assumes (the reissuing of the *Zeitgeist*); it is above all the localization of the appropriation in the "alignment" of technology. Is there a destinal unity such that danger and salvation could be brought together, brushing against one another *beyond the grasp of understanding* and awaiting it? There is here certainly an "eschatology" of History,[91] wherein every appropriation remains assigned to Being itself. One could open oneself to the unthought of metaphysics, allow the possible to be, claim the meditation on the essence of technology to be beneficial, without literally holding thus to the correspondence between provenance and destination, between *a-letheia* and "salvation." This unification is, finally, possible in Heidegger only as the following: *Being confirms History*, a presumption which pushes the questioning beyond itself, beyond perhaps what the hermeneutics of technology demands, and separates the Being of understanding from its articulations.

Is technology really taken up by Heidegger phenomenologically (despite the example of central planning) and ethically (despite the call to meditation)? It is perhaps not enough to admit that "the most dangerous is not technology"[92] (in its factual functioning), nor that one must know to say "no" to the purely technological usage of technological objects[93] in order to permit dwelling in the technological world. The *distance* that Heidegger places between the essence of technology and technology allows this distance to fall into the indifference of its functioning, as if everything had merit there; but the technological world is not so easily unified nor unifiable. The *distance* is the renewal of Platonism: the *Gestell* maintains a certain relation (even remote) to the *eidos*, which is in a fixity of position that the German word connotes—even though modern technology is so dynamic, ruptured and explosive. From the viewpoint of technological reality and technological objects, Heidegger—conforming to Tacitus's adage: *major e longinquo reverentia*—has perhaps taken the theories on the triumph of cybernetics or F. G. Jünger's *Die Perfektion der Technik* too literally.[94] There is no doubt that the "perfection" of a total and self-regulating worldwide planning is the objective of the technological complex; but it is another question to know whether reality follows with its economic and political implications, whether contingency does not continue to play an ineliminable role, above all in this imponderable "factor" called human—individually or as a people. If technological reality "surpasses fiction," it is not always in the direction desired by those involved in worldwide planning or the authors of the most technological scenarios. And this, not merely because of mishaps and failures, but above all because of constitutive reasons that adhere to the limits of technology, to the reversals, negative or positive, that it cannot avoid, to the "zones of indifference," or to the "pockets of resistance" that the technological world tolerates or integrates—but which could turn against it or, at least, destroy its supremacy, thanks to a supplementary "play."

To admit the *Gestell* as an interpretive grill is one thing; to transform it into a descriptive schema is another. Heidegger's thinking on technology can lead to a kind of inverse technicism, as does Ellul's, to the degree where—"salvation" being ontological—we grant Technology the autonomy and the "perfection" that allow it to resolve all its problems. But all Heideggerian hermeneutic knowledge is not so reductive. It seems that Heidegger's thought could *help* us to build a *non-dominating experience* (relatively) of the thing and of the world.

Is it possible to attempt such an approach "in the epoch where power is alone in being powerful?"[95] It would be more than paradoxical, indeed saddening, if the thought of the Simple does not know to respond to us, as one with the learned obscurity of a new Trissotin often pretends to believe. I do not believe this to be the case. I therefore propose several accessible examples of the Heideggerian line of thought.

The thinker is seated in a garden; he contemplates a rose that the wind will perhaps cause to sway. He appreciates and admires the red of this rose. Neither the rose itself nor its color are, he suggests, objects (in the Kantian sense of objects of the physical science of nature), nor objects in the more general sense of representations: "There is therefore a thought and a saying which in no way, are either objectified or carried to representation."[96] Oriented toward the representable and manipulable, are our spirits and our senses still amenable to this differentiation? Are we still capable of thinking this rose, of evoking its tones without immediately transgressing its offered fragility? Has metaphysics rendered us so naively confused, so naturally violent and busy, that every rose and every sign is integrated in a field of objectification, neglected or classified, but not *released* as such into the Open?

We continue the exercise. "Action is everywhere, and nowhere the worlding of the world" [*ein Welten der Welt*]. What does Heidegger want to say? Does not every action presuppose a world? What is explained here is continuously under our eyes, but we do not take account of it: action [*Wirkung*] is effectively praised in order to live (the ideology of work) and in order to live better (the ideologies of political activists and even the organization of leisure). To the national level of worldwide planning and expansion correspond the innumerable authoritative or liberal incentives for individual projects that must participate in a universal dynamism. It is a question of "transforming the world," as Marx already said; the technicians of power—technocrats avowed or scorned—have no other slogan. We, therefore, live in an epoch of the "conception of worldbuilding" [*Weltbild*], where what is important is less and less the world itself (reduced to the "environment" or to deposits of raw materials) than the project [*Bild*] of production-exchange-consumption. "We are still far"—writes Heidegger at the beginning of *Letter on Humanism*—from pondering the essence of action decisively enough." It is certainly evident that the supreme value of the contemporary world is efficiency, when production—in the sense of *poïesis*—is essentially the unfolding of reserved potentialities. The modern human being, the technological human being, obsessed by his or her own dynamism and the "objectives" to be attained, has for "world," in the Heideggerian sense, only the indispensable source of his or her own efficiency; culture, aesthetics, and religious "uncertainty" are the principal phenomena which accompany this new "a-cosmos." But if the world, in its full sense, is absent today, what is it? Heidegger responds in *The Origin of the Work of Art*, "The temple-work, standing there, opens up a world and at the same time sets this world back again on earth, which itself only thus emerges as native ground [*heimatliche Grund*]."[97] This sense of the world, from which modernity infinitely distances itself, Heidegger never ceases to recover, to re-solicit, beginning with the return to Being-in-the-world [*In-der-Welt-sein*] in *Being and Time* up to the evocation, in the lecture "The Thing" and several other texts, of a meeting of the Fourfold: earth, heaven, gods, mortals. A world worthy of this name is a world capable of being *inhabited*: "We do not live here because we have "built," we build and have built so that we can dwell, which is to say that we are *inhabitants* and are *as such*."[98] We arrive here at the horizon that Heidegger intends to substitute for ideologies. Whatever our final judgment may be, we cannot remain insensitive to this: Heidegger in no way leads to abstract scholastics, where one often wants to enclose him (a kind of sterile duel between the Being of being and the truth of Being); philosophy, having left its closet, rediscovers a world more profound and more diverse than metaphysical space, because the crossing of the Fourfold points to discreetly Hellenic references, to the rich orders of the great civilizations that the twentieth-century West discovered at the very moment that it caused them to perish. It is enough to dream of the undoing of our disciplinary architecture of "housing" in a "territory" divided into projects, enterprise zones, and other zones of "development"; what becomes unsettling is the proximity of the errancy which is our lot: the omnipotence of an efficiency that no longer opens to any encounter, any event worthy of this name.

Event: *Ereignis*. The third example that renders Heideggerian thought more compelling concerns the unspecified moment of an appropriation that is completely foreign to metaphysico-technological domination. Just as geometric space obscures the world, the segmented time of projects and plans no longer gives access to temporality as granting. The human being who wishes to master the earth, the concrete figure of the forgetfulness of Being, understands time only as a standard which is always either more precisely delimited or as a limit which is indefinitely surpassed. To this person, the free advent of time has as little importance as the age-old oaks along the freeway: "The more technological that time-keepers became, which is to say, more exact and more profitable in the effectuation of the measure, the less one is provoked to meditate on time in what is proper to it."[99] Now, if it is permitted to evoke much too briefly what is essential in such a dense lecture as *Time and Being*, this thought merits being isolated: "Time is not. There is time."[100] Time is not something which spreads itself out before us, which is at our disposal; tradition tells us that it flies, but this is because we still want to cling to it, to hold it in some shape or form—to maintain it as a possession. And here is where Heidegger tries to plunge to the heart of that which allows the temporalization of time: time gives itself in the triple offering of present, past, and future. The fourth dimension of time is, in fact, the most originary: the offering [*Reichen*] which invisibly temporalizes us. To name this offering *Ereignis*, the appropriating event, is not to capture it; it is to evoke it, to approach it; it is to accede to the dimension of the Open, which is, simultaneously, the dimension of the withdrawal into themselves of the three dimensions that are usually accessible in time.

Heidegger recognizes at the conclusion of *Time and Being* that here we have run up against a limit of expression. This is the wager of isolating in utterances the very gift of time; but is this unutterable essential, perhaps like the God of Hölderlin *offenbar wie der Himmel*, "manifest as the heaven"?[101] What one cannot say, one can at least show—as Wittgenstein argues in the *Tractatus*. But in order to shelter the enigma, it is not sufficient to gloss it in silence: Heidegger went as far as possible in his *transmutation* of language.

Untimely exercises? Certainly and deliberately. In the end, what engages us on this voyage to the country of Heidegger's thought is that from the epoch of Power to the epoch of the "anarchy of catastrophes,"[102] Heidegger proposes radically different ways of thinking of the dominant language-world, its ideologies, and even its counter-ideologies (because Heidegger had never agreed to being included in the "counterculture").

At the foundation of Heideggerian thought is the conviction—always placed in question—that the truth of Being does not surrender to the dominating will of human beings. To be sure, Being allows itself to be exploited, but it is after all humans themselves who exhaust their being by means of willing *themselves* to dominate.

This Heideggerian distance permits the critique not only of Marxism and ecology; it allows a critique of every ideology as a by-product of the metaphysics of representation. But it is not inevitable that Heidegger's thought would be in turn reduced to an ideology, be it that of a "power without violence,"[103] of poetry, or of thinking. It would risk losing its dis-quieting vigor, if its new thematic would be simply juxtaposed to the dominant rationality as an interesting twist, nothing more. It is not in this sense that we came to speak of a desirable "contiguity" between the dominant language-world and reserved languages.[104]

A hermeneutically fertile contiguity can only be "diacritical" in the precise sense that Heidegger gives it here: "To think in a critical manner means: to constantly mark the difference (κρίνειν) between that which demands a proof in order to be justified and that which protests against this in order to be authenticated in the simplicity of regard and of the welcome . . . "[105]

For Heidegger, "the simplicity of a look and of welcome" thoroughly traverses ideologies; born on the side of representation, it goes beyond their systematizations and manipulations. Meditative thinking is, however, not a cover for another risk: to consider that rationality functions in some automatic fashion, and that the κρίνειν does not exert itself at its center. We see, therefore, that Heidegger himself in his posthumously published interview placed, so to speak, *all* the ideologies of the twentieth century at the same level.[106] Now the absence of political-practical analyses of actual forms of domination and control can turn the tables against thinking, very easily rendering its exercise impossible: faced with the technological-ideological "overarmament," can the thought of Being be allowed to remain completely defenseless? There would need to be an entire work of semantical, political and socioeconomic specifications which has still to be completed in order to respond, for example, to the non-ontological and yet vital question: How do we safeguard and promote a civil society, guaranteeing freedoms (at least to the minimum without which thought becomes extinct) in a technological society?

This is the moment when my thinking diverges in a direction that is not completely Heideggerian. Whatever the extent of the divergence, the perspective of a "dynamic" contiguity between different approaches suggests a double opening to the possible. Why would the meditation on the ontological enigma suspend the rational inquiry of a balance or an improvement of the political? Why would rationality be denied the right to acknowledge its own limits and, as a consequence, refuse to impose itself in a founding or engulfing theory of the real in its totality? Is not the triumph of positivism perhaps more fatal than the recoiling of meditative thought on its own resources—in waiting for better times? To allow this ray of light to skim the surface of our somber horizon is to think as much from the rational perspective as from the perspective of the meditation on the enigma that the possible is a growing power, with perhaps a crossover between these dimensions.[107]

The thought of Being can remain vibrant only if it loses neither the impetus nor the space of κρίνειν, even if such a critical vigor can—and ought—to turn itself against its own intuitions and impulses. Thought "ought to think against itself, that which it can only rarely do."[108]

8

Rationality as *Partage*

THE POWERS OF the rational? Having become open to the infinite, thanks to their methodological mobilization and to their diversification, the powers of the rational find themselves forever exposed to this supreme irony: their power is not such that they could *definitively* illuminate their destinal unity, nor that they would be sheltered from reversals which come to flaunt their adequacy. In the expression "the powers of the rational," we know, henceforth, that an antiphrase is hidden. This is also the case regarding the reality of these "powers."

"To think against itself." How could one do better than in the ironic exercise of a dialogue? No language can ultimately enclose the rational; even less, thought as such. But is it perhaps the privilege of the dialogue to yield to detours, to follow digressions, to acknowledge aporias that discourage all other discourses? In the attempt that follows, three voices without face and without place take up the discussion of this new understanding of the Enigma, which we have named "contiguity": they are most often in agreement only about the paradoxically opaque power of a *partage* whose principle wants to be absolutely illuminated. We will discover bit by bit that we already know these voices, and that the contiguity that they enrich is none other than the very place of their exchanges.

Voices: X, Y, Z
* = Silence

Z.- Why envision rationality as *partage*?

X.- At the origin of rationality was an elating discovery of sobriety and rigor, the breakthrough toward truth as source of the world, the accession to invariable principles of nature. Universally practiced for centuries, it rather recently has seen its attraction to the ideal diminish on behalf of its powers: henceforth, we know that its ideal is linked to that of technological civilization in which many have recognized the direct consequence of its successes. Unwarranted identifications?

Y.- A classical type of expression can do nothing but denounce the contradiction in "rationality as *partage*." Rationalism, Bachelard said, in unison with an entire tradition, is "the rule of apodictic values." To consider rationality as a fact or as a destiny is, therefore, to misunderstand the transcendence of axiomatic aims, to confuse levels, to turn the philosopher on his or her ear. Is the objection excessive? Is a fully constituted rationality given to us as a lot or model of life (assuming that one could, in this regard, compare us to souls observed by Er)? I would rather believe that the exercise of rationality remains, above all, an assumed, willed exigency: never a simple destiny. One is not born wise or intelligent as one is born blind or petit-bourgeois: this is to confuse potentiality and actuality.

Z.- However, from the moment when what was a freely accepted ideal (often undertaken with courage, against orthodoxy) becomes an obligation or even a condition of social acceptability, we know that judgment bears less on the goals which remain than on the results already obtained through the work: the constricting appropriations for new generations.

X.- . . . as in every society, but with the advantage of dynamism. We must distinguish between a minimal level where one acknowledges the social diffractions that you evoke (for my part, I would place more ambiguity here), and the higher level where rationality can protect its ideal role. I do not see an incompatibility between the two.

Z.- If I understand you correctly, you admit that rationality can be considered and even judged in view of its minimal constraints, its cumulative effects and its global results, even those that are indirect—and not merely by invoking its norms. But I am no longer in agreement with you when you understand "*partage*" as a brutal and negative facticity: if rationality is our lot, it ought to be entirely so, on the level of its ordinary use as well as on the level of its ideal summit. And destiny is not only a bundle of facts . . .

X.- There is a meeting between rationality and destiny because rationality has attained—despite regressions, reversals, and failures—a point of no return; destiny is, therefore, now lodged in rationalization as an inexpungible necessity and—what is more—a convergence.

Y.- I hope to demonstrate the contrary and the dangers of this seductive theme. Rationality would be rather *anti-partage*: it is that which pulls apart factual positions, opinions; it takes sides. Rationalized or not, is man not forever surrendered to his carnal, temporal, mortal condition, to sickness and to day-to-day problems? And as to his collective destiny marked by technicalization from the perspective of the fight for power, it is so tragically uncertain only because the movement of rationalization has not yet penetrated and transformed this destiny. Also, I would like to know at what level you situate the "convergence" between rationality and destiny.

X.- The immanent tendency to total rationality: to lift it again to its own principle, that is to say, its self-foundation.

Y.- Assuming that one could define rationality as such and discern its specific project, what is striking in the attempts at rational self-foundation is their hyperbolic or metaphysical character, that is to say, their scientific failure. The only philosophical certitude acceptable for a lucid spirit today is negative: it is the recognition of the impossibility of finding the unconditioned principle of knowledge.

Z.- If this mind is truly lucid, it will revive this impossibility in the possible as such, that which is prior to meaning, the destinal gift.

Y.- What do you mean by that?

Z.- That meaning itself is not its ultimate horizon, but that this mind questions that which has rendered possible and continues to render possible this horizon.

Y.- It will be only more radically arrested in its ambition of self-foundation.

X.- It is not certain that going back to an autonomous foundation and the regression to the precondition of foundation completely overlap. They do not respond to the same exigency. Before taking up this difficult juxtaposition, would it not be better to try to respond to my more modest question concerning the orientation of rationality?

Y.- A beneficial limitation that will permit me first to note that an oriented rationality would not be far from a telic rationality. This is a dangerous conception. Why reintroduce finality, an Aristotelian concept that the last great metaphysical syntheses of the previous century (I think above all of Hegel) have attempted to save, one last time, in order to structure the unity of the real and give it the systematic figure of a true totality? The Absolute thus realized is still seductive for more than one person—and not only those rare individuals who accept Hegelian proclamations. Agreed. It is not these people who are going to advance science. Is it even still necessary to speak of Reason? The era of unified, optimistic, and solemn rationalism is over. Only a rational operativity reveals itself in extremely diverse degrees of formalization and experimentation. The real is neither rational nor scientific in itself; it only becomes that partially, thanks to the progress of the positive knowledges. The question that is undoubtedly insolvable is that of the unification of rationality as such.

X.- There are, nonetheless, certain minimal characteristics which define the rationality of a procedure.

Z.- In classical terms, the respect for logical principles . . .

Y.- But one does not have the right to infer the scientificity of the aforementioned procedure (or of whatever theory) from its rationality. On this point, Popper is right to point out that the Kantian theory of knowledge is rational, not scientific, which is contrary to Newtonian mechanics (both rational and scientific). The field of contradiction is immense: scientific experience obliges us to select from all of these compossibles. According to the field of study and at a moment m, "scientific truth" reveals itself as unpredictable in relation to that which rational activity is able, by itself, to foresee in it. As regards the essence of rationality, it is as difficult to define as the "logical form" about which Wittgenstein says—that in order to be able to represent—it is necessary to place us "with the proposition outside of logic, that is to say, outside of the world."

Z.- There is rationality as such, rationality at work in the sciences, rationality applied technologically.

Y.- Agreed, except for the first term—it is too undetermined.

X.- I did not make myself clear when I spoke of the orientation of rationality. I do not situate this orientation at the level that you understand it. It is not a matter of defining a priori the ought-to-be or even simply the being of reason; it is rather a matter of observing the functioning of rationality and of recognizing a general tendency there. This is my program of work: operating a posteriori, to establish a conjecture that permits the disengagement from pure functionalism in order simply to take a look at rationality today.

This is my argument: since, even in the relation of principle to consequence, there is not symmetry, all nontautological reasoning represents an increase of knowledge. The rational milieu is not isotopic, but rather is expanding; this is even more true if one includes experimental discoveries in this movement, as well as the multiple connections that these imply and permit between facts and concepts. We are not absolutely sure that the physical universe is expanding, but for the moment our knowledge is expanding; and every indication is that it will follow its progression, along the direction it is launched, for an indefinite duration. Is this a simple accumulation of information? In one sense, yes, but at the same time, it is a refinement of methods, a maturing of the scientific spirit. Considered globally, this movement is completely different from a reversible relation: it is an overdetermined and negutropic process.

If this perspective is metaphysical, it is in a different sense than that of the idealism of the last century.

Y.- I have difficulty understanding, indeed, how you can speak of rationality founding itself without making it, in one way or another, metaphysical. A truly scientific approach, even in logic, never isolates rationality *as such*. We know that in placing the non-contradiction of arithmetic among the undecidable statements, Gödel goes to the limit of the possible regression toward the foundation—precisely in demonstrating the impossibility of an absolute demonstration. A certain type of philosophical discourse is able to found itself because its specific question is that of self-foundation: it only realizes that which it postulates. This does not hold true for rational scientific discourse.

X.- It goes without saying that I do not aim at a *proven* self-foundation. However, a logical discourse that establishes, even negatively, the conditions of its own coherence appears to me to be confronted with the question of foundation. I no longer understand the self-foundation of rationality in a *strictly* Hegelian sense; if I continue to do philosophy, or metaphysics, it is according to a new orientation that I will call *referential*. Absolute knowledge nevertheless has a referential validity, no more, no less.

Z.- What do you call referential validity?

X.- The best way to explain it is undoubtedly to situate it in relation to Hegelian self-foundation. Reason, which justifies itself in knowledge, manifests and realizes divine power; the work of the spirit accomplishes in humans the living plenitude of that which the tradition thought under the name of God. Such is, in its formal rigor, the Hegelian doctrine. Rationality, for the last time, finds refuge here in a speculative theology: its unity is, literally, absolute.

The rationality at work today is no longer able to make this jointure with theology without regressing. It appears that Hegel *postulated too much*.

"Referential validity" seems to signify that absolute knowledge no longer has anything other than a value of reference, if not a documentary role. But it cannot be simply a matter of knowing one piece of information among others in a series, since knowledge of this information conditions our positioning of the question. What is more, a complete—and not merely formal—understanding of absolute knowledge implies that it is not treated as one formalized "object" among others, but that the understanding had traveled the entire length of its distance and had assumed, in the end, its circularity. We find ourselves, therefore, trapped in the following alternative: either absolute knowledge is reduced to a documentary or formalizable element in a larger metatheoretical totality, but then it is misunderstood as *absolute* knowledge in its specific ontotheological ambitions; or absolute knowledge is understood and developed, but then we are only repeating it. The only way to take up this contradiction is to elaborate it in order to prepare for going beyond it. Let us suppose a new knowledge that poses itself as consciousness and welcoming of this contradiction. This new knowledge would think absolute knowledge both in its formal reduction (as reference) and in the necessity of its repetition (insofar as it is always valid). Now, Hegel never truly conceived spiritual repetition as pure tautology. It is, therefore, not contradictory that the new knowledge *would translate* into a still unaccomplished rational figure, the appeal of speculative unity between effectivity and rationality. The System is both conserved as the best possible philosophical synthesis and placed between parentheses in order to permit its reconfiguration. The detheologizing of the Absolute no longer allows postulating anything more than the minimal unity necessary for the comprehension of the real in its totality, in order to make the connection between the spirit of the world, scientific knowledge, and my knowledge. It is at the interior of these limits that I remain Hegelian, and that I recognize the "validity" of absolute knowledge.

Y.- This minimal unity is always theological; it is always a unifying sign that has been bestowed with a fetishistic substantialization.

X.- Your rejection of unification appears to me to misunderstand a fundamental methodological truth raised by Kant: the task of reason is to "systematize" knowledge, that is to say, to effect the conjunction of concepts according to a principle idea. To deny the possibility of a unified rational problematic is to revert to arbitrarily decapitating knowledge. Are you going so far as to refuse any unity to your discourse?

Y.- The tendency of reason to unify itself does not guarantee any knowledge of experience or even any conformity with this in general . . .

X.- Nevertheless, it is a spontaneous and inevitable tendency that bears witness to the internal structure of rationality legitimately reclaiming its own coherence. I do not see why you would refuse to take this aspiration into account, whereas you find it normal that mathematical statements explain nothing about the reality of experience.

Y.- The connection to sensory experience is neither the sole nor certainly the principal criterion of truth. A logico-mathematical system has an internal validity open to theoretical verification within well-defined limits; the system of rationality that takes itself for the object, on the contrary, gives only a "logic of appearance." Presuming too much of its strengths, it produces a dialectical, that is to say, rigorously nonscientific, usage of concepts.

X.- This is certainly why for rationality its unification constitutes the *philosophical* task. The *focus imaginarius*—the central perspective of Kantian rationality—is not itself reduced to nothing under the pretext that one cannot constitute it into a real object of knowledge. The supreme regulating principle, it allows systematic unity to be problematically *projected*, not dogmatically given.

Z.- Kant has recourse to a mirror image to understand the always particular status of the field of rationality. We see the objects on the surface of the mirror *as if they* were behind that mirror; it is the same for the content of the ideas of reason. Whoever wishes to substantiate these ideas, or even to have objective knowledge of them, is comparable to an automobile driver who would drive without making a distinction between the scenery that he sees in front of him and the scenery reflected in the rearview mirror. Without doubt, it would not be long before . . .

Y.- This reminder and this caution is not addressed to me: I refuse to constitute rationality as such in a system of experience.

X.- You appear to retain from Kantianism only the critique of the metaphysical unification of the rational. This focus does not testify merely to the inevitable character of a misfortune that endlessly returns as do cloudbursts during a Breton vacation. Kant affirms the properly philosophical legitimacy of a problematic procedure concerning the unity of the rational. I, too, believe that there is a horizon of horizons, or rather that philosophy is not complete insofar as it does not problematize the central point.

Y.- Let us admit the correctness of your interpretation of Kant: what does it change in the actual progression of rationality? The problem of unification has been remarkably treated in the transcendental Dialectic, and it remains apt, just as the return to perspective has been marvelously painted by Velasquez and remains for our admiration. Contemporary painting has nevertheless destroyed perspective. Likewise, in rational practice, everything occurs as if one has understood that nothing more was to be gained from the direction of supreme unification and that it was necessary to multiply the centers of reference as modalities of the distribution of forces, to substitute for the old principles of homogeneity the specificity and continuity of counterprinciples of heterogeneity, of irreducible specificities (different from specification as a variety of the homogeneous in inferior species), and of the discontinuity of rational forms.

X.- But must not this substitution, in turn, become systematically constitutive?

Y.- Likewise, must you not correct what you said a few moments ago concerning the "horizon of horizons"? One does not have the "right" to either affirm or deny the *existence* of a horizon of horizons. This level is no longer that of existence. And, since it is a question of level, I will use it to note that the reason that we speak of here is strongly "displaced" in relation to Kantian usage, still dependent on special metaphysics. For Kant, rationality goes beyond experience, whereas, for us, it attests to the ordering of it; this last is understood in a much larger sense, not the sharp cut between the sensible and the supra-sensible (this difference having been relativized).

X.- Even if Kant's model of a jointure of genres, species, and subspecies under a central common perspective is surpassed, it is necessary to maintain at the horizon a point of *possible* unification. Without that, rationality as such would no longer be taken as object. It is quite true that in contemporary terms the problematic of the ideal of pure reason no longer conforms to the ideas of special metaphysics, but rather to the sole perspective of a unification (or, eventually, of a totalization) of rationality. What remains of this problematic—once one has taken away the soul, the world, and God—is nevertheless essential: the question of knowing whether rational discourse offers only a nominal unity or founds its homologies on a unifying principle. After all, among all the possible discourses, why would one not arrive at discovering a regulatory discourse of rationality as such?

Z.- But nothing implies or indicates that this "discourse" could be a Kantian regulator, nor, moreover, that it could be metatheoretical in the sense of a formalization of all discourses that profess to be rational (the very logic of formalization invalidates this latter hypothesis). I would rather see it shattering systematizing or formalizing reductions, penetrating regressions to the foundation that has operated up to the present, in order to install a kind of configuration *other* than pure and simple rational operativity. At any rate, is it not necessary for the philosopher to reserve the rights of a certain form of semantic unity? Without this unity (which again puts into play the initial irreducibility of languages and the so-called natural languages), could we even pose the question of *partage*?

*

Y.- That thought could proceed to the point of conceiving its unity and, eventually, its destiny, I agree. But why attribute this progression to rationality, which is only a qualification of thought, not thought itself? As for me, I have little inclination to postulate a metalanguage through which the essence of rationality (or, short of that, its main components) would be considered as an object of intuition; I recognize, on the one hand, nothing but semantic unity (completely relative to common experience), and on the other hand, formalized languages of the different hypothetical-deductive systems which are actually possible—and these two are "bridged" by the project of scientific objectification.

X.- Through that which science can do, and that which science will be able to do, you postulate a unification of the real. By that you see yourself constrained—despite your denials—to anticipate phenomena (this process has been correctly recognized by Kant as specifically rational) and even pushed to move as far as an auto-anticipation of Meaning, a dangerous, although inevitable, privilege of this philosophical reason that you only assume negatively.

Y.- I refuse both to postulate a real in itself and to speculate on the potentialities of science—a reintroduction of metaphysics into the field of positive analysis. Rationality is certainly a project, but this must always be specified so that it shows itself worthy of its ambitions. In a rigorous rational practice, I never encounter Meaning; the appearance of *a* meaning in the discursive field is itself subject to well-defined conditions of possibility as regards the formal organization of signifying chains.

Moreover, far from a "constituted rationality" dictating to science once and for all the modalities of its progress, it is the inverse which happens: truly scientific work calls into question hypotheses or dogmatic theories that, in one sense, were only too rational. When Lysenko maintained the mutability of genetic capital in view of environmental influences, he had in his favor not only Stalin, but a certain common sense and, in addition, the apparatus of dialectical reason: an impressive combination . . .

If one examines the development of rationality in the last 150 years, that is to say, right around the time of the death of Hegel (who already—must I remind you?—had disparaged the principle of non-contradiction), one sees its criteria diversify and their economy redistributed: the quantification of the probable, the relativizing of the most well tested axioms and foundations, the introduction of the coefficients of incertitude. Is rationality still identifiable? In modern mathematics, there is no longer identity per se: anything can take the place of anything, with the condition that isomorphisms be respected. Is rationality still the truth? The opposition of truth and falsity no longer prevails unconditionally; Quine, following many others, notes that physics embarrassingly demands a rupture of the binary and the "adoption of some form of tri- or *n*-chotomy." Is rationality still the adequation between System and Sense? Since Gödel, Church and Tarski, the undecidability of propositions, which can be neither derived nor refuted, testifies to the irreducible gap subsisting between every formal system and that which it represents. Please pardon these recollections and do not misunderstand their conclusions: they do not lead to a kind of universal relativism (a frequent misinterpretation of the theory of relativity that means the inverse: a progress toward objectivity) after which nothing could be demonstrated or decided, but rather leads to an augmentation of precision and to a refinement of methods, such that every unification (dialectical or not) of the rational field will appear naively regressive, dangerously metaphysical. To reintroduce the absolute into epistemology is to turn up one's nose at a century and a half of progress, and even to return from Comte to Hegel—something that was already dangerous for a judicious and informed mind in 1830! This ought in no way to provoke us to abandon every synthesis; rather it should provoke us to conceive, as geometry permits of spaces to the *n* dimension, a diversified rational field in indefinite expansion.

In view of what I have just stated, I believe it possible and desirable to make a bit more precise where and how rationality is at work when its *configuration* changes according to its points of application. Two examples: in the logico-mathematical domain, the operativity at work is *formalism*; in the field of the human sciences, scientific exigency will be specified in the analyses of the functioning of *discursiveness*.

Formalism: it seems to me that its significance, in fact, spreads to every domain of objects, even to the logico-mathematical domain. Here is why: it allows for the constitution of a language whose alphabet and syntax are, from the outset, perfectly defined and whose deployment—itself mathematically formulated—does not depend absolutely on possible objects of application, but solely on the axiomatic schemas that regulate its functioning. Neither mathematics nor logic will any longer be defined by its objects (numbers or correct argumentation), but from a truly methodological operativity. Why recall this new common ground of contemporary epistemology? Quite simply, in order to establish a foremost account of the contribution of formalism for rationality. Formalism realizes, within well-defined limits, the mastery of language, complete liberation in relation to every unified sense, every substantialization, every objectivism; this is a formalism of the type that was intellectually exhibited (and not merely schematically represented) by the *specimens of operativity* that constitute so many of the purely functional models of rationality. You ask me about rationality as such; I reply, "Look to the work wherein it reveals itself in its purity, no longer in the sense of pure Kantian reason still subject to transcendental focus and to its dialectico-mathematical consequences, but according to the very activity of formalization." Once you have returned rationality to its axiomatic motor, will you be in a position to constitute a rigorous metalanguage of formalization in general? I am not sure of this. But you will at least be headed in a good direction.

The complete mastery of language: I do not claim that it will be realized by human science or by the theory that we have today. However, one works in this direction through the emancipation of *discursivity*, a positive or, eventually, structural study of discursive instances substituting themselves for the intentions and significations secreted by the metaphysics of the transcendental subject. Thus the objectification of regularities, reversals, and discontinuities of literary and historical discourse permits distributing rationality with supple insights ignored until recently by literary criticism and the history of ideas. It is a matter of learning to recognize structured ensembles or systems—but not without the caesuras and ruptures of rhythms created by uncontrollable unpredictabilities and chance events. At a level of objective mastery, inferior to formalism, we see a positive activity constituting itself, rational in its project and in its critical practices, tending to shatter pseudo-systematic unities, in order to construct a discontinuous theory of domains of objects and to exhibit the functioning of diverse kinds of discourse. There again, I do not see what rationality is; the ontologism of definitions is abandoned. Nevertheless, my work roughly isolates the refusal of rationality, attacking the nostalgia for supreme unity or for recuperative interiority. Can we do better in order to assure rational *partage*? Again and always, it is in order to transgress the limit of this *partage* that one elevates the discourse to rationality in its entirety. The extreme limit is attained when one has established and analyzed the discontinuity of the discursive instances to which correspond, in the most formalized spheres, the plurality of hypothetico-deductive systems. At this time, I myself am at the edge of this limit, beyond which the consequence of my discourse will break my discourse.

X.- You claim to attain the maximum coherence in avoiding totalizing systematization. In fact, you displace the question more than you respond to it here.

Y.- I displace it toward what?

X.- Toward a discursive operativity that serves the purpose of your project and that, henceforth, plays the nurturing role of rationality.

Y.- I try, rather, to constitute a structure of reception (minimally designed) which could build an epistemology.

X.- The misunderstanding is due, in large part, to your refusal to construct more than an epistemology (and again! only programmatically) in a place which, in my view, remains that of a properly philosophical reflection.

Y.- My refusal to do more: this is my philosophy.

X.- You have, therefore, an implicit philosophy that you refuse to develop, due to a neopositivist prejudice.

Y.- No, this philosophy remains willfully tacit, as that of the early Wittgenstein: I am able to develop a coherent discourse that is worthy of the exigency of rationality up to a certain limit of generality (for example, in what concerns the formalization of a certain number of objective operations). Beyond that, I find only non-mastery, and I am silent. I say even less about this than the author of the *Tractatus*: I have not pronounced the word "mystical" . . .

X.- The beyond of the limit: it is necessary to name it in one way or another.

Y.- To be sure, but not in such a way that it impedes the possible progression of rational expressivity—which is produced when one definitively falls from the side of the non-rational and pretends to seal the enigmas which remain.

Z.- Your rationalism is open, and you see its merit there. However, the reverse of this opening is that everything that is beyond this limit *will come to be explained or mastered according to the procedures of discursive rationality*. Is this not again an unjustified presupposition, a way of prejudging those blanks in the logico-scientific text that is as improper as explicit metaphysical discourses?

X.- It is the displacement of the philosophical question toward a linguistic strategy whose underlying will remains arbitrary. *Willfully* silent philosophy I have heard: but whence comes this will? You say it is impossible to exhibit *the* rationality as such: one can show only its sense. Even less possible: to show the positivisms configuring the discursive moments. Under the pretext of dismissing metaphysics or even simply the (imperial?) universality of Meaning, you claim to substitute the diversity of the procedures of the production of knowledge for a complete totalization, to substitute the functioning of supposedly neutral machine-systems for the cumulative organization of the System. You *affirm* the diversification of these procedures. Why? You laugh. Not a question to pose? I will pose it again and again. You cannot place yourself *between* the sciences and operate there at the interior of their respective domains in a quietude justified only by these domains. You put on the grey smock of the technician and you wash your hands: how simple it all becomes when one has rejected the discarded clothing of philosophy! In affirming discontinuity, you continue to exhibit a unified figure of knowledge: neither a scientific practice nor a technological arrangement. Your unification is posed as discontinuous. And I sense that you are tempted to proclaim it in the name of the Nietzschean laugh (why, however, invoke an authority, an ancestor?) against an imaginary Hegelianism, caricatured for the needs of the cause? At the risk of disturbing your positivist beatitude, I am led to say to you: Meaning is always at work, even throughout your denials; you will not avoid being the sign of its articulation. At the level of culture that we have attained, you and I testify to a certain kind of organization of intelligibility; by these increasingly homogeneous practices, we realize a formerly unobtained level of exchanges of information; we assist in restricting the network of interferences and connections wherein heretofore unnamed experiences come to be distributed, circulated, exchanged, and universalized.

Meaning: not able to be situated solely from its powers or its technological modes of functioning. Meaning: not a teleological project linked to divine eternity, but rather the space of contact and exchanges that rationalization progressively constitutes. A working? Certainly. A liberation? Perhaps. You have often made reference to the *operativity* of reason: you must recognize that this is defined less by its results or even its principles than by its self-regulating activity. The work of the concept? It now operates in the specific disciplines of which you have spoken; but it can and must be thought in the unity of its effectuation, in that which more than ever remains common to all its advances: its dynamism. This work is undoubtedly more easily grasped at the historico-world level than at the level of scientific discoveries; in the face of the expansion of the technological world, with its transforming capacity, is one not legitimately led to step back in order to recognize a desubstantialization of the sources of power, understanding by this an increase in the potential of production-transformation, at the expense of perceptible and finite determinations which previously were exclusively at play? For example, as Galbraith points out, power passes from the possession of land to the possession of capital, then—more recently—to the holding of a "techno-structure," that is to say, the means of power that are never reduced to "goods" or "money" but which, involving in part the latter, integrates them into a network of designed and organized professional abilities in order to obtain certain industrial, military, and other objectives. On a less global level, but significant for this development of the political economy: the desubstantialization of currencies. It is now universally recognized that a currency is defined by the economic power of the nation that corresponds to it. Now, this power is derived less from the quantity of raw materials that it possesses than from the degree of development of the industrial and commercial network of the nation in question; here again, we see technostructure. In what sense does this work operate? In the sense of detaching itself in relation to contents, on behalf of a self-referential expanding activity.

You tell me that each domain has its logic and its specific methods. I admit this when it is a matter of conducting a strictly scientific or technological study. However, at the point where I am standing, the aim of the philosophic meditation is actually to conceive *that which is*. It cannot do this without analogical and totalizing reasoning. I am, therefore, going to follow my inquiry in asking myself if I observe in other spheres a "phenomenon" that is comparable to what I have just isolated at the historico-world level. It is enough for me to once again take up your analysis of formalization and discursivity: reason is today realized in axiomatization or, at a less rigorous level, in the constitution of a theory of discursivity. According to your own terms, rationality works in one direction: the complete mastery of language. Indeed, if scientific research is enriched by a constant "metatheorization" (the formalization of procedures, reflection on principles, the formalization of this reflection in an indefinite movement), then I argue that at a less "pure"—but no less decisive—level of world history, the common language of technology spreads its tentacles, erodes natural languages, and constitutes a new self-referential space: the worldwide technological community.

I can now show the unity in this analogy: corresponding to a loosening in relation to contents and to signifiers—in stages according to the different levels of contemporary reality—is the constitution of a complex field of exchanges of information. This field—at the level of generality where I envision it (that is to say, at the philosophical level where the analogy becomes pertinent)—appears to me to define itself by *functional self-reference*. It is, therefore, what I propose to designate "meaning today" for want of knowing how to speak of this, and to renounce comprehending that which takes place, before our eyes.

Y.- I have listened patiently to you this time, and have hardly found your discourse absurd. It seems only that in order to support it, it is first necessary to specify your method, the degree of certainty that it claims to attain, and secondly, to situate more precisely in relation to metaphysics—let us say, to Hegelian thought—the "meaning" of which you speak.

X.- The first demand is more easily satisfied than the second, despite the attempt already at work a few moments ago to situate it in relation to Hegel. My method intends to remain truly philosophical: it neither claims a degree of certainty concerning epistemological considerations that you have previously presented (because you were then, discreetly, in the very wake of scientific rationality—although to a lesser degree at the level of the human sciences), nor an intuitive or even religious abandonment of rationality. Its aim is to totalize the rational comprehension of that which is, to attain a sufficient degree of rigor in the analysis in order to subject it to a critical and judicious dialogue, such as our own. You see it: nothing but the most dialectical in the Platonic sense . . . It is a method of rational research, illuminated by the quest for the non-hypothetical, assuming the uncertainties of a nonscientific investigation.

Hoisting again the philosophical flag obliges me to respond to the second question, the key to the rest: What difference is there between the meaning that I speak of and the meaning of Hegelian knowledge? A few moments ago, I recalled Hegelian knowledge; there, meaning could be only the meaning of meaning, absolutely self-referential meaning, the Absolute; it is in relation to this central and final principle that logical spheres, natural orbs, subjective spirit, objective spirit, and absolute spirit are dialectically ordered. I see the following difficult question arise: "If your meaning is no longer absolute, in what manner is it still meaning?"

Y.- It is the question that I was going to raise.

X.- It is problematic at several stages:
1) The New Meaning is no longer able to be exactly similar to a meaning that is absolutely Hegelian. Why?
2) In what way is it still meaning?

3) Is there a relation between the New Meaning and the Absolute? If so, what is this?

Stage 1.- The New Meaning has lost what was once the keystone of the system: divinity. It lost it because this fundamental presupposition henceforth impedes the understanding of that which is. I have already said it: the Hegelian Absolute *presupposes too much* for reason today. To presuppose the complete rationality of that which actually is, linked to the finality of a reconciliation as the immanent aim of the totality of the real (postulated—moreover—understood according to the universality of the dialectical method), is too much for the enlightened understanding. This collides, henceforth, with non-dialectical structures, unabsorbable paradoxes, uncertainties concerning the foundations (and even more the ends) that forbid it to admit that the task fixed in principle (to understand that which actually is), is still possible within the very terms of the Hegelian-like System by renewing its royal axis. I will, therefore, go so far as to concede to you that the System has become cumbersome for a rationality both positive and faithful to the initial demand of the comprehension of the real in its entirety. We have seen the reason for this: the theologico-metaphysical exigencies of the System constrain concrete intelligibility; they often even enter into contradiction with the objective figuration of actual rationality. The Absolute demands too much of us; what it offers in return is no longer completely in touch with the present actuality.

Stage 2.- In what way, however, can the Absolute still have meaning here? Because a Hegelian exigency always remains urgent: to understand the present in its actuality. This understanding still presupposes something, but much less than before: a certain rationality of the real, and no longer the complete rationality of it. Of course, Hegel never wanted to eliminate every chance or all residue of nonsense in the philosophy of History; he merely (but this "merely" is still considerable) claimed that the "true is the Whole," which is to say that, thanks to the totalization of objective figures in History and in the present, one arrives at an exhaustive unity of meaning. If our new approach totalizes, but only to a certain point, without claiming to be exhaustive insofar as it always isolates a meaning, this is because it responds to the exigency of rationality and because it can succeed, to a certain degree, in establishing the features of the present reality. The direction toward which we orient ourselves is now perceptible: there is certainly a "spirit of the times" (another name for meaning); this "spirit"—obviously nonsubstantialized—can be characterized (in the first moment) by its increasing activity of abstraction and organization. I take up again two traits already announced a few moments ago, one quantitative, the other qualitative: the *increase* of information exchanged at all levels, the concentration of potential in a more and more functional *activity*.

It goes without saying that this is more a program of inquiry than a result. You ought to extend to me a little credit, without which my hypothesis will not be able to take a sufficiently precise and convincing configuration.

Y.- My silence is carte blanche. I await my turn.

X.- Let us understand this well: the meaning that I uncover is different from Hegelian meaning because—purely immanent to objectified phenomena—it is connected to no other principle than that of rationality at work. Not totalizing this rationality into an organic system, it has no other claim than to be a grid of intelligibility, an intermediary operative structure between a formal system and the absolute System. In order to respond to the question posed in Stage 2, it is able to be a meaning, because it is capable of regrouping a mass of phenomena, of disengaging there a convergence and nearly an operative law.

Y.- There remains the third stage.

X.- I am coming to that. There would be no relation between the New Meaning and the Absolute if that which is in question had changed completely. On the contrary, we have seen that we are concerned with the same task. What is more, the New Meaning preserves, in the two traits that I have isolated, the characteristics of the Absolute: first, the dynamism integrating contradictions and permitting the attainment of a superior degree of development; secondly, the work of determination that renders the dynamism possible. One must add a last trait that is the most important: self-reference. The Absolute is such because it thinks itself completely and is free from every exterior determination; the New Meaning does not deduce itself from a principle that would remain exterior to the demand of rationality: it realizes this demand in the immanence of the actual present.

Y.- I am perhaps going to turn this last feature against you: does not self-reference permit you to play with words? There is a difference, extremely decisive, between the theological autonomy of a being or a spirit that founds itself, thinking itself completely, and the quite relative autonomy of a machine, an automobile, for example, which depends both on fabrication and on maintenance. Your "new meaning" appears to me to be much closer to the second kind of autonomy than to the first: you yourself, in this regard, have spoken of *functional* self-reference and also of an *operative* structure. The New Meaning, if I have understood it correctly, never detaches itself sufficiently from the phenomenon for which it accounts in order to attain the sovereign autonomy of the Absolute; it offers us only an intellectual autonomy. To return to an example that has already been cited: the technostructure certainly tends to autonomize more and more (in disengaging itself from political power through vertical integration, for example), but is this in the same sense as the Absolute?

X.- Not exactly. But one must not neglect the analogy.

Y.- Nor is it necessary to solicit it.

X.- I have not completely finished my proposal: having indicated what the New Meaning would preserve of identity or analogy with the Absolute, it remains for me to return to that which separates these two configurations and to designate their manner of relation. It is a delicate undertaking. The equation:

$$\text{Absolute} - \text{God} = \text{New Meaning}$$

would be—in my view—formally true. It is evident that the New Meaning is the Absolute de-theologized. But what is the Absolute de-theologized? If God and the Absolute are identical, the equation becomes (A designating the Absolute):

$$A - A = 0$$

The New Meaning would be reduced to nothing, or at least to nothing that is philosophically discernible. I do not hold to this possibility, because this would be a misinterpretation that would interpret the relation between God and Absolute spirit, according to Hegel, as synonymous; it is not God who defines Absolute spirit, but the inverse; God is only the objectification in the religious sphere of the unity whose Absolute spirit is the truth. In this way, the New Meaning is indeed a certain "residue" of the Absolute or, more positively, more dynamically, perhaps a new figure of this same Absolute.

Y.- Be careful! Either you are correct to think that the New Meaning is the Absolute de-theologized, but then, if it is truly de-theologized, in what does the Absolute remain? What is the "residue" if it is not an insignificant? Or you are wrong; the "residue" itself is theological. In this last hypothesis, you have not progressed. You have only displaced the difficulty, you have not left Hegelianism.

In the first case, you say almost the same thing as I do, granting you only a superior degree of generality: systematic unification is no longer possible; one can only go at the very most up to the point of uncertain convergences.

In the second case, I repeat, you are a sheepish Hegelian; you cause this metaphysics to rise from its ashes—without daring openly to take it on.

What is the correct hypothesis?

X.- One thing is clear: the New Meaning is no longer theological in the manner of the onto-theological metaphysics of Hegel himself who "retrieved" the Christian contribution, claiming to assume a harmonious junction with that form of monotheism. In the first equation, I inscribe God, not the theological element in general.

You are correct to point in the direction of the following difficulty: once the God of onto-theology withdraws from the Absolute, is there going to remain a more general theological element that will continue to give the New Meaning a metaphysical character, thereby differentiating it from simple hypotheses of work or theoretical speculations?

Z.- According to Hegel himself, beyond the religious sphere, spirit preserves a divine characteristic: it is *thèïon*. But how is this divinity still worthy of this name?

Y.- The de-theologizing of the Absolute has been undertaken since Hegel's death—undertaken by his disciples—above all, by those on the left, of course. You rediscover the same aporia as the Marxists (although they more often refuse to take it into consideration); has the theological charge not been displaced toward Humanity, the Proletariat? It is not so easy to kill God: every vision of a unitary and a fortiori mobilized world is in secret a substitute, when it is not a hideous caricature. As for me, bare-headed before the great emptiness, I prefer to renounce any kind of unifying synthesis, certain at least of this: I betray nothing of rational exigencies.

X.- I maintain that there is an important ambiguity in the presentation of the New Meaning and that, in requiring me to clarify the relation between the New Meaning and the Absolute, you compel me to support it openly. Am I going to succeed in doing this?

If I remain with what you called the first hypothesis, not only do I de-theologize the Absolute, but I renounce the truly philosophic task: to comprehend the Whole in a unified manner. I hold to an intermediary synthesis, at the very most within the jurisdiction of the philosophy of History. If I gave this impression, it is due to the prudent induction which has guided me in the explanation of the New Meaning, in order to remain rigorous in the exercise of rationality and remain in contact with experience.

If my undertaking remains philosophical, it is inevitable that it continues to concern itself with something like the Absolute, the Divine, that which constitutes the Supreme Being. It is the privilege, the grandeur, and the fragility of philosophy that elevates it to this pinnacle of thinking.

Y.- This is a contradiction! How does this fit with the de-theologized character of the New Meaning?

X.- There is not only a single way to deal with the same question in a comparable horizon. We have spoken quite a bit of Hegel, but not at all of Fichte. The latter sets out from the impossibility of proving the first unconditional principle of knowledge: all of his inquiry will remain situated in this point of departure, that is to say, in the constitution of a speculative synthesis which accepts being constructed on the basis of an impossibility. In an analogous way, as an heir of Fichte (and Kant before him), I come again to the inevitably unified source of rationality, as well as to the self-foundation that I cannot prove, but whose very impossibility is going to continue to condition my inquiry.

Through the New Meaning, the Absolute appears to us under the figure of detheologization: it deconstructs itself as the self-grounding principle in order better to extend itself as exchange, labor, expansion. All this happens as if an enormous energy were liberated from the splintering of the unified divine center, and as if its light were henceforth diffused among the many centers of activity where self-reference exposes itself—hereafter as functional. Functional self-reference is a self-reference relative to precise conditions of activity, a self-reference that can never again completely close upon itself in founding itself. The New Meaning is no longer protected a priori from the meaninglessness, as was the case of the Absolute in Hegel: it signifies that which can never define itself completely (and which therefore remains in part a mark, a sign) and which, nevertheless, rules over the world.

Rationality as *partage* is perhaps this immense constellation that kindles its lights and extends them, not without having understood that beyond the fringe where its powers waver, there is only darkness.

Y.- The frontiers of darkness, I see them pursuing our "chains of reason" at every moment, not in the least warded off by Meaning, be it new!

However, I admit that *if* there could be a unification of experience here, it will operate only through rational work which produces its efficacy only in masking its essence. Is it not necessary to have the wisdom to accept this situation without substantializing a pre-given instance that would play the role of destiny—or even *partage*?

Z.- Is not this "rationalist factualism" plunging us into logical, scientific and technological operations, resulting in the loss of any sense of *partage*? And expanding beyond any limit (like a gas that no longer encounters any resistance) this "vacancy of *partage*" where Hölderlin already saw modern man going astray?

Y.- A rhetorical danger or a real one?

Z.- Please note that I never extolled what was substantialized in the last moment. *Partage* is a limit, but this does not *add itself to* a territory as a line extended at the edge of a terrain; the *partage* of a language, for example, is the bearing of its style. If the sense of *partage* does not inhabit us, no disguise will cover this deficiency.

Y.- Give me at least one example of *partage* in rational terms, without hiding behind a metaphor.

Z.- In Kant, the *partage* of understanding is categorical knowledge.

X.- And that of reason delivered to itself, the contradiction.

Z.- Already a sign of how human reason, delivered to itself, vacillates in a kind of impotent omnipotence.

Y.- Because Kant still looked backward—toward a problematic that metaphysics bequeathed to him and that he still took far too seriously.

X.- You cannot deny that, closer to us, in a contemporary scientific context, lies the problem of the ends of reason: after Planck, Einstein, Lobatchevski, Husserl acknowledges that rationality no longer experiences itself in a unified and harmonious system of the world, but rather in a constant questioning of its foundations. In this sense, the actual *partage* of reason is the crisis.

Y.- But Husserl did not seek salvation beyond the rational.

Z.- Who says to you that I seek *partage* beyond?

Y.- Now I understand even less.

Z.- As it seems that every new affirmation in this domain has a recessive character, let us take a rather negative approach. Rationality is not a lot, comparable to a life of gardening or tyranny; on the contrary, it effaces any ancient destiny, any representable and "embodied" destiny. Does not this sense of *partage* begin with the consciousness of this effacement?

The absence of a sense of *partage* can be discerned, in my view, in a rationality without restraint, deciding everything on the basis of calculations and concepts, without understanding discursive specificities and more: missing the singularities of languages as well as irreducible silences. The rationality of worldwide planning has no other way to make room for the imponderables than to reduce them to marks of indetermination, and, thus, it forecasts solely in a linear expansionistic manner the exploitation of natural resources and the production of the goods of consumption. When we submit the principal questions to it (the right to adore immemorial gods, the freedom to create, to live differently), it finds itself completely disarmed.

A current example of practical nihilism: taking a vacation on a Greek island last summer, I was approached by an American. "I come from New York," she said. "I am a journalist and must take some photographs of old peasants for a magazine. Where can I find them?" She was reassured when she learned that upon rising early the next morning she could certainly encounter the representatives of a way of life that had not yet completely disappeared, people who still traveled perched on their donkeys; she was delighted to be able to catch her plane the next evening after having had a profitable hunt for faces, costumes, and picturesque postures.

Within several hours, I say to myself, the work of centuries will have been frozen in silhouettes and glued to paper. This debut of objectification prepares for the pure and simple destruction of ancient ways of life.

From the strict point of view of our technological reason, no one could fault the journalist. Who, in an analogous case, will still have at the very least a feeling of incongruity? Is our world capable (yet this is hardly the most cruel case) only of this detached exploitation of a condemned, defenseless nobility, scarcely having the right to be respected?

Y.- You have made what is perhaps an excessive effort in order to present yourself as a pedagogue: one will always find superficial examples to caricature a philosophical position.

Z.- At a minor level, I acknowledge it; this intellectual impatience corresponds, however, to the very modern incapacity of self-limitation. Seize all, possess all, understand all . . .

X.- No, the American journalist, she too has a destiny at her pitiful level: the muddled vagabond.

Z.- Undoubtedly, she has a good intellectual capacity. The only thing she lacks is grasping that her absence of limits is her limit.

X.- A destiny is that which one is obliged to accept without being able to com prehend it *at its very heart.*

Z.- Precisely!

X.- Yes, but what is proper to contemporary *partage* (let us not remain, in this regard, with the example of the journalist) is the extreme lucidity that appears there. The question becomes: does this lucidity in any way change the situation? Personally, I think so.

Y.- Finally! Rationality also has its advantages. In what sense? First, it allows for taking a critical distance in relation to the obtained results. Then, relative to what you call *"partage,"* it offers a position of generality, it sets free a problematic. A destiny in the ancient sense would not give itself to thinking: it would possess you completely. "Go forth and the enemy will be conquered," said the oracle. This is not like the "voice" of reason. The silence of critical consciousness is the school of patient lucidity.

Z.- One could disagree with you, arguing that reason also "speaks"—"Two plus two equals four"—and that it results in a much more mechanical constraint than the voice of the oracle. I know that arithmetic is only a small part of mathematics, and that the rational domain is infinitely larger than both this conception and mathematics itself; however, the question is raised as to the type of rationality you are referring to. Is it rationality actually at work, or the ideal Reason whose qualities retreat beyond an increasingly distant line on the horizon?

Y.- In order to respond to you, we must again take up a distinction—already suggested, but not exploited—between the *minimal rationality* necessary to set in motion technological, informational, and other procedures that are those of the present epoch (actually, from this point of view, there is hardly any part of humanity that escapes rationality) and *maximal rationality* (rather than ideal), which is the full deployment of the cognitive faculties at the heart of scientific or philosophical knowledge.

X.- Evidently, it is impossible to claim that humanity as a whole, or even a majority of it, has attained a new height of rationality. It is even doubtful that the statistical minority who are conscious of nature and the importance of rationality truly "carry weight." There is here almost no actual relation between rationality as *partage* imposed at a minimal level (where the turn into the irrational is so often produced) and the elevation to the highest level that contemplative rationality can open. Nevertheless, has one the right to dissociate minimal rationality from ideal rationality? From the moment that monsters and gods were killed by the Enlightenment (in the larger, not the merely historic, sense), it is the whole of the rational domain that becomes determinant. Rationality as developed knowledge is only *a possible*—which will very likely remain such—for the great majority of people, but this does not mean that we must erase its horizon.

Z.- Therein lies the sickness, perhaps incurable, that we suffer. The gods, at times cruel (almost always irrational in their behavior!), were at least guides. On the contrary, the true radiance of reason is too distant for the masses; what remains for them is only technological repercussions and the effects of propaganda. What first befalls them is the destruction of beliefs and ancient attachments. The distinction between minimal rationality and ideal rationality poses the political problem of our times. The masses are exposed only to the destructive effects or the brutal worldwide planning of rationality; the ideal of an "ethic of knowledge" can only remain obscure or too demanding for them. In spite of claiming with Jacques Monod that only objective knowledge—in joining truth and values—proposes to contemporary humanity the ideal that has rendered modernity possible, and which alone is compatible with its development, and, moreover, in spite of praying for the coming of a democratic socialism that conforms to this ideal of coherence and authenticity, you cannot help—despite an enormous educational effort—that this ideal remains abstract and distant for the majority, not able to be politically relayed or translated into a massively efficacious language. For now, not seeing the conclusion, "the ethic of knowledge" is "aristocratic in its sternness," as Althusser has clearly shown. Althusser, however, was content with a Manichean critique of Monod's idealism (because it was based on the automatic opposition between idealism and materialism), pretending to believe that Monod wanted to say that morality leads the world! This critique was all the more unjust insofar as Monod made an attempt, which Marxism had until then been incapable of making, at meditating upon the normative character of the postulate of objectivity and on the fact that an ethics of knowledge is a will to power.

If it is always permissible to hope, then we must recognize that, in imposing itself on everybody, rational *partage* can only reveal the structural inequality between the (creative) exigencies of the ideal and the (social) constraints of the real (the curse of humans in the era of technology).

X.- It was already the reef on which the Platonic ship floundered. All the same, you are wrong to retain only the negative effects of a rationalization, which, after having destroyed irrational monsters, educates and informs. It would be hasty to confuse the profound dynamism of rationalization and its worthy opportunities with Taylorism or rudimentary worldwide planning. Just as it would be politically harmful to neglect or reject democracy under the pretext that it does not eliminate all manipulation. With this account, one could maintain that technology has erased the differences between the advertising campaign for a detergent in a great modern democracy and a campaign to conform in a fascist regime!

Y.- If it came to result in a disjunction of reflection, I would withdraw the distinction by which I have emphasized my concern: there are not two rationalities.

X.- The tension between minimal rationality and maximal rationality can perhaps be reduced. Nothing forbids this.

Z.- "Nothing forbids": here is an example of the forgetting of *partage*, forgetting that rationality is itself a *partage*.

Y.- No. Rationality knows to detain itself: tacit philosophy, we said a few moments ago. The silence imposed on metaphysical rationality only allows reason and nothingness to meet; but in all events: nothing more.

Z.- Silence does not surround reason as the sea does an island. You yourself said that our "chains of reason" are pursued at every moment by the darkness (not, I imagine, the darkness that comes forth to hide a god, but the eternal darkness of cold indifference and anxious silence) . . .

X.- Thought arises in a vertiginous mutism to contemplate sidereal spaces or even, closer to us, great glaciers enveloped in blue nights and advancing one inch per year. If there is a silent philosophy, it is due to a decision of the spirit: "That about which one cannot speak, one ought to remain silent."

Y.- This is a victorious silence that overwhelms gossip.

Z.- More multicolored than a rainbow, silence richly rewards the achievements of art or thought. When it surrounds the *Vespers of the Blessed Virgin Mary*, one would be wrong to separate from this silence, as if its naked density would save us from returning too brusquely to the daily whirlwind, and its halo would come to protect the fragile splendor of Monteverdi's harmonies. But Wittgenstein's silence suffers from the constraint imposed by language itself. Musical, philosophical, or poetical, a language leads to a silence that it deserves.

Y.- Wittgenstein's silence is not as empty as you claim. Extraordinarily masterful (each time that the metaphysical temptation presents itself and, in some way, out of respect for it), his goal is to *show* that which cannot be *said*. Because the famous phrase that was just cited has a reverse, much less well-known, that hides at the heart of the *Tractatus*: "That which *can* be shown, *cannot* be said."

Z.- I agree with you that this silence is not empty. But why this exorbitant privilege accorded to propositions of natural science? The inexpressible comes to paralyze language much too soon.

Y.- This petrification comes to seal a strategy that sought to *neutralize* metaphysical discourse. The appeal to language is therefore not made in a neopositivist spirit; it does not seek to substitute one language for another, but to arouse in each of the two languages—equally inadequate—the sense of the inexpressible. This is an admirable strategy, rarely understood, of a philosophy on guard against its own expressivity, of a rationality whose silence comes to seal rigor.

X.- Is it an admirable impossibility to be silent? Is it the paradox of a discourse that *affirms itself* in delivering itself over to death in a haughty despair?

Z.- This strategy would be, Wittgenstein wrote, "the correct method *of philosophy*" (my emphasis). But, while wholly in agreement with the creative intuition of the first Wittgenstein, according him all the admiration that you wish, I completely disagree that philosophy must make such an imperious appeal to silence (because this operates on the basis of a sterilization of philosophical language and its profound ties to poetry), and that the logical model ought to be restrained in a manner as definite as the tautological and uniquely propositional representation of a finite world . . .

X.- . . . and without sufficient acknowledgment of the fecundity of formalized demonstrativity . . .

Z.- . . . so I am coming to ask myself whether the Wittgenstein of the *Tractatus* does not utilize a congealed and, ultimately, a rather restricted logic to awaken the bad conscience that sleeps in every philosopher. It is, therefore, perhaps sufficient to give an internal critique of this type of rationality in order for us to exit this impasse.

Y.- Why do we need to make an appeal to this problematic? It was a question of knowing how something like consciousness of a *partage* can have a sense for rationality today. Wittgenstein's immense merit consists of delimiting the thinkable and the unthinkable, the expressible and the inexpressible. Is not this self-limitation of rationality the *partage* that you seek?

Z.- This is the self-limitation of a logic unable to be formulated in propositions; but this is not the self-limitation of the rational as such.

Y.- Even so, if the inexpressible is mystical (which, for my part, I did not say), a critical distance is effected in relation to the rational.

Z.- If Wittgenstein counterbalances a congealed logical core with an undetermined *completely other*, without acknowledging the historical relativity of reason, would this be to advance toward this sense of *partage* that I seek? To suppress the question of the enigma under the pretext that it cannot be formulated in logical terms (whose exclusive character has been totally and arbitrarily posed), is this not to take the inverse road from what I wish both to say and show?

Y.- Wittgenstein does not suppress the enigma. He simply asserted this: it does not exist in terms that are able to be logically formulated; the only way to make reference to it without distorting it is through silence.

X.- One need only to reverse Marx's famous phrase to rediscover the same thing. He would have said, "Humanity does not pose insolvable problems."

Z.- Which is, quite obviously, wrong—even absurd.

X.- It is the case that if you despair of rationality, you will not be able to say the *partage* any better than Wittgenstein could treat the enigma.

Z.- Can one say what reserves itself other than poetically?

Ein Zeichen sind wir, deutungslos.[1]

*

Z (to X).- You just made a confession a few moments ago: the New Meaning is no longer protected a priori from meaninglessness; this is a signifier that can never be totally defined, and which remains, in part, a mark or a sign. The difference between the self-foundation of the Absolute and functional self-reference would, therefore, be this: whereas in the first case, the Absolute presupposes itself and proves itself, in the second, it only presupposes itself in order to experience itself in a work where the foundation is perpetually excluded—a discourse as relative to its conditions of appearance as is an economic circuit or a network of information. From a formal point of view, Absolute Meaning was apophantic and expressive (even in speculative propositions); the New Meaning adds question marks. It appears only as heuristic and provisional (a characteristic it shares with diverse phenomena—principally technologies—which it proposes to connote and unify).

X.- For my part, to what did I confess?

Z.- That the superficial polemic on the legitimacy of a discourse of discontinuity makes a place for the ever more perceptible emergence of an agreement between you or, at least, for a convergence in the acceptance of a kind of practical rationalism excluding every metaphysical foundation, every transcendent or even transcendental referent (if this is not purely hypothetical) and postulating that "rationality as *partage*" means: we can expect nothing beyond scientifico-technological work; silence is still preferable to the reintroduction of an allegedly sovereign discourse, a more or less shameful theology . . .

X.- My position is a bit different from what you claim: scientifico-technological work is the sign of the New Meaning that only allows itself to be said philosophically—but, without ever being able to be substantialized. What I call the New Meaning is the *identification* of a work by a discourse which is never completely reduced merely to that which signifies it. I do not accept that this completely relative unification is a theological residue that ought to be eliminated; it is the fragile mutation of a discourse that *was* theological and which is no longer *literally* so.

What remains is no less than rationality able to elevate itself to the thinking of destiny. Consider Leibniz and Hegel. To confine ourselves to the latter, has anyone noticed that the logical element that joins the thought of the spirit and that of nature, at the end of the *Encyclopedia*, is a *Sich-Urteilen*: not merely a reflexive judgment, but a *self-dispensation*? The destiny of reason is self-reference. Having arrived at the end of its route and its Calvary, rationality does not forget its origin; it reminds *itself* that the absolute itself is not without presupposing *itself*; that the finitude of the absolute is the acknowledgment of its closure, its self-limitation, its self-giving. To assume its sovereignty in the structure of a self-referential Same is the *partage* of rationality.

Z.- There has never been argumentation without paradox. First of all, you are correct not to base your argumentation on Leibniz; rather than an elevation of thought to destiny, I find there a rational domination over contingency. Now, I very much acknowledge that Hegelian rationalism preserves it as an ultimate trace of its relation to a destination. But as for a thought of *partage*, as I understand it, what remains from this feeble connection is another story. Hegelian self-reference establishes the Absolute, the adequacy of reason which comprehends the world as its other and is at rest there because this is nothing other than the other side of itself . . . The Limit is always taken up in the interiority of Meaning. On the contrary, in self-limitation, which I endeavor to bring closer, the accent is on "limitation," not on "self."

X.- What would be the Other as limit? The irreducibility of *partage*? The Divine?

Z.- Let us speak of this Divine who has been engulfed in your speculations! Already metamorphosed by Greek metaphysics, captured by Thomas, rejected by Cartesianism, worn out by German Idealism, liquidated by the will to power, what remains of it in the era of technology? Nothing. The New Meaning and the Hegelian Absolute are for me one and the same, just as your divergences, dear friends, appear to me as subtle distinctions of persons of good company faced with this enormous occasion: the Divine has left us. Modernity (the present world, marked by rationality) no longer has access to This Who—Apollo, Yahveh, Christ, or Shiva—could only arise as Unique. All of nonwestern humanity (or the West prior to, or at the margins of, metaphysics and science) thought its relation to the world from this indefinable living connection that one had to name with reverence, love, or terror. It is finished. Why? Because the dynamism of rational mastery has a corrosive force that respects nothing, above all not the Unique; the will to truth is the will to generality: the concept universalizes the Unique, it takes its place—but the product of replacement ought to be henceforth qualified to become a particular "case" in a larger totality or, inversely, to fulfill the function of an articulation that possesses a chain of determination.

Heirs of August Comte, Hegel, Marx, and of the stubborn cohort of the men of the Enlightenment, you shake your head at hearing such a discourse—it causes you pain to listen to it. At times you still manage to believe, with the incurable optimism of the rationalists, to be able to reconcile the movement of Progress with the numerous forces that have been exiled from our earth. When this confidence fails you, you console yourselves by clinging to some divine "residue," which would still inhabit a problematic New Meaning, translucent avatar of the Absolute. In truth, in your heart of hearts, you cannot not know that from the Platonic episteme to the New Meaning, the consequence is clear: it is the single and unique trajectory of the Concept that pursues itself. A few moments ago you included God in an equation; this was no longer Him: this was the god of the philosophers. It is not in an equation, it is in an *alternative* that God (or the Unique) is inscribed vis-à-vis the concept (or meaning).

This is the maximum that one can do, dear friends, in order to translate the untranslatable in the language of the concept itself, in *your* language.

X.- As to the status of the concept, permit me to first of all point out to you that the alternative that you have posed has already been isolated by Kierkegaard in the famous phrase "either . . . or." And before him, had not Hölderlin a presentiment of such a fissure, even before his friend Hegel had completely formulated the System?

We are tied to the pillory as the representatives of the philosophico-scientific movement of the West. If you extend to us this aggressive honor, do you not at the very least place yourself in a delicate situation. Where exactly? You are not a stranger to the concept, since you are able to qualify the Untranslatable in logical terms. How do you escape this corrosive force that you have evoked? The worm does not spare your fruit. From where do you speak in order to throw stones at us?

Y.- And how will rationality, becoming a *partage*, see its course modified, eventually amended?

Z.- I am here, in my turn, called upon to explain myself. This is only just. My outburst surprised even me; certainly it struck something in me, as much as it did in you. If rationality is our *partage*, it spares no one, and scientific practice has taught you that it no longer privileges anyone: every result is to be taken up again, every proposition is to be critiqued. It seems to me, however, that in discussing the relationship between the Hegelian Absolute and what remains of it for us, you distance yourselves from the central perspective: How do we think rationality itself in its unity, in its limits, in its historical signification (or its complete absence of signification)?

In perhaps more fitting terms: it seems to me that you, both of you, take no distance in relation to meaning as such. Your discussion restricts itself, in short, to this interesting, legitimate—but to my taste, still too narrow—question: Can meaning be surpassed in a new philosophic systematic, or is it necessary to confine oneself—for lack of a superior unification—to a negative philosophy, in an epistemological mode? Or better: Can a metatheory of meaning enclose itself upon itself, or must it maintain a simple referential validity?

There are better things to do than to erect or to exacerbate the self-reference of meaning as you did: allow it to appear in its limits. There is no point "in adding on"; but we have to show from where meaning originates, to interrogate it as to its ends. There is no point in apologetically espousing its cause, nor the inverse (in which I would distinguish myself from the great pioneers that you evoked, Hölderlin and Kierkegaard, who sublimely ran up against the impossibility of Reversal: the extreme experience of a completely Other, where our wings blaze in the likeness of Icarus).

This route, opened by taking a critical distance vis-à-vis our tradition, was announced from the beginning of our undertaking; I had said that it consists of going back to what was preliminary to meaning. Is one going to repeat the Kantian essay on a topology of pure reason or—a possibility that you have not explored—to reconstitute phenomenologically the genesis of meaning by starting from the "lifeworld"? These attempts at a transcendental excavation or of critical archeology serve only to *bring back* the practice of rationality in relation to its own foundations. After having made one's way through all the positions internal to rationality, it is desirable to determine the level of specificity of meaning, going back again from objectivity to "demonstrated reason." The principle of reason has an efficacy that renders the constitution of an objective field possible, not the inverse. In going even further backwards, one comprehends the solidarity that is instituted in Greek philosophy between the position of Being as presence and its rationality, its "logicalization." This genealogy of reason allows it to appear, in the light of its presuppositions, as *one* of the utterances of Being—having become exclusive, and it permits it to rediscover the principal play that was more and more masked by the organization of the real.

Let us reformulate the same inquiry from another angle. The self-constitution of meaning leading to the domination of the earth was led to place its own reflexivity at the center of the world, to make of it—literally—the absolute, to place this absolute above man, life above the world (this Hegelian hierarchy was reversed by Lévi-Strauss at the conclusion of *The Origin of Table Manners*); and, finally, this self-constitution of meaning has dialectically determined everything, except its own rationality—which Hegel merely preserves formally. When Hegel said, "That which is, is rational," he believed he had philosophically exorcised hell; he recognized contradiction everywhere, except in this *ne plus ultra* that is the *ultima ratio*, the meaning of meaning—placed beyond the negative or, at least, beyond questioning. Must rationality remain the ultimate, unquestioned instance? We would free many more of its non-dominating possibilities if we were to stop protecting reason itself from the fire of interrogation.

We are in a dominant and dominating world-language, whose internal logic tends to function in a closed circuit: systematization and modelizations, autonomization of technological networks, a more subtle control of reality by "open systems." Is it so negligible to take some distance? To be sure, nothing will be resolved in terms of the dominant world-language, but perhaps we will have begun to free ourselves from the obsession with efficiency and productivity, perhaps we will have liberated certain internal possibilities of rationality—those that have to this point been driven away by self-domination?

Is this rationality truly possible, capable of conceiving the world, of comprehending itself, and, nevertheless, of admitting that time, language, and the divine are not always able to be annexed to the self-constitution of meaning? Is it possible to scrutinize this Inconceivable/Imponderable, which the concern for the systematic domination of the world has up to now driven away? My answer is yes, given two conditions: first, that the specificity of language be strictly respected (the style of meditating thinking is not to be animated by the conceptual impatience of knowledge; that which is held in reserve in the Imponderable cannot always be approached according to the laws of calculus). The second condition, itself two-sided: that one frees rational exigency from the adhesion to a universal necessity (the ultimate mythology of rationality) and frees meditative exigency from the certitude of the accomplishment of metaphysics (a Hegelian relic in Heidegger). This implies that the "world-becoming" of metaphysics in technology ceases to be decoded as a fatality in a unique sense and that it does not relocate the destiny of rationality exclusively in calculative thought. The examination, at once critical and meditative, of the structural homology between the auto-nomy of the language-world of the absolute and the auto-matization of the world-language of technology ought to permit of releasing the *chances* of knowledge and to free thinking in a world that is increasingly technological.

Meditative thought must guard against merely being satisfied with the factual, almost mute, admission of an imponderable; it ought to be conveyed by an inquiry comparable to that which a great thinker recently attempted: the discovery of an instance that concerns us more intimately than the presence of the present—the gift itself. This gift, anterior to all being, all objectivity, all presence, is the "there is" or even the deployment of a time in its illuminating withdrawal. This welcome of time is not reduced to the passive reception of pure time; it delivers—in the same time as temporal experience in general—the opening to that which time reserves as epoch. I am delivered to the upsurge of time; I am, more fundamentally still, required to respond to a destinal situation: an incarnate temporality whose face utters heritage. The gift is destinal because it delivers, with itself, every tradition that masks it: it delivers precisely that rationality whose self-referential tendency tended until now to obliterate every origin, every "non-grounded" sending.

Hypothesis, wager, hope: rational experience can open itself to this gift that conditions it and, doing this, carries out a mutation that our time desperately needs. It is true that occidental thought, in its dominant movement, has tended toward the self-constitution of knowledge, without considering destiny or the preliminary ontological site. But the future is open: an enriched direction for rationality is offered there, a neglected direction that would be capable of again enriching the experience of meaning confronting its destiny.

Have I responded to your concerns? Have I clarified the place from which I speak? It cannot be elsewhere: the locus is meaning, ceasing to turn on itself, and finally delimited as *partage*.

X.- To be frank, I see two principal openings in what you have just said: the first (even if it comes, in part, *in fine*), very much inspired by Heidegger; the second, wanting to be original. Do you agree?

Z.- For the moment, yes.

X.- As to the first, would you admit that it tries to clear a third way of thinking between rationalism (or metaphysics in the sense that Heidegger understood it) and the limited reversals to which you have alluded?

Z.- Indeed, why not?

X.- Then could you explain why the second opening does not take its direction from meditative thinking in the Heideggerian sense, whether the turn you take justifies starting from Heidegger or is it opposed to him?

Z.- More than ever, meditative thinking is necessary; I attempt to take a step that is not totally explicit in Heidegger, although it proceeds from his analysis of metaphysics. Heidegger thought that metaphysical rationality is a *Geschick*, a "sending" that we cannot purely and simply place to the side; according to him, this sending proceeds from itself, traverses its own resources in cybernetics and technology. *Thinking*, henceforth, has nothing more to do than to draw out a most secret destiny, that of the truth of being, which, reserved for the moment, is capable of being illuminated one day and, perhaps, of forming an epoch. For me, there is here a regrettable methodological incoherence. Certainly, Being requires thinking to be "authentic," but it finds itself paralyzed in regard to the exercise of rationality which no longer—it seems—has anything to do other than to record the ineluctable. On this point, I put into question the Heideggerian thesis of the accomplishment of metaphysics. It is difficult to admit that metaphysical rationality is our destiny, while denying that nothing can be modified through our manner of assuming it. We would have only to respond to the injunction of Being afterwards, at a later time—a totally indeterminate time.

X.- Would you want to bring together these two modes of thinking that Heidegger separates into dichotomies?

Z.- These two modes of thinking are unable not to be brought together.

X.- I am very much in agreement with your critique of Heidegger insofar as, in my view, his "third path" is an impasse. If, as he himself concedes, one is not able to overcome metaphysics, but then one must come back towards it—in order to renew it.

Y.- Pardon me for interrupting your dialogue in which I would have difficulty choosing sides.

Before referring to the small luminous point that captured my attention in the preceding exposition on the "destinal gift," I insist upon a clarification. My request will undoubtedly appear feeble and paradoxical. Feeble, because it concerns a question of terminology; paradoxical because my opposition is itself susceptible to being overcome—as soon as this difficulty is cleared up in the manner that I hope.

Let us understand ourselves clearly: the "rational core" (to take up an expression of Marx, which applies much more strongly to scientific rationality than to its Hegelian dialectical form) *cannot be a destiny.* I believed I had made this clear from the beginning; obviously I have not done so strenuously enough. Is it necessary to recall that this rationality is not a lot that imposes itself obscurely, but is a meaning that illuminates? That this is not a constraint that suddenly paralyzes with impotent terror, but the necessity of an order that requires my intelligence and my will in order to quiet and satisfy them; that it leads to a taking up of action thanks to a mastery of its pre-givens, and not to a simple accepting—even heroically—of the factual? Briefly, rationalization fights the powers of destiny—an assertion that has for me an abiding and eternal truth, but I could support it with numerous examples drawn from History.

Z.- I do not understand "destinal" in the sense of a *fatum* that is massively imposed without allowing any margin of flexibility for a proper response.

X.- If that is the case, it is perhaps more precise not to speak any further of destiny vis-à-vis rationality, but only of *partage.* I assume that you clearly had this idea in mind . . .

Z.- Yes.

X.- All the difficulties are not thereby erased. Because even if rationality in itself *ought* to be more than a destiny, its technological results reduce it to a destiny for the masses who benefit from it, or submit to its "perfections," without any consciousness of the exigencies which have allowed for this progress and which are at the source of all understanding.

Z.- It remains that thinkers—in the larger sense—can rise to knowledge of *partage*.

Y.- From blind technicization, endured as a destiny, to rationalization accepted and desired as something more: if this is your "program," I am in agreement.

I come, therefore, to the convergence that I believed I discerned between us regarding this that you call "partage." Although rationality itself, in its constitution, cannot be a *fatum*, it is illusory to wish to escape not only to rationality such as it presents itself today, but still further, to a growing rationalization at all levels.

This perspective is difficult to accept only if we cling to old metaphysical conceptions or to a fictive idea of nature. We fall back on the side of destiny when we are mistaken about rationality itself.

We are at the dawn of an era whose instances will be new and even more uprooted. Henceforth, everything will come from the rational and scientific project. The source of Being will be mediated, artificial (in a nonpejorative sense), directly universal. The "natural" will change even more. The relation to local links and rootedness in natural languages will become effective from the height of the concept, like the plunge of an eagle swooping from on high. No object, no particular content (even the earth) will be determinant for the development of this rationalization.

In this expansion of knowledge, we will take up, whether we like it or not, our residence.

Have I betrayed you?

Z.- You have less betrayed than surprised me. I did not anticipate this prophecy from you. Regarding the rationalization of the world, how can we not agree with you? Already more than foreshadowed in our time, this movement can only be amplified and imposed on us *like a destiny*. It is certainly the case that everything will depend on the *manner* according to which humanity will assume the inevitable. There can be some surprises there. If rationalization is passively accepted as an indispensable collective resignation in favor of the most efficient organization, we will know the worst of destinies: enslavement and tyranny. If, on the contrary, rationalization is experienced as a call, as a new source of creativity that our renewed energies can channel, then *perhaps* it is a new dawn that awaits the world, even more radiant than its Greek model, an epoch of intelligent and heroic accomplishment, and also of enthusiasm in the face of irruptions of novelties, in the firmament of the earth as well as in that of the spirit?

There is the destiny of Ajax and that of Ulysses. Two different lots, but also two ways of responding to adversity. Ajax, or stubbornness, leading to folly; Ulysses, or steadfastness, before the divine ruse . . .

X.- You have just cited two heroic destinies: these individual examples can only with difficulty lend support to the illustration and the comprehension of that which awaits humanity as a whole. The heroism of the masses is a myth. Heroes of reason, there have been some, there will be still more; but they only announce the Meaning that will be imposed in any case.

Y.- We come up against the difficulty of knowing what margins of freedom rational *partage* allows us. When I spoke of a growing rationalization "from which all will come," I wanted to isolate the portion of the *inevitable*. If rationality becomes a *partage*, it is because it imposes something that we cannot refuse.

Z.- You go much further than many rationalists.

Y.- I do not believe that. You yourself evoke a caricatured image of the rationalists. Take Monod: he no longer saw—from the side of the "Kingdom"—any issue for humanity other than the pursuit of scientific progress; at this level no real choice is possible, insofar as the other alternative is to fall back again into the "shadows," that is to say, into the inconceivable on this side of History. If you like, science tells us, "Me or Chaos," as did de Gaulle in 1965.

Z.- And perhaps we will have both, as was the case in France after 1965!

Y.- Be serious! No one today is opposed to the pursuit of scientific progress, except for a few hippies who, moreover, deserted the world of technology more than that of knowledge.

X.- In our time, we do not see clearly how the cause of science is distinguished from that of technology.

Z.- We see it even less because science is manipulated by technocratic ideology to its own ends.

Y.- Reason itself is impotent against this deviation. It is up to the thinkers to protest!

Z.- If the collusion is installed de facto (which is already the case) and reinforced, will it not become excusable to prefer the fall into prescientific chaos over the scientific (allegedly) organization of chaos?

X.- The worst politics is sometimes understandable; it does not make it any less appalling.

Y.- Why do you not choose it?

*

Z.- At the very moment when we appeared to come together, we have perhaps never been so far apart. You would prefer to see me "on the other side," that is to say, on the side of the antirationalists or the mystics whose case reassures you because their absurdity exiles them to the side of this chaos in relation to which you situate yourself and reinforce your certitudes. I disturb you in maintaining a foot in each camp.

Y.- Do not transform yourself so quickly into a victim of misunderstanding.

X.- Rather, explain why our agreement on rational *partage* was deceptive.

Z.- Because once you have established the inevitable, you subscribe to the progress of rationality—nothing more.

X.- That is not true. I think that through the philosophical conception of that which actually is, one can elevate rationality above its scientific and technological performance, disclosing it to itself (in the self-referential sense just explained).

Z.- Once one has conceived that which is, it remains to think that which is held in reserve . . .

X.- Which is to say?

Z.- . . . to think the prior conditions and the not-said that condition the figure of efficiency.

X.- This is a marginal task with regard to the first.

Y.- I believe I understand you when you claim "to think that which is held in reserve" and to uncover, in short, the philosophical experience of that which it has up to this point masked: you credit language with possibilities of monstration that authorize you—as you believe—to encroach on the unknown, to displace the frontier toward a kind of West that is all the more exhilarating because it is less determined. There is complete disagreement between us: no guarantee of objectivity supports your "demonstrations"; I prefer to keep silent before that which is held in reserve, to allow the artists to evoke or paint "that which slips away" (I think of Michaux).

Z.- Why refuse thinking the possibility of becoming, in its turn, an art? Why paralyze a priori language in decreeing that rigor ends at the frontier of the sciences? True art has another kind of rigor.

Y.- Let us say that, until now, I have not found this rigor in philosophical thinking.

X.- Philosophy ought to mediate the lessons of art, not assimilate them.

Z.- Its task, nevertheless: to unfold the scope of language, to discriminate the possible, to make the link between divergent—but nevertheless congruous—orientations.

X.- You want to think "that which holds itself in reserve," but do you intend to make the connection between the work of rationality and unthought signs?

Z.- It is toward this jointure that thought directs everything.

X.- You appear to ignore that scientific work is increasingly interdisciplinary.

Z.- What I presuppose is not integrated in already constituted disciplines, since they—and even philosophy to a large extent—have no regard for what holds itself in reserve: the destinal gift and the signs that it bestows on us.

X.- You dream of a philosophy that is comprehensive enough to integrate Heidegger's contribution in such a way as to enrich both his thought and rationality itself.

Z.- I cannot deny it.

X.- A beautiful mirage. Why would the dominant rational work open itself to that which rejects it?

Z.- In the sense that I understand the meditation on rationality and its meeting with the unthought, the work of the rational is never rejected. The perspective, henceforth, widens: letting-be overtakes rationality (something Heidegger never took into consideration). The place of the meeting would no longer be dominating rationality and excessive willing. This would be the deployment of a meaning that requires only the amount of will (to truth) that is indispensable to its activity, but which lets itself be.

X.- There would be a nonexclusively dominating rationality?

Z.- Exactly.

Y.- Could you be more precise?

Z.- There are already—in the work of the rational such as it is permitted to be constituted at the heart of actual scientific practices—irreducible uncertainties, fundamentally undecidable moments, impassable paradoxes, many nonsystematized articulations, uncertain regularities.

Y.- These moments are inscribed in a project of the whole that is itself dominating, since it aims for a definitive understanding.

Z.- It is not a question of desiring the absence of all domination or even of denying the dominance of rationality. Beyond the moments mentioned, one can envision a truly philosophical experience susceptible to the founding of a rationality that would be a predominance of spirit, rather than its complete domination. I can explain myself: a predominance is essentially a characteristic, decisive *among other elements*. Is it utopian to seek *a* rationality delivered from the empire of the will to power (which is to say, a rationality for whom domination would not be the sole nor even perhaps the principle end), endowed with a lucidity that would permit it to grasp itself as *partage*, admit its proper limitation, and open humanity to other possibles?

Y.- I admire the subtlety of your *distinguo*, all the while asking myself about the limits within which it makes sense.

Z.- It is a question of designating a self-limitation that allows rationality to no longer masquerade as the autocrat of life. Can the New Meaning perhaps better respond to this expectation than the Hegelian Absolute?

Y.- Does not the contemporary scientific spirit give endless examples of this "nondominating dominance," broken, on the contrary, by all forms of totalizing metaphysics, including positivism?

X.- Nevertheless, you have admitted that rationalization *imposes itself* as an inevitable fate. It imposes itself *absolutely*.

Y.- In adopting your own terminology, you made an unwarranted passage from the truth of the New Meaning to the "spirit of the world."

X.- Do you wish to say that I excessively technicize science? I am simply starting from what exists today.

Z.- If one admits of a nondominating part of rationality, then that which is imposed absolutely is not entirely rationality or even the better part of it.

X.- I ask only that we recognize this marvelous rational advantage. I merely argue that for one of you it is already acquired by the truly scientific spirit, and that the other sees only a new philosophical method capable of obtaining it—the latter not knowing how the spirit of the times gives birth to, or at least, permits this method.

Z.- This thought I am espousing wants to be that of the simultaneity and the opening of the possibles. The place of the Spirit in this time: carved in foreign galaxies, which brush against and exclude. A sidereal silence (or background noise) among thoughts, disciplines, groups: this is not only true at the interior of the scientific sphere, but also between that sphere and that which challenges it. Distance, indifference, ossified Babelism.

That some spirits, faced with these wounds that numb, with this desolate spiritual landscape, do not lose hope or contact with viewpoints most foreign to each other, is not only a sign of their quality, but an indication of the profound exigency of that which I also agree to call the spirit of the times. "We live in several worlds": when Lévi-Strauss wrote these words at the end of *Triste Tropiques*, he wanted to say—in a way that was fundamentally quite Hegelian (but without misunderstanding the difference that subsists between a vision of History surrendered to entropy and another enclosing absolute sense)—that the person of progress guards in himself or herself the stigmas of what his or her most distant ancestors have been: we are that which humanity was and has made of us. A continuity of meaning and a homogeneity of discourse are each time presupposed, whether truth is understood as a "progressive expansion of meaning"—as Lévi-Strauss also said—or a process integrating contradictions. We must capture this "plurality of worlds" at the base of our living memory, but equally in the present of our shattered culture: truth, History, and even meaning are splintering. This splintering also includes our world.

"We live in several worlds": perhaps now we can better discern what this signifies for the most acute and most attentive minds, those who still know how to *listen*: we are, certainly, beings who are conscious of having traversed the essential stages on the path of comprehension and of the rational organization of things and people, realizing within themselves, through their scientific and technical practices, the maximum possibility of exchanges of information, integrating them in a network which is more and more woven of disruptions. Perhaps we are *also* these unpredictable and secret beings who detach themselves from all this, and whose existence is revealed, attached like a madrepore to a reef—invisible from the surface: the free arrival of things and beings, the call of the Near to be named, the evocation of the unspeakable; beings exposed at the same time to the unfolding of the absurd, themselves seized by the furor to demonstrate the nonsensical, to exhibit the ridiculous.

To think this situation without minimizing the extreme givens is a difficult task: impossible? How can we know this in advance? The wager: Lévi-Strauss himself is not far from taking the wager when he allows (never hazardous occurrences) his own structures to be traversed by a disillusioned look at the powers and limits of that which we call meaning; thus, his extreme rationalization of myths—even when it appears to reduce them to our logic—transports us toward the relativity of all meaning—a movement where the "explanation" is no more spared from relativity than myth itself.

If "we live in several worlds," then there is not one single manner of experiencing rationality as *partage*. Whether the point of support is close to the center of gravity of reason or whether it benefits from the fact that meaning and language do not unfold at once at the limits of scientificity, it is a matter of establishing this cohabitation that the young Nietzsche saw between "intuitive man" and "rational man," between Dionysius and Apollo. In the sanctuary of "Know Thyself," a place was equally made for the god of obscure madness. But when the luminous lyre made its music too monotonous, the Inspired was already exiled far from mortals.

*

X.- A partially non-dominating rationality, a plurality of language-worlds: I see the assumed junction of your discourse drifting towards an acceptance of distortions and shatterings that meaning has always had to fight against in order to constitute itself. And how can you still speak of a "spirit of the world" if this too is non-dominating?

Z.- Why think of the future of meaning from the perspective of its past? And have you also not acknowledged in the *New* Meaning a nonabsolute character?

Y.- At any rate, an endeavor of this kind, despite its efforts at synthesis, remains a limited discourse, less close to Hegel than to his adversaries.

Z.- To make rationality a *partage* is to return from the claim of universality toward a more determined, and at the same time *unique*, horizon (although potentially non-exclusive); it is perhaps to reintroduce singularity where it had been the most methodologically expelled. That there is today a dominating "spirit of the world," I do not deny: it is the point of departure of my analysis. I argue that the New Meaning is at work, not that it exhausts all sense and demonstrations; I accept the necessary knowledge of this which is, but I try to approach what holds itself in reserve under the figure of the efficacious: science and thought have an avant-garde and vigilant vocation—something you fail to recognize.

X.- It is only fitting to ask whether that which you yourself call avant-garde does not correspond to a reappearance—or a survival—of savage thought, in such a way that partially nondominating rationality should be interpreted as the inevitably regressive part of all rationality?

Y.- Do you understand savage thought in Lévi-Strauss's sense?

X. Of course. This author having been evoked a little abusively, I attempt to recall that in delimiting the respective methods of savage thought and scientifico-technological thought, Lévi-Strauss does not advance so far as has been claimed concerning the admission of this reserved dimension, which would be the "destinal gift."

Z.- He has, however, something to teach us about the cohabitation of scientific and savage thought, which he prefers, in the end, to call thought in a savage state, distinguishing it from "cultivated or domesticated thought which seeks to obtain a result." This last expression emphasizes, in a very significant manner, the dominant character of scientifico-technological thought, while leaving intact the possibility of a coexistence or a co-penetration between the two thoughts—compared to plant species, wild or not, which intermingle and are sometimes crossed.

X.- Does this summary not turn against you, if it forbids scientifico-technological thought any exit from the kind of domination that is imparted to it by its specificity? Rationality is certainly partially nondominating, but not in the sense that you have argued; its availability continues thanks to the savage state of thought—like a nature park in our upside-down environment. Nothing assures that this savage state of thought is not more threatened than the natural reserves threatened by an evolution that operates against them.

Y.- Unless partially non-dominating rationality, far from being reduced to the savage state of thought, promises something more? Such as a superior consciousness of the new exigencies of a coexistence . . . a new consciousness . . . epistemological . . . ?

Z.- I acknowledge that the example of the nature park is significant and can be turned against me. You will object to me: nothing has changed in scientific domination; simply, a "ruse" of the technological system manages the reserves or the pockets which are simply pacifying parentheses for "sensitive spirits"—nothing more. I also recognize that my position does not completely agree with Lévi-Strauss; his "problem" is not domination; it is not even rationality as such, even less destiny. For him (especially in *The Savage Mind*), it is a matter of establishing the status of one kind of knowledge in relation to another, of showing that thought at the savage state is, in its own way, already scientific, and, in addition, always alive under diverse metamorphoses.

Nevertheless, how is the evocation of Lévi-Strauss justified? It is, insofar as the author of *Tristes Tropiques*, undertook, in his own way, a recognition of the Other, which, although apparently reductive (by its formalism) and opposed to the effort to listen that I believe necessary, represents a perhaps unprecedented undertaking of cohabitation between heterogeneous approaches to the world. Moreover, what is philosophically important in Lévi-Strauss is less the objective ensemble of the scientific results than the lucid and modest consciousness of the limits of *our* rationality, a consciousness which emerges at certain privileged moments, when "the man of study," in some manner, passes to confession. I thus rediscover an outline of this new epistemological consciousness which we just took into account.

X.- You admit that your recourse to Lévi-Strauss had something unexpected: his work marks and accentuates the reabsorption of History into the synchronic, enclosing in a definitive and privileged formalism a rationality whose operativity recognizes no "transcendence," not even that of a commonplace *partage*.

Now, when you speak of a partially non-dominating rationality, do you understand a rational and voluntarily non-dominating state of thought appearing as rational *after the fact*? The latter case corresponds to the so-called savage thought that we call primitive. It does not seem to me that there was identity between the two states. Can a thought of the savage state, captured in art or in the margins of the dominant world, have the same economy as the savage thought of the "primitive," insofar as this concerns the relation to the cosmos and natural order as well as to that which concerns its own organization?

Z.- Perhaps, through me, you question Lévi-Strauss himself? Ambiguity exists in his work: the thought of the savage state is the *bricolage* of contemporary artistic thought, as well as of so-called primitive thought. But this analogy of methods hides a profound difference in the socio-political and destinal condition of the two spheres where this thought in its savage state unfolds. In our society, this is surrounded, pursued, condemned: we can ask whether Lévi-Strauss himself manages to prevent himself from a pejorative judgment when he rediscovers it at work in the *Critique of Dialectical Reason*. In closed and stable societies, whose last representative our technology encircles, this thought alone held sway: a considerable dissimilarity that comes to rob these peoples of their future; their ancient destiny is no longer the unique or the last recourse. Our *partage* becomes their destiny.

X.- Do you understand the passage from destiny to *partage* as an exit out of the exclusive? When Montezuma awaited the return of Quetzalcoatl, was the future not entirely hanging on the passage of Venus to the meridian? There was no other perspective on the horizon than this spot on the sun, which a jade statuette immediately commemorated . . . Destiny in its great book was sovereign.

Y.- Was everything for Montezuma linked to an ironclad necessity? This is doubtful, since certain Aztecs did not accept surrendering to Cortez as the God they had awaited.

X.- The divergence would bear on the identification of Quetzalcoatl, not on the perspective of his return. The same is true for the identification by the Jews of Christ as the Messiah. In both cases, the destinal element is ineffaceable, it is the awaited return of God or his privileged representation: it is inconceivable that one is not prepared for this.

Z.- Are we dominated or possessed by rationality as Montezuma and his subjects by the conflict between the winged Serpent and the smoking Mirror? *Partage* is not, fundamentally, less exclusive than destiny. If we cannot escape rationality, it is because the inevitable pursues us. The so-called primitive people are now subjected to *our* exclusivity.

Partage and destiny give a name to a profound and supreme constraint which is not within the power of humans to lift. On the one hand, *partage* is not reduced to the mechanistic realization of events fixed in advance by a blind force; it is not bound at the level of ontic determinations; and it is not the *visatergo* that governs the unfolding of rationality. Above all: the sensitive string that is touched by *partage*, which already in Greece achieved the summits of tragic lucidity, this is not merely the execution of certain facts and gestures, it is the sense of the Open.

Y.- Which is to say?

Z.- Paradoxically, the encounter with the ineluctable that, more than ever, opens the space where humans play out their fragile language: the exaltation of an irreplaceable response.

X.- Tragedy was fixed within the limits of a ruthless precision, whereas our *partage* increases the possibles. Is this empty mask the ultimate Ruse of destiny?

Z.- The sign of this destinal power is the image that one makes of the future. There is no longer any God or any singular trait that is attached to that which is to come: only this mythical future depersonalized by the concept.

Y.- The idea of the future more than anything which is able to come . . .

X.- This call of the possible causes us to forget the interior of the limits within which we move.

Y.- The limits of the universal!

X.- With a reflexive, piercing consciousness that the Greeks did not have. If we are unable not to choose the path of rationality, we must call this path necessary— although it is not dictated by some goddess of Reason. What role shall *partage* be assigned? Freedom results from an understanding of Necessity: for the first time, humanity is compelled to *free* itself. Only rationality is truly necessary and, consequently, only it can free us: when this truth imposes itself on the world, then—as Hegel understood—the Absolute realizes itself and rules *without partage*.

Z.- But you yourself just showed that Hegelian rationality arrives at consciousness from its self-dispensation.

X.- Rationality comes to think destiny and even its own presupposition; but ought it go so far as to admit another, preliminary instance?

Y.- This *partage* is different from a destiny in the strict sense; it is also different from Greek *partage*. I understand this "anti-*partage partage*" in the following way: rationality is our future, a mixture of logico-scientifico-technological constraints and unpredictable circumstances.

Z.- There is not a contradiction between the play of possibles and the constraint of *partage*. The former results from the latter.

Y.- Mallarmé, the poet of the conjunction between "obscure disaster" and hyperbolic, lucid rationality allows us to maintain in an entirely different manner that—whatever the "complete account being developed" from the perspective of knowledge—thought will not cease having to confront chances that cannot be abolished, or even, always restated, outside of the grasp of languages, the enigma of Being. Oedipus is dead, the Sphinx silently scoffs at us until the end of the centuries.

Z.- How can you deny that every destiny dupes man? Er does not believe his eyes when he views most of the souls, including those who have drawn a number offering them the largest selection, hurrying without reflection to the most treacherously alluring packages of life. Souls believe themselves to be free, they are: the philosophical view, through a dream, perceives the framework of necessity in which the thoughtless who are freed are going to rediscover an order—as, in our time, the statistician sees social atoms obeying, despite themselves, the law of the greatest number.

Nevertheless, never as much as today has the spirit had such a complete view of the possibles and at least the illusion that nothing is definitively established.

Y.- Who can know the degree to which this is an illusion?

X.- In spite of our differences, it seems that you would share it. Perhaps I have not covered them? While admitting this incertitude, I hardly see by what miracle Western rationality today would turn itself around and reveal itself to be—even poetically—non-dominating. Why should the dash toward mastery suddenly be broken? I admit that one calls Necessity cruel, without disguising it as Progress, and I do not admit that one can refuse it. In brief, Z's theory appears to me to be a useless subterfuge.

Z.- It is not a theory; rather, it is a perspective of thought. Those who affirm necessity impress more than the pioneers of questions. I seek only not to kill the possible, while taking the side of the ineluctable. From this side, the pursuit of rationalization has been recognized; it would be a dead end for the thought of *partage*, to overlook (or believe to have overcome) rational power. From the side of the possible, it seems that a rationality that had passed the test of "questioning" would no longer be strictly identical to its initial state.

Y.- Is it enough for it to avoid scientism, technicism?

Z. A rationality that would benefit from the sense of its "finitude" could perhaps do even better.

X.- It would be a question of a superior consciousness that would not change everything to scientifico-technological practices.

Z.- How do you know? If ground-breaking thoughts take another course, how would practices not be, sooner or later, modified? Perhaps one day scientific discoveries will be admired as the mountains . . .

I mean to say: not as the records of a narcissistic human will, but as signs of the Enigma.

Y.- Do you not reintroduce, thanks to the destinal, a supreme interiority, an instance of shelter?

Z.- An interiority open to the interiority of *partage*, a shelter continually enveloped by the Inconceivable/Imponderable. . . . If dominating rationality thus attains *its* reserve, can it only renew its ancient powers?

X.- Beings of knowledge, are we still capable of allowing ourselves to be ballasted by a singular *partage*? Although up to this point, a person's destiny was a house, a sky, a well-defined daily life, we are now almost without tangible attachments, indifferent to place, time, captivated only by the unknown that can be formulated.

Z.- Do not be so quick to generalize! Certain people try to make the connection . . .

X.- Humanity has always been tempted by the impossible.

Z.- But we must not despair. People first believed that the impossible was the technological mastery of the world; they now fear being unable to escape from their success.

Y.- Undoubtedly, but is the human being still the interlocutor? Is it not technology, or rather Power, as an autonomous Complex, that has attained its point of no return?

X.- This is a diagnostic question. If the destiny of the power of science has crossed the threshold of the irreversible, is that not the end of the matter?

Z.- If . . .

PART III

Of Rationality and the Enigma

To you, hardy seekers, explorers . . . to you alone I will recount the enigma that I have *seen*.

—Nietzsche

"INEPTITUDE IS THE desire to conclude," writes Flaubert.[1] How can we not agree with this, above all as concerns a subject such as this one, where it is less a matter of definitively concluding a piece of research than a matter of assessing an initial investigation—in order to take it up again later? It is at least still permissible to question. First of all, this dialogue. To admit the part of staging in no way invalidates the exchange which takes place there. If it goes without saying that Z is closest to the position supported in this book (in pushing ever further the recognition of a "non-dominating rationality" at work in contemporary logic and science), why was the author delivered to this uncertain, but above all *regressive*, joust, in which much, undoubtedly too much, was conceded (if only through the attention given to their theses) to the "critical rationalist" Y, and to the "neo-Hegelian" X? It was in order to carry out a diacritical recapitulation of the previously studied positions, but above all, finally to demonstrate the exercise of a *contiguity* that is effectively at work between different thoughts—nevertheless confronted with the same rational *partage*—and to give *even here* (non-programmatically) the example of a rational course which does not exclude the understanding of Being (the originary enigma continually restated, indefinitely retranslated): to not forever defer the exploration of the rational possible, and to show, through the resources of the dialogue, the play (interpretative, but also perhaps effective—in the relation of humanity to the world) that subsists between the phases of potentialization—and beyond all potentialization!

Now, we must again interrogate this increase of the enigma that weighs down the secret of Being: the enigma of the destiny of the power of rationality. Then, we must slip into the Labyrinth, to lie in wait, to overtake some of the threads which the *understanding of Being* weaves there. We must seek to the very end this possible reserve in the power of the rational, which is also obscured by it in the jointure of a *partage*.

The destiny of power or the second enigma

Marx writes of communism in the *1844 Manuscripts:* "It is the resolute enigma of History and is known as that solution."[2] Every road taken in our careful inquiry comes to contradict this claim, whether it is expressed in the name of communism, capitalism, or even absolute Spirit. The enigma of History will never be resolved; this is at least one certainty which the rationalist can attain in terms of a philosophical genealogy of its power. As Heidegger said concerning the epoch of metaphysics, these phases form a "free following"[3]; their succession and their connection assume a necessary character only retrospectively. The burden of *partage* that is our lot resides less in the intrinsic necessity of the process of the whole than in the irreversible network of potentializations that we feel ourselves constrained to assume.

No "solution" is therefore proposed to History—which reveals itself here as the increase of power, becoming massive and almost irresistible in its ultimate phase. Because this history, envisioned from the angle of potentialization, began furtively in Greece with the incommensurable discovery of an absolutely new dimension—necessary and universal—of the intelligence of things, this will be, henceforth, the archetypical Rationality (Phase II), without which science as such would not be constituted, and techniques would have remained tenuous, always more or less artisanal. This leap of an absolutely new power into the unknown is all the more enigmatic because it remains latent and, while taking the most decisive step which alone perhaps truly broke the link with the immemorial, it only rendered the methodological accession to power possible (in Phase III).

To render possible: this appears quite negligible in regard to battles and treaties, and even in consideration of the effective exercise of a technology and of a mode of production. There are those who will say that I wanted to rewrite History, but starting from its skeleton, the accumulation and the metamorphosis of the Possible. There is no point in rewriting History, but only in slightly increasing our understanding, starting from a perspective that I believe to be privileged. Naiveté? Were the "events" of anecdotal history not reduced to those timely encounters, awkwardly reassembled in chronicles or crudely accounted for by Providence? In the end, Hegel and Marx arrived with a "philosophical history" which would have us believe that Spirit or Humanity were absolute principles of a necessary process. For us, this is naiveté. The Spirit does not achieve sovereignty in History by turning destiny into Necessity. The inverse is true. Rationality potentializes without a responding sovereign moment, and it finds itself engaged with its own *destiny of power* that it is tempted to repudiate: astonished in the face of the consequences of an originary enigmatic event. Does Humanity give itself only problems that it "is able to resolve"? Recently, it is the inverse. Humanity is thrown, in spite of itself, in a course toward power that it has not chosen and that it controls less and less. In order that History can be definitively illuminated as "humanism accomplished," it was necessary that power, from the beginning, was the measure of man. The dream of harmony that rationalism applies to historical expansion is a mirage continuously belied by our critical genealogy of power. No more than the pseudo–"Asian mode of production" can explain the structure and rites of the Chinese Empire or the caste system in India does the modest state of Hellenic technology and modes of production illuminate the fantastic breakthrough toward episteme: the unpredictable appearance of a new place of potentialization. And if, at the dawn of Modern Times, potentialization and capitalization appear at times to confound their projections, this conjunction is in no way explained by purely economic factors, and its explication is reduced, not increased, by the systematic reference to class struggle. Innumerable revolts carried out against machines, nostalgic aristocrats attached to their ancient privileges or their land, the quasi-visceral conservatism of peasants, of artisans, and of the petit bourgeoisie—the great majority of humanity *suffers* the potentialization in Phase III. Likewise, still recently the socialization of the means of production in the East had neither abolished assembly-line production—product of the "capitalist" division of work—nor prevented the Party and the State apparatus from placing themselves unconditionally at the service of scientifico-technological imperatives, priorities of Phase IV. We see the telling result. Humanity *is not able to resolve* the problems that it has not posed, but which are *imposed* on it by a potentialization that surpasses it (as early as Phase II and increasingly thereafter).

Here, therefore, is the restraint imposed on rationality scrutinizing the torrent of History: to renounce excessively systematic harmonizations, to admit of irreducible ruptures (between phases of potentializations), to accept this fundamental (and yet secondary in relation to the ontological) enigma: the rational has a *destiny of power* subject to the Law of Reversal (from the rational into the irrational), and, therefore, no one is able to say definitively what the key is, no one is able ultimately to measure meaning. Is it so scandalous to claim this? Must not science continually revise and diversify its "archetypes," to deny that an "absolute structure"—not to offend Raymond Abellio—applies equally to the movements of heavenly bodies, chemical elements, or blood groups? Thus, *mutatis mutandis*, the theory of potentializations does not claim to *resolve* the enigma of history, but only to *circumscribe* it. To the objection that this theory would not be sufficiently explanatory, I respond that every science of History is a mirage (everything that presents itself as such is only bad metaphysics) and that the philosophies of History explain too much, more than they are able.

This genealogy of rationality is not metaphysical in the sense of special metaphysics, in the sense that it is not supported by any fundamental principle of the justification of the real. What remains for it of general metaphysics is only the desire for intelligibility. Still, does this will to illumination not unconditionally investigate the "truth of truths"? It is suspended on a hermeneutic of contiguity between the rational language-world and the reserved language-worlds. The reception of the ontological difference is not invalidated here; quite the contrary, the differentiation (internal to rationality) between phases of potentialization is associated with the difference between potentialization and the possible (thus one can translate the ontological difference in terms of our problematic).

We will better understand the aforementioned by casting a retrospective glance at the road traveled in this book. The destiny of power had first appeared as an objection to the purity of the rational ideal. A first lesson of the inquiry was: *all ration-*

ality potentializes. There is not, for us, a non-dominating rationality. A second lesson came almost immediately to give nuance to this non-illusory statement: *rationality potentializes unequally*. This marked the founding of the theory of potentialization: a common measure is revealed only between Phases III and IV; *compared* to Phase IV, Phase I is relatively "non-dominating": the real there is never subjected to a universal and necessary rule. Is Being, therefore, power through and through, as metaphysics would have it, and is its less dominating figure able to be obtained only through a kind of privileged culture of Phase I? In preserving a radical difference between potentialization and possibilization, I do not prejudge this question. Taking up again the ontological enigma *as such*, I leave open the question of the *power of Being*. Being is. To let Being be—contrary to all metaphysics—is not to think it a priori or retrospectively as idea, energy, subjectivity, will, etc. Between the possibility of Being and the power of the rational plays forever the play of Difference.

If this book has placed the accent on the internal differentiations of the rational and not on the possibilities of Being, the motive for this is simple, and can be explained in a twofold manner: the very theme of this inquiry was not Being, but the power of the rational, first of all apprehended in its most impressive and irreversible manifestations. Next, and above all, to admit that *rationality is our partage* is to understand that the ontological enigma is not proposed purely to us, but *redoubled* in and by the second enigma of the history of power (in Heideggerian terms: the withdrawal of Being as dispensation of the figures of Western metaphysics).

One of the innovations of this inquiry: to try to think this double enigma without breaking the rational thread; to presuppose, therefore, that at the heart of the possible, there is released the possibility of an "examination of the conscience" of the rational facing its destiny of power; in still other words, that the rational is able to trace its genealogy without enclosing itself again on this self-grounding sufficiency that gives itself free rein in metaphysics. I found it necessary to gamble. What is at stake now clearly shows itself: this which holds the two enigmas together, this which renders the thought of *partage* possible, is the possible held in reserve at the foundation of the power of the rational—the intelligence of the enigma. I do not claim that this intelligence makes or will make an epoch, but it *permits us to think* the destiny of power as enigma.

Itaque dici potest Omne possibile Existiturire. This thesis of Leibniz was founded on an anterior thesis: *Ens necessarium est Existificans*.[4] How is this recollection relative to our concerns? It is in that our hypothesis rediscovers one of Leibniz's intuitions: every possible aspires to existence, the possible does not cease to deploy itself. But the difference is important: one no longer affirms with the certitude proper to a great metaphysician, that existence depends on necessity. To admit *partage* is to let this enigmatic edge (all the same, double) be inculcated at the limit of rationality (and metaphysics): on this side of all linear time (Being) in History, or beyond it (the deployment of the figures of power: the potentialization of the rational).

This intelligence of the possible is no longer that of a sovereign calculating God. It is that of a powerful-impotent rational augmenting its possibles while grasping its limits.

Toward an understanding of the enigma

Whether it was a matter of History or of Being, the question of their enigmatic character did not directly pose itself *as such*. I have identified History as destiny of power, an explosion of potentializations, and finally as a *second enigma*, in terms of a genealogical and critical inquiry that has, at the same time, confirmed our presupposition concerning the insufficiency of all rationalisms (formalisms or positivisms) and asserted the impossibility of a metaphysical self-justification of rationality. From its perspective, Being itself suggests, in a preliminary sense, that it is always more originary than all being, (or every conceptualization of it), that it is taken into consideration and inscribed in language only by a thinking that is already laden with every experience of the world and that is attentive, above all, to gathering that which escapes its canonical forms.

In both cases, rationality does not spontaneously move toward the enigmatic. Its first movement consists rather of repulsing it, circumscribing it, reducing it. The organization of the Relational in a logic, the constitution of epistemological networks only serve to reinforce this tendency that is completely characteristic of rationality in its "natural metaphysics." To problematize every enigma in order to *resolve* it is the immanent project of science, under all its possible forms. Without doubt, contemporary science renounces the attempt to embrace everything in a unique structure or a universal formula, but it is only resigned to this, restrained and forced; we know that Einstein was never completely consoled not to have found the unitary theory of the universe. We have viewed it in relation to the realities and ambitions of the "new science": the admission of the Complexity is not the fact of a conversion in science in a purely meditative and "non-dominating" sense, but results from its very success. Only a science that has attained a certain degree of precision can be preoccupied with numbers, or with statistically evaluating the irreducible imprecisions. Henceforth, science will be able to know the price that it must pay to realize its procedures and to proceed with its measures,[5] an informational cost that is much more significant to knowledge than is the financial cost. This knowledge, diversifying itself in "complexifying" itself, increases the area of deployment and the efficacy of science. Should we confuse this extension with an authentic understanding of the Enigma?

I would risk giving excessive credit to rationality as such if I acknowledge in it the power of elevating itself, spontaneously and through itself, to this understanding. The neoscientific claim of the unconditionals of the "new science" would see itself thus announced and even supported a priori by an originary potentialization that would have the double privilege of an increase of power without failing to recognize the Open. But in the name of what and by what right is the rational consecrated and made absolute? This rationalism, so subtle whatever the formulation, would again succumb to the temptation that was fatal to all its brothers: to exclude the Principle of limitations that it recognizes in reality, to withdraw rationality from the play of its own determinations—in short, to reformulate a metaphysical principle. All my efforts will consist of denouncing this shuffling of the cards, to make the stakes appear clearly, to oppose to rationality its destiny of power in order to obligate it to be taken up as *potentializing*.

Envisioned as an "examination of the conscience" of rationality, our critical genealogy realized a self-determination of the power of the rational. Knowing itself therefore as powerful and having no "shame," does not the rational gather itself in a critically superior moment, a reflexive reference that renders possible and focuses the difference between phases of potentialization? Thus, in the most contemporary terms, "the command of gold, the sacred common of reason" dear to Plato would be actualized.[6] This new revindication of philosophical knowledge in terms of unity[7] could be argued in the following terms: since every coherent questioning of the essential is established only at the expense of a self-reference of a discourse (this is, here again, the price to be paid), is it important to maintain a caesura between the "rational" and "thought"? It is always thought, more or less intelligent, more or less meditative, that synthesizes and appraises the results of human work and, furthermore, which, in the crucible of actuality, adapts and enlarges the results of science in order to make a new culture. Does Heidegger himself not remark that *Vernunft* comes from *vernehmen*,[8] that reason carries in its wake a complete perception of Being-in-the-world?

This type of evidence will seduce, therefore, to the very end of thought. Evidence that I believe to be in error and that I must expose in order that the play can appear. Is this negligible and even sophistic? What difference is there between an "open rationalism" (with which I am not satisfied) and the thought that I seek? Earlier, I noted the difference between the phases of potentialization and the "possibilization" of Being. Relative to the ontological enigma itself, the stakes can thus be formulated: does the whole of Being free itself in the Relational or does it hold itself in reserve?

The question does not allow half measures. Indeed, if the Enigma of Being is only one unknown (or indetermination) among others, it can be considered solvable in the end. But more subtly, it can be classified as definitively insolvable *among other impossibilities* to which science is resigned. An objective fact among others, were it negative. In the manner of the undemonstrable in mathematics and of the undecidable in logic, the analysis of Being seems to render futile and useless every return to ontology and even more radically, every approach to Being as enigma.[9]

But the question is precisely this: What allows us to assume that a linguistic or formal analysis can be applied, without anything further, to Being *as such* (and not simply to Being as copula, analyzable in its grammatical functions)? To recognize

Being as *question* (in the sense that Heidegger distinguishes it from every "problem"), is, therefore, truly the preliminary question par excellence, that which recalls the precedence of Being over all its parts.

In familiar terms: there must first be Being, so that we can attempt to formulate a judgment—adequate or not—about it. In more elaborate terms, (that the whole of this inquiry undoubtedly makes more comprehensible): without the unfolding of the possible, there is no potentialization. The possible in its "quiet force" reserves Being; potentializations give structure to being.

Why this pure and simple reaffirmation of the ontological difference, when the rich diversity of rational *partage* has been explored, when the danger of a formalism that threatens the distinction between Being/beings has been pointed out? Precisely because this reiteration is neither pure, nor simple, nor innocent. The acquired experience ought to allow once again—but more explicitly—for the augmentation of world-languages (rational potentializations, ontological possibles), and to show that they are not separated by impervious divisions, that their play among the disciplined forms of language is not set once and for all, like figures in a classical ballet. The differentiations that the critical distance of thought imposed ought not to obscure, even in virtue of the ontological primacy of the possible, that the linguistic occurrences of Differences are unpredictable. The Difference can be said in many ways (and multiplied in differentiating itself from the phases of potentialization). No thinker can ultimately possess all the keys to the understanding of the Enigma; no language is the definitive depository.

The rational and the possible

Techno-scientific power is revealed in a cumulative process that is ever more accelerated, and no one can predict its outcome or arrest. This "statement" has been elaborated by a critical genealogy of rationality that has engendered the rationality of an increasing methodological order, the refutation of speculation, and an inventory of its certitudes (of which the most contemporary version thus includes the demonstration of undecidables, the establishment of imponderables). This rationality has no ontological response other than the *effectuation* of power, no limit other than the explosive (or implosive) punctuality of its effects. "To effect in order to effect" becomes its explicit or implicit motto. Everything must yield to this.

The rationality of the Labyrinth ignores (or suspends) this movement. It does not comprehend in order to act or in order to effect. It neither finalizes nor capitalizes intellectual activity. It has no plans of conquering the world or of utilizing beings. Its creative provinces do not objectify themselves in any definitive method; it does not exhibit them, it leaves them coiled in the depths of the spirit. Its temporality is not linear; it does not sustain any cumulative process; it precipitates the cycles of an incessant, although unpredictable, Return. Its lucidity is as extreme as its skepticism; there is no motto that is not fraught with irony. There, understanding is friendly to image. No surprise is excluded, no miracle forbidden.

A cruel paradox wherein humanity perhaps attaches itself to death: it is not in the Labyrinth that our contemporaries are enclosed, but in a course in which they do not know the outcomes any more than did Breughel's blind. At such a point, the rationality of power captivates the dominant obsessions, so that the escape toward the Labyrinth risks becoming impossible—even in fiction, including science fiction.

Is the translation (and the "fissure" at the interior of rationality) of the Heideggerian difference between "calculative" and "meditative" thought going to return us to the divergence of rationalities, vis-à-vis some small dichotomy between the technological Apparatus and the ontological Secret? Are the Labyrinth and the Secret merely one and the same? Is it not the final triumph of the Labyrinth to lead to this question?

From the moment that such questions are posed, the language of thought reunites with the possible. Undoubtedly, indeed, the "victory" of the rationality of the Labyrinth is affirmed, in that there is no definitive response, no triumph of reason. If the enigma is preserved, why would the Secret be violated? If the present world were as available as the white page upon which I write, no access would be prohibited, no "forbidden sense" would be final, and the "amicable struggle" between thinkers and poets (and scientists) could be given free reign. But, as clearly as the thought of death, I recall the *partage* of power, this irrationality of the quasi-exclusive domination of *one* dimension of rationality.

If it was so important to tie up the threads in Power and beyond it, at the heart of potentializations and the source of the possible, it was deliberately to confront this *partage*, marking its omnipotence and limits, breaking thus with the tyranny of a world-language claiming exclusivity; it was to shatter the claim of Power that consisted of capturing the whole of the rational. Rationality as *partage* is in no way to be confused with the pseudo-fatality of Power. It holds in reserve a possible that is not absolutely reduced to the present and future effects of power.

This "result" is encouraging (as is all renewal with the possible), but it proves to be double-edged. If it is vain to wish to speak *in the name* of the rational or to wish to exhaust it, this beneficial disillusion comes to contest not only Power, but all sovereignty of Sense, all exclusive diction.

To salute the great poets, Mallarmé, Borges, is not to recognize in them a unique privilege concerning the poetry of the enigma. To mythologize the understanding of logical paradoxes is a path that has been up to now unprecedented, but out of which it would not be necessary to make a new system.[10] Likewise, the Heideggerian advance gains nothing by being transformed into prophecy. *Wege nicht Werke*: pathways, not works, Heidegger writes at the beginning of the *Complete Works*. Multiple are the possible paths of thought in language; diverse as well are the faces of Coherence and the "nuances" at the heart of thought.

At any moment, thought is possible, but its coherence is not achieved. The border between thought as such and rationality would be traced if we could admit that all the possibilities of thought have a priori a rationally determined status. Consider this phrase: "Every angel is terrifying." Is it necessary to assign a rational status to this thought of Rilke? If I suggest that this short sentence is an ordinary expression, I have respected neither its context nor its unique meaning. If I take account of its poetic "charge," I renounce speaking of it in completely rational terms. Certainly, it appears logical to acknowledge that such a "statement" is "poetic"; but, if poetry is not clearly defined, is it thus rational to name the difficulty rather than resolve it? As for denying poetry or assigning its "status" at the bottom of the ladder of rationality, this is always possible. Is this, however, the most rational attitude before the polysemy of a language? In this way, every rationality quickly collides with the eventuality of critiques that are rationally better founded, more reflective, and more subtle than the theses presented. But are these critiques themselves protected from more sophisticated critiques, and so on? Do they not even risk looking irrational in relation to a rationality at a higher level? The will to the complete rationalization of life and thought collides with Reversal; reason rediscovers this law within itself, in thought, as well as beyond itself, in the vast world and its history.

Reversal

It is perhaps not too late to give an account of the Reversal that has often been put in question. What do I want to say when I claim that the will to complete rationalization turns into irrationality and that, generally speaking, every applied rationality is exposed to the betrayal of its initial ideal? It is not a matter of contesting the rationality of the rational or the fact of irrationalities, completely identifiable and often irreducible absurdities; it is absolutely not a matter of claiming that the rational and the irrational are *equivalent*, nor that, through some kind of magical operation, what was intrinsically rational yesterday will no longer be so tomorrow. The sum of the angles of a triangle does not cease to be, once and for all, in itself, equal to two right angles. Under these conditions, does it make sense to speak of the Reversal of the rational?

To answer this, I will methodically depart from the most incontestable rationality: the logic of inference and its formal procedures. Is the reversal not practiced there? Quine notes, in effect, from the beginning of his *Methods of Logic*,[11] that the conjunction is *commutative*, which is to say that the conjunction of pq is equivalent to qp. But this reversal, a fundamental operation of logic and mathematics, operates under the form of permutation at the interior of an equation or within the limits of a relation of equivalence. It is narrowly defined in the most rigorous formal terms— nothing directly comparable to the Reversal that we attempt to define.

Another form of reversal is rhetoric: it is a matter of a *chiasmus*. In its tautological form ("white bonnet" and "bonnet white"), it is equivalent to the preceding case: permutation. It is at least formally equivalent to it, but its rhetorical function goes beyond its logical form, because the utilization of such a *chiasmus* is subordinated to the project of reducing or polemically denying an assumed antithesis. For a more decisive reason, the increase of terms in a discourse exceeds simple permutation, when it allows for slight variations (A-B/B-A), whether or not this is with an eye to elegance.[12] To deny an antithesis is to still practice an antithesis, and this certainly allows for nuances, irony playing an essential role here. The context is, therefore, decisive in determining the true role of the *chiasmus* in a discourse: analytical, decorative, slightly ironic, profoundly polemical, etc.

Can we approach the sense of Reversal? No: the *chiasmus* remains rhetorical, even when we envision a much more general and fundamental reversal, intervening in every sphere of human activity—psychological life, social, political, etc. Nevertheless, perhaps we come a little closer to the goal, thanks to this bias: the importance of context. The living discourse cannot be purely formal and, consequently, "the reversal for or against," dear to Pascal, cannot have here a univocal sense. Is it by accident that this reversal is going to be carried out par excellence, first of all by Pascal in the moral and the political (as well as in their respective relations)? Since sophistry, the close ties between rhetorical antithesis and the moral and the political have been recognized, analyzed, extolled and beseeched. The *logos*, the human soul, the City: so many sensitive "places," in innumerable echoes, of the figures of reversal and of revolutions, more or less easily "representable."

One will, however, object: the *chiasmus* and the antithesis are perfectly recognized figures of discourse; they have nothing of the irrational in themselves, whatever the extremely variable and, at times, aberrant circumstances under which they occur. That which is reversed by the antithesis is not rationality in general; it is a specific position, a specific proposition. The antithesis discursively controls the violence of the shock between adversaries; formulating the object of litigation, it contributes to reducing the gap and rationalizing it. Better said: logic comes to the aid of rhetoric in giving its status to the demonstration of the falsehood of a proposition or thesis. *Refutation* is the negative and, above all, argumentative side of demonstration. It can be substituted for demonstration since this constitutes a petition of principle (hence, after Aristotle, it is the principle of contradiction).[13] The contradictor, if rigorous, dreams of only one thing: to exhibit the internal contradiction by which the adversary's thesis destroys itself. Commenting on the Aristotelian definition and distinction on this matter, P. Aubenque notes that the ἔλεγχος is not merely the "syllogism of contradiction," but more specifically still: "a refutation of the adversary through himself."[14]

No more than antithesis does refutation effect a breach in rationality. It raises again, on the contrary, the objection that has already been presented against all hasty reconciliations with the Reversal of the rational: the refutation reinforces even more the logic of a discourse that successfully restricts the adverse thesis to a *self-refutation*, thus gaining a genuinely rational triumph over personal conflicts as well as over the passions. However, if it is true that argumentation lives on rationality, it finds itself particularly exposed to the Sophist danger: "what makes the Sophist, according to Aristotle, is not his art, it is his intention (*proaïresis*)."[15] This is to say that rationality with the right to the argumentative field does not shelter this field from effective irrationalities: the appearance of rationality is all the more dear to the Sophist because it is the best that money can buy. When the play of arguments can itself also give rise to illusion, *effective* rationality ought to take into account the *proaïresis* of each disputant, which is to say a "supplement" difficult to predict, because the form—correct or not—of the arguments hides even more than it reveals. As was stated earlier in regard to the chiasmus, the *context* of the refutation is decisive and, in the same way, the formal character of reasoning is extended.

Consequently, this Reversal that we are trying to track does not seem to be absent from either rhetoric or dialectic: its silhouette stands out as soon as rationality is no longer reduced to its pure form. Rationality, effective and total, is not acquired directly, but, from beginning to end, it sees itself jeopardized with potential irrationalities; it is a matter, therefore, of a polysemous context delivered to a hermeneutic deciphering in which intentions, finalities, and effective investments, from the most diverse exterior sources, weight the discourse. Even the principle that Aristotle considered "the most firm of all,"[16] the principle of contradiction, is not perhaps as insulated as Aristotle thought. To the "physicians," to Heraclitans, to Mégeriques, etc., Aristotle retorts that, if the principle of contradiction is not demonstrable, this is not a weakness: one cannot demonstrate everything, and, above all, to want to demonstrate the principle of contradiction resorts to supporting it with the very principle to be demonstrated, which is a petition of principle.[17] What remains is that the supreme principle that safeguards rationality escapes, through its chief characteristic, to purely formal criteria of reasoning.

Is this overflow of the rational, in both directions from its core of formalism, sufficient to admit of a Reversal of the rational into the irrational? This would certainly go too far. Morever, is it still necessary to recall and to clarify that which we truly have in view through the Reversal? It is time to play a straightforward game with respect to this question.

In attempting to formulate a law of reversal, nonformal but, nevertheless, universal, do we not rediscover—in fact—the Hegelian dialectic? A law at once ontological and rational, the dialectic puts an end to the principal condition of all coherent thought, the Aristotelian exclusion of contradiction; it places itself there and devotes itself to its very soul. "All things are in themselves contradictory,"[18] proclaims Hegel, insisting on the fact that contradiction is not the product of an exterior reflection, a contingent lack that is reducible, but rather, is inscribed in life itself, as its innermost capacity: "Something is only living to the degree that it contains in itself contradiction and it is entirely this force that never ceases to take it up."[19] Rather than recalling the examples that teem under Hegel's pen[20] (since reality in all its aspects and its degrees is composed only of contradictions posed and surmounted), it is important to grasp that contradiction is principally immanent to identity itself: the tautology A = A contradicts itself, since it can be posed as such only in differentiating itself from non-A. Or, "difference in general is already contradiction in itself."[21] The principle of contradiction is, therefore, only the "negative form" of the principle of identity.[22]

The speculative dialectic does not respond at all to the question raised from the outset of this book: its uneasiness does not even touch it. Hegel never envisions the power of the rational as a reversal, eventually harrowing or strange, of rationality itself. Hegel does not take into consideration the Reversal of the rational into the irrational, since for him everything is *superiorly* (speculatively) rational. What the dialectic concedes on the one hand—the omnipresence of contradiction—it takes back with the other, because every contradiction is resolved and reconciled in the Absolute. The System has certainly been enriched from all the profundity of onto-theology; it is no less the metaphysical apogee of rationalism.

In fact, the Reversal that I have in view has nothing of a universal and systematic law that would be applied uniformly and gradually from logic to the philosophy of spirit. When Hegel presupposes the complete rationality of the real, the Reversal is thinkable only if being, life, and the Spirit exceed rationality. But to what degree? It is precisely this question to which it is impossible to respond from the customary or metaphysical exercise of rationality: this would be a petition of principle nipping the question in the bud. If the Reversal is neither a rational operation (as permutation) nor a systematic figure in the Hegelian sense, nor even a metaphysical foundation (because it enunciates no option on the definitive truth of Being and never sees a reconciliation or onto-theological "last word"), then what is left? A presumption of thought.

Seemingly fragile and of unprecedented status, this assumption has, however, been rendered inevitable by the insufficiency of rationalist responses, the too frequent absence of a questioning of the effectuation of the rational and its shadowy face, and the apparent impertinence of the confrontation of the rational with its power. We have seen that Weber started down this path, and his reflection on the modern process of rationalization caused him to inquire into the "inexpressible process" that vitiates all totalizing rationalization. He never formulated a formal rule, and for good reason.

If the Reversal sanctions the very *limit* of rationality, and potentializes to the point where it is no longer recognizable in its effectuation of power, it can be outlined neither formally nor completely, even less quantifiable (as if a *threshold* was assignable to it from a definite degree of rationalization). But this limit is not, however, experienced in any less cruel manner in the most concrete spheres of contemporary life.

A very precise example allows us to understand this, and serves as a springboard to take us closer to the Reversal, in its sites and in its phases of predilection. The rationalization of industrial work has been conceived and specified by the famous engineer Frederick Winslow Taylor in his *Principles of Scientific Management* (1911). Even though "management" has undergone numerous developments (especially in the techno-commercial domain, thanks to information systems) to the point of becoming an autonomous discipline, Taylor has formulated its principles: to analyze each element of work in order to replace the antiquated "handmade" with simplified and precise gestures, to select and to train scientifically workers in the function of precise tasks that they would have to accomplish (and whose rate of production will have been calculable); to instigate a constant technological cooperation between "management" technicians and the workers themselves.[23] The principle rule of Taylorism is the perfectly consistent optimization of work. At no moment did Taylor envision that this beautiful worldwide planning of work, optimistic and optimizing (in every sense), would lead—in a way that was unforeseeable for its designers—to those situations tragically the reverse of enlightened objectives thus established: "Harmony, not discord. Cooperation, not individualism. Maximum production, instead of limited production. The development of each person to the maximum efficiency and prosperity."[24] A "manager" today will note that Taylor still remains too idealistic in his good intentions, and that the numerous problems of "management" are not systematically inventoried and analyzed by industrial sociology. Without contradicting this, I must, however, note that the objectives of industrialization globally remain the same as Taylor's and that they form a whole, as was foreseen. If Japan today is so "performative," is it not because it literally applies the aforementioned principles? But it is enough to take up these objectives one by one in order to appreciate their ambivalence and the reversals that they are full of? Harmony? Except for strikes, often provoked by rationalization itself, which have been ongoing since 1911! Cooperation? The expansion of the workplace goes together with an unprecedented individual atomization, a retreat into self, urban insecurity, etc. Maximum production? To a certain degree, but at the cost of an unbelievable gap between developed and underdeveloped nations.

Taylor's rationalization is the *index* of the general quest of the actual scientifico-technological world: the search for the highest performance in terms of *input-output*.[25] This optimization of performance is essentially *programmatic*; we have seen that Research and Development is the indispensable driving force. Contemporary techno-science can be characterized as a vast *program of optimization* (its motto: expansion), in which every radical Reversal is in principle excluded. Of course, one admits that there are "problems" (or "contradictions"), but this is in order to better avoid the question of the rise of the autonomy of Power as such.

This example and its principal lessons show that Reversal is precisely that which rationalization is not able (nor does it desire) to think: its limits, its sanction. Reversal is the sign of all that opposes the will of complete rationalization toward the marks of its unilateralism. Prowling around the edges of the program of optimization are the specters of Reversal: catastrophes, crises, the apocalyptic are always possible.

Although these indications make futile an enumeration of the "sites" where Reversal is produced, they allow us to understand that it can occur *wherever* rationality is no longer limited to its strictly formal usage, that is to say, essentially as soon as a politic—in the large sense—is in play. To take up again the example of Taylorism: the severe critique of his psychological, social, and human results does not impugn the core of this foremost rationality. Whatever the future destiny of Taylorism, the segmentation and optimization of work that Taylor realized remain perfectly coherent, and are not invalidated by any turn. Reversal does not bring about the *formal* in Taylor's rationalization. What, then, is its concern? It apprehends the relation between the project and its effectuation in the practical field where "human freedoms" are confronted and united. As with every methodological resolution of a problem since Descartes, Taylor *distills* the difficulties. Consider a laborer laying bricks, calculating the number of bricks handled per day, per hour, etc.; what typology is going to best respond to the "economic principle"? It is clear that what we call Reversal does not intervene at the *interior* of these rational operations; it does not even directly contest the program (every program of rationalization, moreover, in its projections and operations, providing something other than the "problems" relating to the program itself) *inconceivable*. Reversal stands out whenever the very terms of Taylor's problems are replaced in their "context" of the whole; according to Taylor, the "means-end" relation was enclosed in the problematic of the pure and simple profitability of work. According to his critics, including ourselves, this problematic is confronted with its social, psychological, moral, and even ontological "cost." The question becomes at once more global and more radical; it allows for multiple formulations, but these must, in one way or another, displace, radicalize and undoubtedly reverse Taylor's problem: "What has one truly gained in submitting all of human work to the principle of profitability? Why has this principle become the sole regulator of *all* practice? How is it that this rationalization created a new situation, not foreseen by its initial program?"

Reversal does not contradict *rationality* head-on; it scoffs at *rationalization*. In Weber's terms, it is at this moment that *Zweckrationalität* reveals its contradictions and limits. What is it that reverses? Not the rational of rationality: more appropriately, what we have called the *excessively rational* [*surrationnel*], that is to say, the project of methodological rationalization. However, is this project—whether it is grasped at the most "noble" level (Descartes) or at the basest level (Taylor) of technology—less rational *after* than it was *before*? Does excessive rationality reverse itself in accomplishing itself? Or is the problem only displaced?

If Descartes and Taylor were both confronted with the results of their "practical philosophy" that issue from the method, those of the "rationalization" of work, would they acknowledge themselves to be forcibly "defeated"? If the same presuppositions are maintained, rational potentialization can only reiterate its demands, restate its objectives. Unseen contradictions, problems, crises, reversals—so many facts "manageable" by an excess of rationality. We have seen that the unbelievable increase of risks encountered at every level of technicalization has caused us to renounce neither "technological progression" nor the pursuit of over-armament.

It is, therefore, not enough to claim that the conditions of Reversal appear as soon as we "globalize" the initial problems. Certainly, the intervention of different "contexts" reveals the complexities of the problem, their density, and their psychological, moral, and political fabric. One example: the henceforth famous report of the Club of Rome on the "limits of growth" remains technological, but it *signals*—at the very interior of the technological sphere and in the most quantified terms—a possible and perhaps desirable reversal of the drive that until now has been almost exclusively toward economic expansion. What is important in this regard is the subject of a major technological risk, the danger of a world war, or more fundamentally, of our destiny of power, always returning to this essential methodological difficulty touched on by Weber's questioning of rationalization, pursued by our analyses of the phases of rational potentialization. How does one *think* this power of the rational non-irrationally, whose rationality itself has such difficulty taking a true measure?

By the very virtue of the presupposition of a "contiguity" between different language-worlds, I have had to respond to this exigency on two fronts: at the interior of rationality and at its limit. *At the interior*: in distinguishing the phases of potentialization, we understand that the power of the rational can decisively come up against Reversal in neither Phase I nor Phase II, nor even in the initial impetus of Phase III (the methodological *project*), but only when the effects of power in Phase III become massive enough that one begins to be obligated to pose the question of power (this occurs in the nineteenth century with Marx and Nietzsche within the limits already mentioned); we understand, finally, that Phase IV, the era of extreme risks, is the ultra-sensitive zone where rationality itself must take into account not only its ideal (*Wertrationalität*), but also the deep-seated ambivalence of its accomplishments (in the displacements and sudden changes of *Zweckrationalität*.) *At the limit of the rational*, thought yields to evidence. The real is rational in neither its eruption nor its entirety; rational potentializations do not convey the full extent of it. Rationalism, in refusing to think the limit of all rationality, naively believes that Reversal is avoidable; it must, then, sooner or later, fall to the "exterior of itself"—to transpose a Hegelian expression. The irony of History will come to belie its claim to formalize the nonformalizable. What will come to reverse itself, therefore, is not the (formal) rationality of the rational: it is its *practical* significance, its availability, and its ontological reference. As in paranoia, logic will perhaps save us: in the midst of the wreckage of all sense. Through its Stalinesque destiny, dialectical rationality itself recognized its most massive and cruel Reversal. Henceforth, in Phase IV, the potentialization of the rational is undoubtedly burdened with a Reversal of every ultimate rationality. It is indeed this vertiginous turn that the *Dialektik der Auf-klärung* attempted to explain, in terms that were too reactive and without taking sufficient critical distance vis-à-vis the potentialization of the rational.

Thought by an understanding open to the ontological enigma, Reversal becomes this *limit* that all coherence and all relation cannot eliminate from their field: the constitutive limit of the power of the rational.

The non-will and the possible

If all essential thought ought to be able to think "against itself," it is because no truth is protected from the Law of Reversal, whenever it leaves the strict limits of the formal. Now, what fundamentally prohibits an ideology from inquiring of its own metalanguage and its profound repercussions, is its inflexibility, despite its apparent self-criticisms, concerning the theses that constitute it once and for all. And what reinforces this inflexibility is the transformation of thought into militant incitement, into a mobilizing reference for the will. Every essential breakthrough in the domain of thought is stalked, now more than ever, by this operation of the *reduction to ideology*; Marcuse was a consenting victim of this, Habermas only partially freed himself, and even the tentative Heideggerian is threatened.[26] Almost no one escapes this groundswell: the mobilization of the will in the contemporary phase of rationalization. This will surprise only those who misunderstand the multiplicity of the forms invested by this mobilization, which will go so far as to install itself in nonviolent ideologies. It transgresses the superficial, psychological manifestations of militarism (having placed its "funds" in security, as Heidegger saw it, the Complex of Power can allow an apparently non-negligible margin of play for individual subjectivities); even in the case of nonviolence, it subordinates action to a supposedly efficient end. In fact, neither Marxism, nor ecology, nor nonviolence, nor Marcuse, nor Habermas goes so far as to pose the radical question of the limits of all willful action in the face of a process which has all the appearances of a "sending of Being." Max Weber is certainly the only one along with Heidegger, and after Nietzsche, to have grasped the *destinal* aspect of the process of rationalization, its irreversible and almost fatal character, transcending projects and counterprojects.[27] But paradoxically, even though conquering industrialization (in Phase III) seemed to call for an explicit and even triumphant mobilization of the will to power, the actual phase of potentialization appears to imply that the mobilization of the will is *acquired in advance*, which reinforces the plausibility of this Heideggerian intuition: the metaphysical foundation of the new epoch is no longer strictly the will to power, it is the *will to will*.

Heidegger has, therefore, in my opinion, washed the horizon of every ideological necrosis, pushed the diacritical force to this "statement," so radically demobilizing for every militant zealot and perhaps discouraging even for every will (a challenge launched at Zarathustra). No will can have reason, neither from the Complex of Power in its massive (and dynamic) facticity, nor above all from the relation (ontologically sealed) by which being is given directly, irrevocably, under the figure of calculation and at the site of manipulative organization. One is far from the Frankfurt school and its commendable, but perhaps still naive, attempt at reversing a "bad" domination into a rational power mastered in and through a dialogue or democratic "interaction." The increase of power does not seem to be, in itself, a difficulty for Habermas: the hypertrophy of the "complexity" appears to him to be compensated by an increase of critical vigilance. Who is not an activist in this fashion? Even the hippies believed that the abandonment of the System could seize upon this and, in the end, lead, step by step, to a kind of general work strike: again a will to efficiency, metamorphosed, displaced, simulated.

Does nothing more remain except to leave the will to will to itself? Yes. But even that is not *nothing*: the possible, for Heidegger, is a new understanding of Being, that Being is able only to withhold itself, from itself. To prepare this understanding. ... The rest does not depend on us. Thus subsists a fragile will maintained in the non-will: the flame of the attentive thinking under the breath of the possible.

Already, the preceding "explications" with Heidegger's thought have shown that it is at this extreme point, where my steps glide into his steps, that a divergence is produced (the task is not to appreciate this divergence or its value that I have only to propose)[28]; if the possible alone can save us from the fatigue of the will (and its vanities), why disregard this rational-irrational possible of a "dynamic" contiguity, that is to say, a still potentializing contiguity, where rationality and the ontological enigma would no longer be excluded? It is not rational mastery alone that has rendered this conjunction difficult and improbable: the master can be astonished before his own masterpiece and even feel the mad desire to think, to admire, at the height of mastery. The greatest artists, the most profound scientists (artists of rationality) have felt this enthusiasm.[29] This mastery knows itself to be exposed to the Law of Reversal; rationality is not made from its own unconditioned essence, has not transformed its own power in the ultimate horizon, has not invested everything in the ever more exclusive and fateful increase of this form of power.

To accept necessity to the point of liberating the possible in its gratuitousness: Nietzsche is perhaps not far from thinking the eventuality of this, when he writes, for example in *Human, All Too Human*: "Therefore a superior civilization must give man a double brain, two brain chambers, as it were, one to experience science, the other to experience nonscience. Lying next to one another, without confusion, separable, self-contained: our health demands this."[30] This citation is taken from the fragment "The Future of Science." Nietzsche departs from the considerable difference that reveals the economy of scientific satisfaction: a considerable "source of pleasure" in its inventive exercise, it becomes banal and boring when one "apprehends its results." For Nietzsche, what is questionable here is not the value of science in and for itself, but its civilizing effect, or even its socially powerful effects. The idealist optimism of a Renan is contradicted by neither a reasoning nor an appeal to historical facts: it is—mockery!—the most elementary psychology that manifests the *reversal* of the energetic power of science. A genealogy of science proves to be as necessary as a genealogy of morals, and even more, in this time marked by science (and uniquely by it): the scientist, he too, despite his nobility, is *constituted* only from instinctual humus.[31]

What is of importance for the present understanding of this text is not only the logic of reversal from Platonism that supports it (the intelligible measured by the sensible that is attached to it). To return science to its energetic economy does not go back to condemning it in the name of the will to power, as a reader who had been persuaded in advance of Nietzschean "irrationalism" would be tempted to do. Above all, it is important to grasp that the Nietzschean diagnostic immediately leads to a *treatment* (the irony of this medicalization of philosophy was in some ways transmitted by the *artistic* side of this superior medicine): to separate into "two brain chambers" that which gives pleasure ("that which is nonscience": the passions, religion, metaphysics, and above all, *art*) and the *regulator* (science). Thus Nietzsche perceived not only that our civilization was completely dedicated to science (the other "factors" becoming more and more "decorative"), but that it ran the risk of seeing itself—the supreme irony—returned, from the summit of the triumph of reason to the irrational, returning to the barbaric (even though science believed itself infinitely distanced from this). Reversal stalking a rationality that believes itself to be the complete reason of life: it, nevertheless, must give an account of itself to life. The "irrationality" that Nietzsche proposes is the only possible salvation for a rationality that tends to lose its vital connections. Not an irrational by default; an irrational through the increase of rationality—the superior irony, faithful to the lessons of the Greeks, in its care to correct one principle though another, the Dionysian by the Apollonian.

A superior irony: it is necessary to save science from its own successes and, in order to do that, to prohibit it from an entire part of our mental activity. Supreme potentialization of this *retention* of the power of science: placed in data banks and reserved for *regulatory* ends. To distance science from the sources of sensual, passionate excitement is to forbid it to exhaust them; it is, moreover, to nourish pleasure and to stimulate the will-to-live. Our thinkers of tomorrow propose only still more "rational" "solutions" to the "problems" of industrial civilization: the systematic planning of urban space, the environment, culture, and leisure. The Nietzschean "response" is the inverse, infinitely more demanding: to impose this *ironic reserve* on rationality itself, which will allow it to take up the Law of Reversal, and for the lack of which it will suffer in desolation and derision ("illusion, error, the chimera, all associated with pleasure, are going to reclaim, step by step, the territory they formerly tended."): an *art of life*, which is the ironic reply to the implacability of rational *partage*, is indeed the *contiguity* between art and science, life and the rational. This *juxtaposition*, proclaimed by Nietzsche in order to attain "health" and preserve the chances of a superior civilization, can only be dynamic, moving with the times, as does life; a new source of power and mastery—for every civilization worthy of being called vital. To play one against the other (art against science) in order to promote one *with* the other—who would pretend that this task is easy, that it can become the daily bread of humanity overnight?

Who, above all, would claim that this new distribution of the cards of thought guarantees "success"? The epoch that unfolds—and no one can be certain of its duration—is one in which *nothing is enough*, neither the non-will nor the will to power, nor even the appeal to the rational. But, if thinking cannot directly transform the world, nor by itself alone avert the dangers that permeate the Complex of Power, if it must even renounce the efficiency that Phase IV championed, it can free the creative will for *something other* than the systematic planning that dominates the planet. Paradoxically, the quasi-autonomous, if not automatic, functioning of the Complex of Power can facilitate this liberation of the will, and it is undoubtedly in the field of the social imaginary[32] that thinking can best be associated with a "dynamic contiguity" between the dominant and the reserve. Just as the freedom of indifference is the lowest degree of freedom in Descartes, so automatization, as Simondon remarks,[33] does not express the highest possibilities of technology. Every "automatic" (or artificially deemed necessary) conception of the recourses presented to humanity can only be challenged. Only the superior irony of art—demiurge of thinking and its imaginary powers—seems to offer, through complicity with the possible, the power of a future worthy of the name.

Eirôneia: irony as superior feint, studied dissimulation, is never the privilege of Socrates, in its style or in its content. In its style, it can lead, within and beyond the dialectic limit, to a canny intelligence, sister of a poem.[34] In its content, it can traverse the *eidos* in order to rediscover the fraternity of the vital night. *Eirôneia*: in accepting its enclosure, rationality is not renounced. How would one lose anything essential in claiming that the Enigma comes to confront intelligibility, and that it knows to sustain its mocking glance?

There are more things

If Borges, in titling a story,[35] gives an indirect homage to the famous speech of Hamlet, I am certainly allowed to be freely inspired by both Shakespeare and Borges: there are more things in the rational and the irrational than in all of Power; there are more things in thinking than in rationality; there are more things in Being and the possible than in thinking.

Instead of denying it, rationality thus stands before the possible, as much before Non-sense as before Sense. It admits that the Chinese boxes where the Enigma was sheltered are opened. It exploits the immense and unpredictable "common frontier," carved as the very contours of language, that it maintains with the thought of Being and the possible.

But if there is not an incompatibility of principle between the exercise of rationality and the sense of the enigma, we know that the singular concern for coherence and the aim of the effects of power do not open this possibility; quite the contrary, they block it. One names "thinking" this "supplement" through which rationality suspends its purely rational exercise, no longer satisfied with its own results and raising the challenge of the Enigma. This is the gap that is opened and measured, and speaks in a privileged manner in the *critical moment*.

In critiquing itself in a thematic manner, philosophical rationality is obligated to think against itself, to master or to take up its own contradictions; it is at the service of the Enigma. However, Kant's great example teaches how critical lucidity is a difficult conquest: no longer does every curiosity merit the status of enigma; not every inquiry attains the dignity of critique elaborated by reason. Why had this critical undertaking become necessary and beneficial? This is because, as far as one can support the odds of evoking it so briefly, pure reason delivered to itself had lost the measure of its theoretical capacity and its practical legitimacy. The delimitation of the conditions of the possibility of knowledge, of the respective powers of understanding and reason, and of the practical import of pure reason, therefore, permitted humans to reinsert themselves in their autonomy. Rationality, provided that it submits to its own tribunal, rediscovered all its rights to be the *templum* of humanity. "Practical by itself alone,"[36] reason was universally legislative, and, presenting man with the moral law, proposed the only legitimate model of the "sanctity of his own will."[37]

If Kant had "brought back from the holy mountain the law that is vigorous," according to the words of Hölderlin,[38] is this heroic work invalidated today under the pretext that rationality has been the object (and, in part, the complicit subject) of a gigantic *inveigling* on the behalf of Power? To claim this would be to yield to the least justified historicism, to sacrifice oneself blindly to today's dominant forces, and to break the golden thread that connects thinking to meaning.

The critical field does not close with Kant; it opens starting from him: the passage of two centuries is sufficient for the relation of the rational to power and to its own possible to be completely redistributed, at least *de facto*. A "critique of historical reason" is, therefore, imposed by virtue of this urgent task. Others before me have foreseen this, and have tried their hand at this task. In this regard, I have been able to appreciate Habermas's efforts.

But this *urgent state* of thinking is not at all imposed in order that we cease to list the Kantian critical experience among the resources of the rational *dynamic*. The purity of "you ought" has the value of a sign transcending all instrumentalizations, all strategic devices. It calls the rational to its vocation of freedom and to an absolute autonomy, unconditioned in relation to all potentializations. This reflection on the moral law is not yet the pure respect that Kant demanded, but it reconnects with a possible that does not seem mortgaged by the will to power: under the starry sky, a mark of the Enigma.

Nevertheless, we now know how Power is powerful and that it is not enough for reason to present itself as such, in its purity, in order to be a recourse (even less for the will to be haughty in a choice). If an increase of rationality could save us from catastrophe, who would hesitate? This is why the Possible was sought after, and the will separated from the most imperious forms of potentialization in order to rediscover its freshness.

Wager on the possible? Without doubt: the possible *as such*; but this ultimate choice of thinking is exposed to a misunderstanding that we have continually denounced: the pseudo-evidence of a strategic mastery of the possible imposed as the *only* rational possible. We would not have taken Power seriously if it had been only an exterior obstacle, a megamachine for which it would have been enough to leave alone. Power intends to take possession of the possible, to master it for game theory, to organize it through worldwide planning, to make it profitable through Research and Development. Even if it does not completely achieve this (because the process of creation thwarts every prospective and short-circuits the phases of potentialization), it establishes institutions, means of control and incitement, and viewpoints that function as so many predeterminations and prejudgments of every new idea, of every discovery.

The Possible itself, therefore, risks being ensnared. Is it not Research, in Phase IV, that develops a religion of it? Are we not imitating Power while pretending to free ourselves of it—a condition much less enviable than that achieved from reversals and revolution? When Hölderlin analyzes the opposition between Antigone and Creon, he remarks: "The rational form which tragically develops here is political, and more precisely, Republican insofar as between Creon and Antigone, between the formal and the counter-formal, it is by excess that equilibrium is maintained equally."[39] Our *partage*, alas, has gone beyond the simplicity of this tragedy: between Power and our possible there is no longer a mediating *Vernunftform* that takes form, and no equilibrium is in sight. Lacking a face and incalculable, the process of Phase IV does not allow for opposing a "counter-formal" to a "formal."

The more that humanity potentializes, the more it must cultivate a principle of selection of the possible: to protect that which is held in reserve to the very degree of its immense powers. Did not the Chinese and the Greeks ensure that they would refrain from allowing their inspired discoveries to tilt toward pure and simple effectuation? Perhaps Phase IV should be characterized as a potentialization that is completely unleashed? Indeed, every process of power implies an organization that assures its very deployment. Mechanisms or regulations exist today (whether they concern bank credit or armament surveillance) that, despite their insufficiencies, prevent the catastrophic rupture of a world system. The distress in the face of this situation is ambiguous: it is often reduced to asking whether these mechanisms will be effective enough or to complaining about their need for improvement. Other things are in play, however, and their shadow has constantly followed this inquiry: *the quest for a measure for the possible itself.*

The understanding of the Enigma would be a possible among millions of others and an arbitrary wager of thinking if it were not precisely *this which is cruelly missing* in the process of rationalization. Insofar as it is a *measure,* it does everything to disconcert our contemporaries prepared to accept only the dominant model of rationality. It is as if, as Pascal has noted, "everything which is incomprehensible does not cease to be."[40] If our condition and our history were not paradoxical, they would fit more easily into the heart of the rational and, consequently, their developments would be all the more predictable. Even the rational does not discern, by itself, that its own genealogy is in defiance of every definitive explanation of history and every mastery of the future. The rational *in itself* thinks neither Being nor History: Hegel, like every rationalist, attributes too much to it in making it an absolute subject.[41] The measure of the possible: only thought qua thought (rational-irrational, definable-indefinable) can help rationality approach it and language welcome it. Thus occurs the understanding of the Enigma, the fruit of a long maturation within it of thought and of dialogue, and also within it, between its various forms, it is always confronted with the unthought, the unthinkable.

If it is necessary, one last time, to show the stakes, let us recall how the *rationalities of reference* proceed from actual "efficiency." There is no mystery in their being set-into-work: their authors clearly announce their presuppositions. Hence, Brillouin describes the theory of information: "Science commences there where the signification of words is narrowly restricted. . . . Our statistical definition of information is only based upon rarity. If a situation is rare, it contains information. . . . "[42] From this comes the statistical measure of information from the famous Boltzmann-Planck formula concerning the evolution of a system toward the most probable state. Thus, von Neumann explains, from his perspective, that game theory is applicable only to an individual who wishes to obtain (in the case of a customer) a *maximum* of satisfaction or (in the case of an entrepreneur) a *maximum* of profits: "The individual who seeks to obtain these respective *maxima* is . . . said to react "rationally."[43] *Money* will, therefore, be the middle term that will allow one to quantify these problems of *maxima.* Thus, von Bertallanfy explicitly presents general systems theory as "a scientific study of . . . totalities that, not so long ago, were considered as metaphysical notions surpassing the limits of science." These dynamic interactions of origins and different levels (in particular in the economic, biological, and social domains) are henceforth integrated into the scientific edifice, thanks to a mathematical formalization that erupts from the study of the phenomena of self-regulation.[44]

It is not important here to decide once and for all which of these theories is the most operative in today's techno-science, to know whether Frank Ramsay[45] has gone beyond von Neumann, or Ashby[46] beyond von Bertalanfy. What is important for this proposal: *all* of these theories through which contemporary science significantly increases its field of application (to the point of posing the problem of self-totalization), only make us see, if not resist, the initial presuppositions of all scientific progress. This methodological focus is possible only if we start by limiting the sign to a sum of information, by limiting human beings to agents of the maximization (or optimization) of exchange, and by limiting all vital economic and social spheres to elements that can be homogenized. We rediscover here the conclusions of our critical study of the idea of the "new science": the rigorous and fruitful development of science is in proportion to the agreed-upon sacrifices made in this immediate moment and throughout its long progression to delimit its objects and to clarify, quantify, and control its procedures. The extension of fields of study to domains that up to now have been considered "taboo" and the changes of rank[47] in no way modify this fundamental epistemological situation. It is not necessary to credit science with more than it can grasp, under the pretext that it will henceforth embrace everything.

What must this rationality do in order to discover, in addition, this understanding of the Enigma that it lacks? The myth of the "new science" was foreshadowed in Bergson's response: scientific rationality must *invert* its movement in order to coincide, once again, with the vital impetus from which it has been separated. Although it fails actually to rediscover all the "depth of field" that implicates life itself in its relation to Being and to language, I believe Bergson's view promises too much, as a certain attractive neo-scientism does today: making us hope for a kind of miraculous conversion of science, it allows us to believe, even as it defends itself, that the inversion of "natural metaphysics" would permit an *intuitive expansion* of science, the almost direct passage of the rational to the mystical,[48] or, at least, to a poetization of nature and existence.[49]

Neither in thinking itself nor in the face of an institutionalized and "autonomized" techno-scientific Complex do we await "expansion" or a miracle. We cannot avoid the most cruel and perhaps most irreducible hiatuses. And nothing promises a reversal of the epoch, under the pretext that the possible still appears, streaking the horizon. The *partage* is difficult, despite the encounters that thought allows and the toothing-stones that thought sets!

The understanding of the Enigma, the possibility open to thought in this era of intensive rationalization, is neither a panacea nor even a beneficial "extension" of rationality. In this regard, one will perhaps be, once again, surprised that we attend so much to the role of thinking *as such*, particularly in our denouncing the ease with which rationalism is affirmed—from Hegel to the adherents of the New Science. This differentiation is all the more surprising because this thinking *as such* proves to be imponderable, almost indefinable. Is it necessary to sacrifice so much to an evanescent moment? Without doubt, Heidegger suggests a more originary and encompassing approach to thinking: welcoming, gathering, acknowledging (*Denken* as *Danken*). But how do we reconcile this ultimate concession to Heidegger with inquiry, not found in his work, into an understanding—or even a rationality—of the Enigma?

Pascal writes: "One would not know how to make a small thought issue from all the assembled bodies. This is impossible and of another order. One cannot draw a movement of true charity from any body or spirit; this is impossible, and of another supernatural order."[50] What ought to capture our attention in this famous thought is less the *content* of the three orders (flesh, spirit, charity) than the very idea of an infinite (or infinitely infinite) *distance* between these *radically* different orders. A distance understood, moreover, phenomenologically through its absence, starting from the most tangible "fragment," that of bodies (kings, the wealthy, captains do not have a sense of other grandeurs).[51] Pascal does not say that the spirit is nothing in itself, but that it is nothing when judged by another order. Relativism? This is evoked by the *representation* of one distance by another, a representation that sets rationality to work logically (the analogy of proportion: bodies are to spirits as these are to charity), and mathematically (the relation between zero, the infinite, and the infinitely infinite) in order to make reason understand that "there is an infinity of things which surpass it."[52]

There are thus not only three orders. There is a fourth without which neither distance nor representation would be thinkable: it is the order of "accuracy" ["*just-esse*"],[53] that is to say, thinking that recognizes the proportions and submits to them. This order of things, the philosophical order, is the "heart" in the broad sense that Pascal understands, and about which it does not suffice to say that it is not confused with sentiment, that it undertakes an essential relation with calculation. It is necessary also to grasp that the heart is the "metalanguage" of the three orders of "reality." And it is necessary to grasp all that the heart can grasp.

Thinking *as such*, in the sense that I understand it, is from the order of the *heart*, not from the second order. It is not reduced to the fact that one thinks, or even that one thinks with exactitude or virtuosity, but that one thinks the very thought (in my terms, that one meditates on the Enigma that is thought). In this repeated astonishment, deepened and rethought, that which is sheltered is as much imponderable to reason reasoning as is charity. Transposing Pascal: from all the bodies and spirits, we do not know how to pursue a movement of true meditation on the Enigma. This formulation, more phenomenological and circumspect than that of the *Pensées*, endeavors to approach the same mystery.

Detechnicizing the future

"Technology decides everything."[54] If it were necessary to have a symbolic reference summarizing the radical desymbolization with which systematic technicalization threatens human freedom, this phrase of Stalin would offer it. Cynical in that it casts aside the communist ideal, this slogan is all the more so by its lock on the future, its implacable will to efficiency that disarms not only all resistance, but even every slight inclination of choice. If "technology decides everything," then there is nothing which is not already programmed, forecast, "solvable." Revolt is useless. Hesitation, deliberation, anticipation, hope are unthinkable. To be sure, one will object that this speech is not interpreted in its larger and more terrifying sense; in this presumed neutralization of every end, we cling to the crude wink of the autocrat, astonished by the only finality actually pursued by Joseph Stalin: power. The totalitarian madness wears the most efficient mask, because it is the most neutral; this is not the ruse of reason, but its counterfeit, an allegedly realistic grimace of the most nihilistic will to power ever formulated and "applied."

Although one has to take into account some degree of cunning, once replaced in the logic and history of Stalinism, this sentence is the *compendium* of techno-discourse that claims the immanent "finality" of the systematic technicalization toward which Phase IV tends. We recall "Gabor's rule": "Everything that is technologically feasible ought to be realized, whether this realization be judged morally good or condemnable."[55] Does this not say the same thing: nothing escapes technology, which replaces all other criteria, and, in particular, morality? Nevertheless, upon examination, the identification would not be correct. Gabor's rule has a formal character (itself technological) that can be inserted into Stalin's discourse, to be used by him as far as he wishes to go; but it is—in fact—less totalitarian than the latter. In only expressing an opinion about the "technological feasible," it implicitly admits of realities that are "non-technologically feasible." A general rule permitting the formalization of technological progress, it does not, therefore, have the same character of "techno-discourse" as Stalin's phrase.

Beyond the particular case of Stalinism, which at any rate cannot be repeated *stricto sensu*, it is the future that interests and calls to us. The ultimate phase of the potentialization of the rational outlines a trajectory which, to be sure, still orders a space of play with other potentializing phases and allows a distinction between rules that are properly technological and techno-discourses; but we cannot deny that it radically engulfs the future: in programming it (in its effective operations and its organization of Research and Development) and in mystifying it (in different strata of techno-discourse). Stalin's phrase was terroristic in itself, through his refusal of all uncertainty. All evidence shows that the effective process of the technicalization of the world remains actually much more complex, because it is still open. If a spokesperson of the "technostructure" were to intervene in this inquiry, he or she would not fail to observe that autocracy and terror are not necessarily the most rational or the most efficient forms of rationalization: in any case they represent only its stages, relayed by more sophisticated procedures of mobilization and control at the heart of a general process of development. If there is here ample material for discussion among economists, political scientists, specialists of developing countries, etc., the question *for us* is not properly technology: the question returns to and assumes the reading resulting from the antinomy of the power of the rational, in light of the inquiry into the phases of potentialization.

In the antinomy of the power of the rational, the future is at stake: ought or ought not everything that is technologically feasible be realized?[56] The technological thesis takes possession of the future; its operative rationality refuses all interruption. In its most radical form (Stalin's phrase as techno-discourse par excellence), it will consist of asking whether it is "technologically feasible," rejecting every other language-world.

In order to pose and to think the antinomy, it is necessary that an antithesis— confronting unconditional technicalization—give it life; therefore, reaffirming the possibility of a free access to the future, the capacity of self-determination (or of respecting that of the other) again becomes the *absolute precondition* of all techno- logical rationalization and now a radical possibility for interrupting the functioning of this rationality.

My first attempt ended in a despairing statement: the impotence of pure prac- tical rationality in the face of the technological thesis (at its foundation exploiting operative rationality). The genealogy of potentialization permitted the discovery of a play at the heart of rationality, as well as the discovery that the technological Ap- paratus is only one phase of this. Renewing the possible at the very heart of ration- ality, it allowed for the unfolding of a fundamental range of recourse: not only the *partial dismantling* of the technological apparatus, but the *rational exploitations* of other phases of potentialization. To do battle at the very heart of the scientific insti- tution and the University, in order that the epistemological phase (II) protects its autonomous rights, is to refuse the imperialism of the technological System. To fight, therefore, in order that the farmers and artisans protect or discover their absolute dignity, their irreplaceable beauty,[57] is to help hold open the play of potentialization. If the technological techno-discourse is not interrupted *at the very heart of the ra- tional*, we are led to believe that its program is developed in every possible way, that it takes possession of both the rational and the opening of time: the abuse of power and the force of an arrogant Power that no longer allows anything independent, no longer respects anything, and neutralizes the sacred.

The sacred is originally *apart*.[58] Thus far only humanity has been confronted with this *partage*, now the link between meaning and time. To potentialize *beyond time*, such is the outrageous pretension of Phase IV. But the possible is given tran- scendentally in a contrasting eruption, a triple dissemination of its *ek-stases*, where the future has the privilege of offering the advent of time itself. The future is the sacred in time, this inviolate part—independent—however, the giver of presence and of withdrawal.

The power of the rational, open to the possible, allows this *partage* to flourish. It unfolds and defines itself in three *sacred parts*: the ethical, the political, the poetic.

These three parts are the fragile sites of the resistance to complete technicali- zation, the crests where humanity still preserves the future *as such*: its *partage*.

The categorical imperative remains, in its purity, the only possible *rational an- tithesis* to the expansion of the scientifico-technological power of the rational. Its limit can appear only from the *form*, the absolute limit—of the absolute form. In this antinomy, the rational itself finds itself divided between its effectivity (the logic of its power) and its most cherished possible (self-determination as the infinite prin- ciple of freedom). The ethical imperative has nothing in itself that yields before

Power or enters into its play. It reserves, in principle, a rational core that escapes to a Reversal.

The political (understood not in the sense of politics, but as the confronting of the possible by a people concerned with life or death) is also a sacred part, in the sense that here the people assume a future. The essential political decision is rational-irrational; it knows itself to be exposed to Reversal, and thus, on this point is different from the purity of the Imperative. It is partly linked to power, in that it cannot completely refuse to be taken up without relinquishing its fundamental rights, and in this way differs from the Poetic, touched with gentleness. Particularly in this era of complete rationalization, it sees itself assisted, enclosed, and suffo- cated by the techno-structure apparatus, reduced even in nuclear deterrence to a lapse of time that is continuously recalculated.[59] But insofar as there will be hu- mans worthy of the name, the honor of a people faced with the future will prevail against techno-discourse.

The Poetic is this dimension that, in the arts and even beyond them, carries every language and every monstration to their extreme limits. Dancing before the arch of the rational, it is sense exceeding sense, it speaks of the unutterable, hears what cannot be heard, the very future founded on humanity in the reserve of its possibles. The Poetic traverses contradictions and reversals; it is the part that origi- nally and destinally exceeds the power of the rational.

Experience opened to these possibilities is named *contiguity*. Carrying the power of the rational to its limits, it responds—"hope from despair"—to this co-de- mand of inconceivables. The irrational is there, but it now necessarily maintains the pure form of practical rationality beyond all the reversals (predictable, foiled or not) of a technological rationality that can also be ensnared in its failures, break- downs, etc. The incalculable is there, but we ought not excuse ourselves from ac- counting for it—with it, not on it—measuring time from ourselves, always our ad- versary. The unspeakable is there, but to speak what can be spoken and in the ruin of all the temples to maintain the letter is a sacred task.

There is no need to invoke our certain death. Finitude is inscribed in the struc- ture of life, in the fragile destiny of the planet as in all beings—magnificent or ri- diculous—who play out here their ephemeral chance. Incorruptible reason sees its supratemporal constancy confronted with this very fragile *partage*—bound with the menacing *partage* of power. As if "eternity," scoffed at by its counterfeits, were able to surrender only to those who are excluded from it.

Faced with this double finitude, thought can be a resource, if the surplus of power does not block the path of life, allowing it again to lift the miraculous weight of its improbability. Reason does not any more offer a direct recourse to our anxiety than it offered its power to the curiosities and impatience of our ancestors. But much would already have been done, if the true power of the rational were to redis- cover the richness of its phases, if humanity were to play at all these levels, if its possible were to offer variations, at the edge of the Enigma, there very near.

"One would be indeed well served," Pascal writes, if one were to see the "connections" of all things.[60] The rational "indeed well served" no longer has the omnipotence that its sovereign metaphysical self-reference accorded to it, nor the omni-efficiency imposed by its techno-scientific version. Disjointed in phases, confronted with Reversal, delivered to the ontological and destinal Difference, the power of the rational is, nevertheless, more than ever the power of a particularly fragile and still not completely despairing possible: our future.

Notes

Preface

1. Friedrich Nietzsche, *Sämtliche Werke*. Vol. 12, *Nachgelassene Fragmente, Herbst 1885 bis Herbst 1887*, ed. Giorgio Colli and Mazzino Montinari (Munich: DTV, 1980).

2. Jean Ladrière, *Les Enjeux de la rationalité* (Paris: Aubier-Montaigne, 1977), p. 63.

3. Georges Waysand, "Szilard et Majorana," in *Les Pouvoirs de la science*, coll. Dominique Janicaud (Paris: Vrin, 1987), p. 220.

4. See Karl Popper, "Toward a rationalist theory of tradition," *Conjectures and Refutations* (New York: Basic Books, 1967.)

5. See his interview in *Le Monde* (29 September 1992), where he argues "No technological revolution is able to exhaust a language."

6. I have developed and clarified the dialogue with this current of thought in *A nouveau la philosophie* (Paris: Albin Michel, 1991), pp. 141-49.

Introduction

1. Jean Ladrière, *Les Enjeux* (Paris: Aubier-Montaigne), 1977, p. 28.

2. Willard V. O. Quine, *Methods of Logic*, 4th ed. (Cambridge: Harvard University Press, 1982).

3. "Logic, like any science, has as its business the pursuit of truth. What are true are certain statements; and the pursuit of truth is the endeavor to sort out the true statements from the others, which are false." Ibid., p. 1.

4. See Ladrière, *Les Enjeux*, pp. 39-41.

5. Jon Elster, *Ulysses and the Sirens* (Cambridge: Cambridge University Press, 1979), p. 1.

6. Ibid., pp. 28-35. See the excellent critique of sociobiology by Claude Lévi-Strauss, *Commentaire*, no. 15, Fall 1981, pp. 365-72.

7. As early as in *La Structure du comportement* (see n. 2 of the Introduction, Maurice Merleau-Ponty, *La Structure du comportement* (Paris: P.U.F., 1942), pp. 2-3. *The Structure of Behavior*, trans. Alden L. Fisher (Boston: Beacon Press, 1963), pp. 9-10.

8. John von Neumann and Oskar Morgenstern, *Theory of Games and Economic Behavior* (New York: Science Editions, 1964), p. 9.

9. Ibid., p. 8.

10. Ibid., p. 9.

11. See Donald Davidson, "Psychology as Philosophy" in Jonathan Glover (ed.), *The Philosophy of Mind* (Oxford: Oxford University Press, 1976], pp. 175-96; Elster, *Ulysses and the Sirens*, pp. 154-155.

12. This failure manifests itself particularly in analytical philosophy's recoil upon itself, which most often condemns it to bickering. It limits itself, for example, to the "practical reasoning"

of a subject, to the respective adjustment of "propositional attitudes" and its supposed intentions. (See the discussion of a thesis by Donald Davidson—"How Is the Weakness of the Will Possible?"—by David Charles, "Rationality and Irrationality," *The Aristotelian Society* [London: The Society, 1982–1983], pp. 191–212).

13. See chapter 5.

14. André Lalande, *La Raison et les normes* (Paris: Hachette, 1948).

15. Ibid., p. 6.

16. Ibid., p. 228.

17. Edmund Husserl, *Die Krisis der europaischen Wissenschaften und die transzendentale Phänomenologie*, 2d ed. (The Hague: Nijhoff, 1962), p. 347. *The Crisis of European Sciences and Transcendental Phenomenology*, trans. David Carr (Evanston: Northwestern University Press, 1970) p. 299.

18. In France the most eminent representatives have been Bachelard and Eric Weil.

19. See William Herbert Newton-Smith, *The Rationality of Science* (Boston: Routledge and Kegan Paul, 1981).

20. Ladrière, *Les Enjeux*, p. 70.

21. Ibid., p. 71.

22. But nothing proves the contrary either.

23. Bachelard used this term positively, which is clearly not the case for me. (See Gaston Bachelard, *La Philosophie du non* (Paris: P.U.F., 1940), p. 138.

24. See paragraph 37.

25. Martin Heidegger, *Vorträge und Aufsätze* (Pfullingen: Neske, 1954), p. 208. "Logos (Heraclitus, Fragment B 50)," in *Early Greek Thinking*, trans. David Farrell Krell and Frank A. Capuzzi (New York: Harper and Row, 1975), p. 60. [Hereafter all English pagination will be given after the original.]

26. Martin Heidegger, *Lettre sur l'humanisme*, trans. Roger Munier (Paris: Aubier, 1957), p. 247. "Letter on Humanism," in *Basic Writings*, ed. David Farrell Krell (New York: Harper and Row, 1977), p. 227.

27. Martin Heidegger, *Holzwege* (Frankfurt: Klostermann, 1957), p. 247. "The Word of Nietzsche: God is Dead," in *The Question concerning Technology and Other Essays*, trans. William Lovitt (New York: Harper and Row, 1977), p. 112.

28. Martin Heidegger, *Phänomenologie und Theologie* (Frankfort: Klostermann, 1970), p. 40.

29. Martin Heidegger, *Der Satz vom Grund* (Pfullingen: Neske, 1957), p. 205. *The Principle of Reason*, trans. Reginald Lilly (Bloomington: Indiana University Press, 1991), p. 125.

30. Ibid., p. 205/126.

31. Ibid., p. 205/125.

32. Heidegger, *Lettre sur l'humanisme*, p. 100/221.

33. Heidegger, *Der Satz vom Grund*, p. 196/120.

34. Heidegger, *Lettre sur l'humanisme*, p. 76/211.

35. Ibid., p. 74/205.

36. Martin Heidegger, "Zeit und Sein," in *Zur Sache des Denkens* (Tübingen: Niemeyer, 1969), p. 25. *On Time and Being*, trans. Joan Stambaugh (New York: Harper and Row, 1972), p. 2.

37. See Dominique Janicaud, *A nouveau la philosophie* (Paris: Albin Michel, 1991), pp. 122–34.

38. Heidegger, *Phänomenologie und Theologie*, p. 41.

39. See Chapter 7.

40. To which we shall return several times, above all *in fine* (Part III).

41. Theodore F. Geraets (ed.), *Rationality to-day* (Editions de l'Université d'Ottawa, 1979), pp. 46–48.

42. Auguste Comte, *Discours sur l'esprit positif* (Paris: Garnier, 1909), p. 305.

43. See Robert Lenoble, in *Histoire de la science*, vol. 5 of *Encyclopédie de la Pléiade*, ed. Maurice Daumas (Paris: Gallimard, 1957), p. 426.

1. Facing the Incalculable

1. Cornelius Castoriadis, "Développement et rationalité," in, *Le Mythe du développement*, ed. Candido Mendes (Paris, Seuil, 1977), p. 223. "Reflections on 'Rationality' and 'Development'," in *Philosophy, Politics, Autonomy: Essays in Political Philosophy*, ed. David Ames Curtis (New York: Oxford University Press, 1991], p. 193. [This essay was originally given in English.]

2. Patrick Lagadec, *La Civilisation du risque* (Paris: Seuil, 1981), p. 65. See also by the same author, *Le Risque technologique majeur* (Paris: Pergamon Press, 1981); *Major Technological Risk*, trans. H. Ostwald (Pergamon Press, 1982).

3. Lagadec, *La Civilisation du risque*, p. 173.

4. See ibid., p. 63.

5. Ibid., p. 202.

6. To take a single domain, that of underwater capabilities, nuclear submarines, undetectable and with an almost infinite radius, can launch independent nuclear warheads from a range of 5,000 to 10,000 kilometers. A French nuclear submarine armed with M20 missiles has a total striking force equivalent to eight times that of Hiroshima. This has already been surpassed. See *L'Express*, 15-21 July, 1983, pp. 33-43.

7. Castoriadis, "Reflections," in *Philosophy, Politics, Autonomy*, p. 193.

8. See below and chapter 3.

9. On the *extension* (of time and space) that assumes total power, see Elias Canetti, *Crowds and Power*, trans. Carol Stewart (New York: Farrar Straus Giroux, 1973), pp. 281 ff.

10. See chapter 8.

11. Jean Ladrière, *Les Limitations internes des formalismes* (Louvain: Nauwelaerts, 1957), p. 68.

12. See Jacques Payen, "La production d'énergie," in *Les Techniques de la civilisation industrielle*, ed. Maurice Daumas (Paris: P.U.F., 1978-79), p. 38.

13. Definition from Paul Robert, *Dictionnaire alphabétique et analogique de la langue française*. (Paris: Société de Nouveau Littré, 1969).

14. Max Weber, *Economy and Society*, trans. A. M. Henderson and Talcott Parsons (New York: Macmillan, 1964) p. 152.

15. "If the participants accept the imputation x as superior to y in order to resolve questions of distribution, then x is superior, *dominating y*." John von Neumann and Oskar Morgenstern, *Theory of Games and Economic Behavior* (Princeton: Princeton University Press, 1944), p. 37.

16. Weber, *Economy and Society*, p. 153. This is an assertion that contemporary sociology would no longer accept, in spite of its debt to Weber. See the discussion by Niklas Luhmann, "Zweck-Herrschaft-System," in *Politische Planning*, by Niklas Luhmann (Opladen: Westdeutscher Verlag, 1971), pp. 90-112.

17. Niklas Luhmann, *Trust and Power* trans. Howard Davis, John Raffan, and Kathryn Rooney (Chichester: John Wiley, 1979), pp. 107-108, 153-54.

18. See, in particular, chapter 7.

19. Konrad Lorenz, *On Aggression*, trans. Marjorie Kerr Wilson (New York: Harcourt, Brace and World, 1966), p. 24. [English translation altered.]

20. Ibid., 174 ff.

21. George Steiner, *The Portage to San Cristobal of A. H.* (New York: Simon and Schuster, 1981).

22. Lewis Mumford, *The Pentagon of Power*. Vol. 2 of *The Myth of the Machine* (New York: Harcourt Brace Jovanovich, 1970), pp. 248-53.

23. Max Weber, *The Protestant Ethic and the Spirit of Capitalism*, trans. Talcott Parsons, (New York: Scribner, 1958), p. 13.

24. Castoriadis, "Reflections," in *Philosophy, Politics, Autonomy*, p. 184.

25. See Chapter 7.

26. Max Weber, *Le Savant et le politique*, trans. J. Freund (Paris: Plon, 1959), p. 70.

27. Ibid.

28. Ibid.

29. In this regard, see Weber's discussion wherein he himself refuses to confine himself any more to an ideological antithesis of Marxism than to a purely mechanistic search for "causes" (*The Protestant Ethic*, pp. 90-92).

30. See Ibid., p. 26.

31. Karl Löwith, *Max Weber and Karl Marx*, trans. Hans Fantel, ed. Tom Bottomore and William Outhwaite (London: Allen and Unwin), 1982, p. 41.

32. For example, Weber, *Le Savant et le politique*, pp. 85, 86.

33. Weber, *The Protestant Ethic*, p. 181.

34. Weber, *Le Savant et le politique*, p. 68.

35. See ibid., p. 70; Weber, *The Protestant Ethic*, p. 26.

36. See Weber, *The Protestant Ethic*, p. 25.

37. See Weber, *Le Savant et le politique*, p. 96.

38. Raymond Aron, *La Sociologie allemande contemporaine* (Paris: P.U.F., 1950), p. 152.

39. Ibid., p. 153.

40. Löwith, *Max Weber and Karl Marx*, p. 47.

41. Ibid, p. 48. Cf. ibid., pp. 41-53.

42. According to Löwith's expression, ibid., p. 49.

43. See chapter 7.

44. Cited by Löwith, *Max Weber and Karl Marx*, p. 53.

45. See Part III.

46. See, in particular, Aristotle, in *The Basic Works of Aristotle, Posterior Analytics*, ed. Richard McKeon (New York: Random House, 1941), p. 152.

47. Löwith, *Max Weber and Karl Marx*, p. 51.

48. Georg W. F. Hegel, *Wissenschaft der Logik* ed. Georg Lasson (Hamburg: Meiner, 1934), vol. 2, pp. 87-88.

49. Castoriadis, "Reflections," in *Politics, Philosophy, Autonomy*, p. 191.

50. Ibid., p. 186.

51. Heidegger, *Holzwege*, p. 88. "The Age of the World Picture," in *The Question concerning Technology*, trans. William Lovitt (New York: Harper and Row, 1977), p. 135.

52. Martin Heidegger, *Die Technik und die Kehre* (Pfullingen: Neske, 1962), p. 47. "The Turning," in *The Question concerning Technology*, p. 49.

2. Power in Its Effects

1. The 15th letter in Voltaire, *Lettres philosophiques*, vol. 2, ed. Gustave Lanson (Paris: Droz, 1937), p. 27.

2. *Rationem vero harum gravitatem proprietatum a phoenomenis nondum potui deducere.* Ibid., p. 40.

3. See Gilbert Simondon, *Du mode d'existence des objets techniques* (Paris: Aubier, 1969), pp. 67-68.

4. Heidegger, *Vorträge und Aufsätze*, p. 13; *The Question Concerning Technology*, p. 4.

5 Aristotle, *Metaphysics*, in *The Basic Works of Aristotle*, ed. Richard McKeon, p. 826.

6. See Part III "Detechnicizing the Future."

7. Ladrière, *Les Enjeux*, p. 71.

8. Michel Serres, *Hermès*. Vol. 3, *La Traduction* (Paris: Ed. de Minuit, 1974), p. 78.

9. See Jacques Ellul, *Le Système technicien* (Paris: Calmann-Levy, 1977). *The Technological System*, trans. Joachim Neugroschel (New York: Continuum, 1980).

10. See Wassily Leontief, "L'économie du monde en l'an 2000," in *Pour la Science*, Nov. 1980, pp. 138-53; Donella H. Meadows, "Rapport sur les limites de la croissance," in *Halte à la croissance?* (Paris: Fayard, 1972), pp. 212-13.

11. Heidegger, *Holzwege*, p. 137. *Hegel's Concept of Experience*, trans. Kenley Royce Dove (New York: Octagon, 1983), p. 62. Here I again take up several aspects of a lecture given in November 1982 at the University of Poitiers on "Phénoménologie, conscience naturelle et monde technique," which appears in *Phénoménologie et métaphysique* (Paris: P.U.F., 1984), pp. 105-24. One would do well to consult this text for a more complete and detailed commentary.

12. Heidegger, *Holzwege*, p. 137. *Hegel's Concept*, p. 62.

13. Ibid., p. 136/61.

14. Among many other passages, I cite the following: "The name 'technology' is understood here in such an essential way that its meaning coincides with the term 'completed metaphysics'. (Heidegger, *Vorträge und Aufsätze*, p. 80. "Overcoming Metaphysics," in *The End of Philosophy*, trans. Joan Stambaugh (New York: Harper and Row, 1973), p. 93.)

15. This internal naturalness in the speculative is the necessary appearance of this absolute, which manifests itself, in particular, in the role of the term N at the heart of the final syllogism of the *Encyclopedia*. See my contribution, "Savoir et nature chez Hegel," *Annales de la Faculté des Lettres de Nice*, no. 32, (Paris: Les Belles-Lettres, 1977), 49-57.

16. Heidegger, *Holzwege*, p. 137. *Hegel's Concept*, p. 62.

3. The Ruses of Technicism

1. Karl Marx, *Capital*, volume 1, trans. Ben Fowkes (New York: Vintage Books, 1976), p. 607.

2. The distance between developed and underdeveloped countries continues to expand: between 1945 and 1965, it doubled (from 1 to 40, instead of 1 to 20). In 1975, the average annual income per inhabitant of the United States surpassed $6,500; in Mali, it was about $70, for a discrepancy of 1 to 93. (See *Grand Atlas de l'histoire mondiale* [Paris: Albin Michel, 1978], p. 294).

3. Jacques Ellul, *La Technique: ou, L'Enjeu du siècle* (Paris: Colin, 1954), p. 79. *The Technological Society*, trans. John Wilkinson (New York: Knopf, 1965), p. 85. [English translation altered.]

4. Ladrière, *Les Enjeux*, p. 71.

5. Simondon, *Objets techniques*, p. 126.

6. Ladrière, *Les Enjeux*, p. 59.

7. Aristotle, *Nichomachean Ethics*, in *The Basic Works of Aristotle*, ed. Richard McKeon, p.

8. See above.

9. This is contrary to what an overly hasty reading of Fr. G. Jünger would claim to infer. The autonomization of technology has only begun: see Friedrich Georg Jünger, *Die Perfektion der Technik* (Frankfurt: Klostermann, 1953), p. 360.

10. Following Ellul, *Le Système technicien*, pp. 26-27. *The Technological System*, p. 19.

11. Frederick W. Taylor, *Scientific Management* (New York: Harper and Row, 1911), pp. 7, 13, 140.

12. Michel Ghertman, *La Prise de décision* (Paris: P.U.F., 1981), p. 346.

13. Jürgen Habermas, *Technik und Wissenschaft als "Ideologie"* (Frankfurt: Suhrkamp, 1968). "Technology and Science as 'Ideology,'" in *Toward a Rational Society: Student Protest, Science, and Politics*, trans. Jeremy J. Shapiro (Boston: Beacon Press, 1970), pp. 81-122.

14. Burrhus Frederic Skinner, *Walden Two* (Toronto: Macmillan, 1948), pp. 106, 109.

15. As we will see in relation to the "seizing of power" by technology over language. See chapter 4.

16. Ellul, *Le Système technicien*, p. 195/177.

17. It is, therefore, a constitutive characteristic of the new phase of potentialization (named "Phase IV"; see chapter 5.).

18. Arnold Gehlen, "Der Mensch und die Technik," *Die Seele im technischen Zeitalter* (Hamburg: Rowohlt, 1957), p. 7.

19. Ibid., p. 8.

20. Ibid., p. 9.

21. See ibid., pp. 13–17.

22. Ibid., p. 21.

23. See ibid., pp. 20–21.

24. As Gehlen did: ibid., p. 11.

25. Ibid., p. 12.

26. We will return to a more extensive treatment of this subject in chapter 5.

27. Alexandre Koyré, *Études d'histoire de la pensée scientifique* (Paris: P.U.F., 1966), p. 189.

28. According to Ellul's expression, *La Technique*, p. 82 ff/85 ff.

29. Marx, *Randglossen zum Programm der Deutschen Arbeiterpartei* (Berlin: Vereinigung Internationaler, 1922), p. 22.

30. See Kostas Axelos, *Marx, penseur de la technique* (Paris: Ed. de Minuit, 1961).

31. Cornelius Castoriadis, *Les Carrefours du labyrinthe* (Paris: Seuil, 1978), p. 227. *Crossroads in the Labyrinth*, trans. Kate Soper and Martin H. Ryle, (Cambridge: MIT Press, 1984), p. 235.

32. Karl Marx, *Economic and Philosophic Manuscripts of 1844*, trans. Martin Milligan (Moscow: Foreign Language Publishing House, 1961), p. 39.

33. Marx, *Capital*, 1:493 n.4.

34. See ibid., 1:444 n.7.

35. Ibid.

36. See ibid., 1:493 n.4.

37. See Castoriadis, *Crossroads*, p. 242.

38. Cited by Castoriadis, ibid., p. 241.

39. Karl Marx, "Principles of a Critique of Political Economy,"

40. The "production of new needs is the first historical act" (Karl Marx and Friedrich Engels, *The German Ideology*, 8th ed., ed. R. Pascal (New York: International Press, 1960), p. 17).

41. Karl Marx, "General Introduction to the Critique of Political Economy," in *A Contribution to the Critique of Political Economy*, trans. N. I. Stone (Chicago: Charles H. Kerr, 1904), pp. 306–307.

42. See Marx, *Capital*, 1:503.

43. Ibid., 506.

44. The skill of the worker is rendered useless in mechanical work and "vanishes as an infinitesimal quantity in the face of science, the gigantic natural forces, and the mass of social labour embodied in the system of machinery which, together with those three forces, constitutes the power of the 'master'." Ibid., 549.

45. Ibid., p. 526.

46. Ibid., p. 546.

47. See ibid., pp. 527–28.

48. See ibid., p. 562.

49. Karl Marx, *Grundrisse: Foundations of the Critique of Political Economy*, trans. Martin Nicolaus (Harmondsworth: Penguin, 1973).

50. Radovan Richta, *Civilization at the Crossroads*, trans. Marian Slingova (White Plains, N.Y.: International Arts and Sciences Press, 1969).

51. Ibid., p. 286.

52. Ibid., p. 28.

53. Ibid., p. 26.

54. Ibid., p. 87.

55. See ibid., pp. 30, 90.

56. See ibid. A chart of the concentration of sulfuric anhydride in industrial zones is not enough to make up this deficiency.

57. See ibid., 208-209.

58. Ibid., p. 209.

59. Benjamin Coriat, *Science, technique et capital* (Paris: Seuil, 1976).

60. Ibid., p. 29.

61. See ibid., pp. 44-45.

62. Coriat is not far from recognizing this. Ibid., p. 60.

63. Ibid., p. 51.

64. We find in Coriat an example, among others. Ibid., p. 50.

65. Ibid., p. 176.

66. Ibid., pp. 137-39.

4. Jonas in the System

1. Francis Bacon, *Essays, Advancement of Learning, New Atlantis and Other Pieces*, ed. Richard F. Jones (New York: Doubleday, Doran 1937), p. 464.

2. Marx, *Capital*, 1:544.

3. Marx and Engels, *The German Ideology*, p. 24.

4. Ibid.

5. Ibid.

6. See primarily Talcott Parsons, *The System of Modern Societies* (Englewood Cliffs, N.J.: Prentice-Hall, 1971).

7. Ellul, *Le Système technicien*, p. 90. *The Technological System*, p. 79.

8. Ibid., p. 91. n.2/79 n.4.

9. Ibid., p. 88/77.

10. Ibid., p. 100/91.

11. This is to a large extent the case with Jean-Claude Beaune, who, while accepting the "principle of totality," applies it in a solely *functional* manner. See "Variations sur la définition du terme *Technologie*," *Les Études Philosophiques*, no. 2 (1976): 195. Cf. the same author, *La Technologie*, coll. "Dossiers," (Paris: P.U.F., 1972).

12. Jacques Ellul, "La technique considérée en tant que système," *Les Études philosophiques*, no. 2 (1976): p. 148.

13. Ibid., p. 151.

14. Philippe Roqueplo, *Penser la technique* (Paris: Seuil, 1983), pp. 15, 28, 127, 99.

15. Ibid., pp. 27, 128.

16. Ibid., p. 34.

17. Ibid., p. 73.

18. Ibid., pp. 129, 144.

19. Ibid., p. 147. We respect the typography of the author.

20. Ibid., p. 118.

21. See chapter 3.

22. See Part III.

23. Ellul, *Le Système technicien*, p. 151/138. Cf. Ellul, *La Technique*, p. 120–34/133–47.

24. As Mumford has noted (see Ellul, *Le Système technicien*, p. 163 n. 2/150 n. 22.

25. Ibid., pp. 158–165/145–150; also, *La technique*, pp. 107–20/116–33.

26. Ellul, *Le Système technicien*, p. 173/158; *La Technique*, pp. 88–102/94–111.

27. The phenomenon is not known to be limited to the case of E. Morin, on whom Ellul concentrated perhaps too much of his wrath (*Le Système technicien*, pp. 219–24/199–204.

28. Ibid., p. 229/209.

29. Ellul, *La Technique*, p. 83/89.

30. Ellul, *Le Système technicien*, p. 231/211. (Emphasis added.)

31. Ibid., p. 232/212.

32. Ibid., p. 244/224.

33. Ibid., p. 256/235.

34. Ibid., p. 259/238.

35. The formulation in the first book was clearer. See *La Technique*, p. 75 ff/79 ff,. and particularly, page 84: "There is, thus, no freedom of choice." [Translation modified.]

36. Ellul, *Le Système technicien*, p. 265/243.

37. Ibid.

38. Ibid., p. 99/107. (My emphasis.)

39. Ibid., "La technique considérée en tant que système," p. 166. "Technology Considered as System," p. 183. [Translation modified.]

40. Ibid., p. 165/182. *Le Système Technicien*, p. 361/325.

41. Ellul, *La Technique*, p. 165/182.

42. Ibid.

43. Ibid.

44. Ellul, *Le Système technicien*, p. 326/296.

45. Ellul, *La Technique*, p. 389/428. [Translation modified.]

46. See note concerning Illich, in Ellul, *Le Système technicien*, p. 328/297 n. 13.

47. See Ellul, *La Technique*, pp. 107–120/116–33; *Le Système technicien*, pp. 186–218/167–98.

48. Cited by Ellul, *Le Système technicien*, p. 194 n. 1/176 n. 6.

49. See Simondon, *Objets techniques*, p. 93.

A first version of the following pages appeared in *Man and World*, no. 16, (1983): 349–64, under the title, "La Technique et le langage."

50. Ellul, *La Technique*, p. 120/132. [Translation modified.]

51. We are more uncertain than Ellul about this point (cf. *La technique*, pp. 119–20/131–32).

52. Simon Nora and Alain Minc, *L'Informatisation de la société* (Paris: La Documentation française, 1978), p. 26.

53. Ibid., p. 11.

54. Ibid., pp. 51, 53.

55. See ibid., p. 120.

56. Nora and Minc are content to allow McLuhan's discourse to blend with their own: the "information marketplace" on page 124 refers to the "global village."

57. Nora and Minc, *L'Informatisation de la société*, p. 117.

58. Ibid., p. 15.

59. Ibid., p. 61.

60. Ibid., p. 60.

61. See Ellul, *La Technique*, p. 115/126. [Translation modified]; and *Le Système technicien*, p. 208/189: "The truth is that the universality of the technological system *causes the rupture* of the human world for a long time, and not its unification."

62. "M. McLuhan Denounces the Action of the Media," *Le Monde*, 16 December 1978.

63. See Simondon, *Objets techniques*, pp. 19–23 and passim.

64. Jean Baudrillard, *Le Système des objets* (Paris: Gallimard, 1968), p. 16.

65. It is important not to confuse this. The intervention of informational languages is evidently a very recent stage of technicalization, whereas mathematical language has, for three centuries, forged the new conceptual instruments of modernity. P. Dubarle, however, points out a considerable change: "Whereas one could previously say that every mathematical object was only of *a certain type* of intellectual theme proposed to rationality among others, it clearly appears nowadays that, for man, mathematical language has become an inescapable instrument for *all* rationality: any claim to rationality must include a mathematical language." (Dominique Dubarle and André Doz, *Logique et dialectique* [Paris: Larousse, 1972], p. 59.)

66. Simondon, *Objets techniques*, p. 138.

67. Baudrillard, *Le Système des objets*, p. 270. (My emphasis.)

68. Ellul, *Le Système technicien, p. 195/177.*

69. See Baudrillard, *Le Système des objets*, p. 76.

70. Nobert Wiener, *Cybernetics and Society* (Garden City, N.Y.: Doubleday, 1954), p. 16.

71. See, among others, Heidegger, *Der Satz vom Grund*, pp. 200-203/122–24; interview granted to *Der Spiegel*, no. 23 (1976): 212. Translated by William J. Richardson in *Heidegger, the Man and the Thinker*, ed. Thomas Sheehan (Chicago: Precedent, 1981), p. 59.

72. Wiener, *Cybernetics and Society*, pp. 17–18.

73. Heidegger, *Der Satz vom Grund*, p. 203/124.

74. Wiener, *Cybernetics and Society*, pp. 74–75.

75. Maurice Merleau-Ponty, *La Prose du monde* (Paris: Gallimard, 1969], p. 7. *The Prose of the World*, trans. John O'Neill, ed. Claude Lefort (Evanston: Northwestern University Press, 1973), p. 3.

76. On the limits of this practice and of artificial intelligence in general, see the excellent work of Hubert L. Dreyfus, *What Computers Can't Do* (New York: Harper and Row, 1972).

77. *Der Spiegel*, p. 209. Sheehan (ed.), *Heidegger*, p. 57.

78. This is a work that takes a prospective and normative view. Dennis Gabor, *Innovations: Scientific, Technological, and Social* (London: Oxford University Press, 1970).

79. See the examples given by Gilbert Hottois, *Le Signe et la technique* (Paris: Aubier, 1984), pp. 90–91, 95–98.

80. On this point, I am in complete agreement with Hottois, but I hesitate to recognize an "obscure transcendence," a beautiful, but too Manichean, expression (Ibid., pp. 152–64).

81. See Ellul, *Le Système technicien*, p. 231/211. Cf. Dennis Gabor, *Inventing the Future* (Harmondsworth: Penguin, 1964), pp. 140, 162. Oppenheimer explains this rule with significant candor: "When one perceives a thing which appears technically sweet, one seizes it, realizes it and questions it; one will know what to make of it only later, after one has achieved technological success." (Cited by Robert Jungk, *Brighter Than a Thousand Suns* [New York: Harcourt Brace, 1958], p. 296.)

82. See Hottois, *Le Signe*, p. 171. Cf. Hans Jonas, *The Imperative of Responsibility: In Search of an Ethics for the Technological Age*, trans. Hans Jonas, with the collaboration of David Herr (Chicago: University Press of Chicago, 1984), p. 11.

5. The Phases of Potentialization

1. Aristotle, *Metaphysics*, Θ, 8, 1050a.

2. See André Lalande, *Vocabulaire technique et critique de la philosophie*, 7th ed., rev. and enl. (Paris: P.U.F., 1956), p. 1212.

3. *Omne autem quod recipit aliquid ab alio, est in potentia respectu illius*: Saint Thomas Aquinas, *De ente et essentia*, trad. C. Capelle (Paris: Vrin, 1956), p. 59.

4. Otherwise said, it is my critical perspective that discovers potentializations and measures their scope, but this does not imply that my point of view is subjective or that it claims to interpret an absolute metaphysical truth (in the Hegelian sense).

5. In what I call Phase II.

6. Ladrière, *Les Enjeux*, p. 71.

7. But which testifies to the homology, realized in and by techno-science, between the scientific operation and the technological operation. Jean Ladrière formalizes the five characteristics in *Les Enjeux*, pp. 39-43, 64-65.

8. See below.

9. Thus, if Marxism is so weak in prehistoric or ethnological anthropology, it is because it has no conceptual instrument truly appropriate to these fields of study (throughout the centuries, pre-capitalistic humanity is assumed to have struggled in processes as problematic as the "mode of Asiatic production").

10. One sees that I understand the notion of phase in a sense different from that of Simondon (cf. "La notion de phase appliquée au devenir: la technicité comme phase," *Objets techniques*, pp. 159-62 and passim). Certainly, phase for Simondon is an "aspect" that is a "phase only in relation to others," each phase having a genesis that is not subject to something like dialectical necessity, and finally, "no phase as such possesses complete truth or reality." All of these characteristics allow for a rapprochement. Nevertheless, Simondon assumes a system of phases "in equilibrium with its neutral point"; effectively, to confine ourselves to this example, science is—according to him—the mediation between technology and religion. I do not seek, in my endeavor, such an inclusive theory. Even if I can concede to Simondon that technicity results in a "dephasing" of magical thought, I have not made a decision here as to the pertinence of such a concept. We situate ourselves at the *interior* of the sphere where the notion of the effect of power can make sense. The phenomenology of the effects of power are, therefore, restricted to the recognition of the minimal unity of a mode of potentialization of type I, which to a great degree belongs to Simondon's "technicity," but within immanence and without the generalization of an ontological content.

11. At the interior of the limits that I believe must be assigned to phenomenology.

12. To a certain degree, I will compensate for this brevity by a return to explicative developments, described in other sections of this text.

13. See Marshall Sahlins, *Stone Age Economics* (Chicago: Aldine-Atherton, 1972).

14. See Galbraith on the eminent value of the earth at the first (substantialist) stage of the political economy: John Kenneth Galbraith, *The New Industrial State* 2d ed., rev. (Boston: Houghton-Mifflin, 1971), pp. 50-53.

15. Marcel Detienne and Jean-Pierre Vernant have collected many convincing arguments in *Les Ruses de l'intelligence. La Mètis des Grecs* (Paris: Flammarion, 1974). *Cunning Intelligence in Greek Culture and Society*, trans. Janet Lloyd (Atlantic Highlands, N.J.: Humanities Press, 1978).

16. See Claude Lévi-Strauss, *La Pensée sauvage* (Paris: Plon, 1962). *The Savage Mind* (Chicago: University of Chicago Press, [1963], 1973).

17. See above.

18. In the pages that follow, I give an illustration of this phase from an analysis of the philosophical foundations of Euclid's *Elements*.

19. Although I do not agree with the entirety of his thesis, I refer here to Reiner Schürmann, *Heidegger on Being and Acting: From Principles to Anarchy*, translated from the French by Christine-Marie Gros in collaboration with the author (Bloomington: Indiana University Press, 1987).

20. We will return to this in Part III.

21. See Hubert L. Dreyfus, "De la techné à la technique: le statut ambigue de l'ustensilité dans *L'Etre et le Temps*," in *Martin Heidegger*, dir. Michel Haar (Paris: L'Herne, 1983), pp. 292-301.

22. See Simondon, *Objets techniques*, pp. 65-67 and passim.

23. See Maurice Daumas (ed.), *Histoire générale des techniques*, 5 vols. (Paris, P.U.F., 1968) 3:x.

24. See ibid., 4:349.

25. Ibid., p. 582.

26. Ibid., p. 352.

27. Ibid., p. 366.

28. Ibid., p. 493. (My emphasis.)

29. See Jungk, *Brighter Than a Thousand Suns*, p. 194.

30. There are many examples of this: there is no other explanation for the stimulation of "creativity" in the United States. It is well-known, moreover, how the aforementioned creativity of the "great intellectual" hippies was utilized by electronics companies.

31. One can ask whether it is due to Phase IV itself, or to the fact that it is not completely imposed.

32. Lévi-Strauss, *La Pensée sauvage*, p. 357. *The Savage Mind*, p. 269.

33. Heidegger, *Sein und Zeit*, Section 15, p. 69; Cf. ibid., Section 22, p. 102. *Being and Time*, p. 98, p. 136.

34. See Daumas (ed.), *Histoire générale des techniques*, 2:xvi.

35. See ibid., 4:viii.

36. Ibid., 3:xix; 4:322.

37. Ibid., 4:314.

38. Ibid., p. 353.

39. Ibid., p. 42.

40. Ibid., 3:xix.

41. Ibid., p. 39.

42. René Taton (ed.), *Histoire générale des sciences.* Vol. 2, *La science moderne (de 1450 à 1800)* (Paris: P.U.F., 1958), p. 515.

43. Ibid., pp. 517-18.

44. See Daumas (ed.), *Histoire générale des techniques*, 3:40.

45. Pointed out by Daumas, concerning the increase of naval traffic at the end of the 18th century, ibid., 3:xi.

46. Doing this, I do not ignore that I come to the end of a tradition that he shapes. But my project is not historical: it is phenomenological and genealogical.

47. See Euclid, *Extraits des Éléments*, ed. Charles Mugler (Paris: Gauthier-Villars, 1967), pp. 13-14.

48. On the "tyranny" of Euclid, see Jean-Jacques Salomon, *Science et politique* (Paris: Seuil, 1970), p. 133.

49. Husserl, *The Origin of Geometry*, in *The Crisis of European Sciences*, p. 357.

50. Ibid., p. 375.

51. Ibid., p. 378.

52. See ibid., pp. 360-61.

53. Husserl, *The Crisis of European Sciences*, p. 340.

54. Husserl, *The Origin of Geometry*, p. 368.

55. Euclid, *Elements*, I.

56. G. W. F. Hegel, *Encyclopedia*, Section 256 (New York: Philosophical Library, 1959).

57. Euclid, *Elements*, definition 7.

58. Descartes, *Discours de la méthode*, ed. E. Gilson (Paris: Vrin, 1947), pp. 17-18. *Discourse on Method*, vol. 1, trans. Elizabeth S. Haldane and G. R. T. Ross, (Cambridge University Press, 1911 (1981)), p. 90.

59. Certain translations are frankly obscure; hence, that of Georges Kayas (Paris: Editions du C. N. R. S., 1978): "A straight line is a line whose extension between any of its points is equal to the distance between these points," vol. 1, p. 1.

60. Euclid, *Elements*, ed. Mugler, p. 31.

61. Plato, *Gorgias*, trans. Terence Irwin (New York: Oxford University Press, 1979), in *The Basic Works*, ed. McKeon, 473a.

62. Aristotle, *Metaphysics*, Δ, 5, 1015b; cf. also Γ, 4, 1007a.

63. According to him, the postulate "All right angles are equal among themselves" is not necessary.

64. Aristotle, *Posterior Analytics*, in *The Basic Works*, ed. McKeon, A, 72b.

65. See Koyré, *Études d'histoire*, p. 180.

66. Ibid., p. 64

67. Ibid., p. 72.

68. *The Works of Francis Bacon*, 14 vols., ed. James Spedding, Robert Ellis, and Douglas Heath (Stuttgart, Frommann: 1963 [reprint of 1857-1874]), 1:159, aphorism III: "*Scientia et potentia humana in idem coincidunt . . .* "

69. Ibid., p. 36. See also ibid., p. 154: "*opere naturam vincere.*"

70. "Nature is conquered only when one obeys it," aphorism III. On this metaphor see ("The chaste matron rules man in obeying him"), ibid., p. 158.

71. Aphorism III: "*quod in contemplatione instar causae est, id in operatione instar regulae est.*"

72. *Melior et certior intellectus adoperatio,* aphorism XVIII.

73. *Homo, Naturae minister et interpres . . .* , aphorism I.

74. *Itaque ordo quoque demonstrandi plane invertitur,* ibid., p. 136. (The reason why the order of demonstration is completely inverted . . .)

75. René Descartes to Mersenne, 9 October 1638, in *Oeuvres*, ed. Charles E. Adam and Paul Tannery (Paris: Joseph Gilbert, 1950).

76. Koyré, *Études d'histoire*, p. 152. On this perspective, see Heidegger, *Die Frage nach dem Ding* (Tubingen: Niemeyer, 1962), p. 61. *What Is a Thing?* trans. W. B. Barton, Jr. and Vera Deutsch (South Bend, Ind.: Gateway Editions, 1967) p. 54.

77. Husserl, *The Crisis of European Sciences*, p. 33.

78. See the Galilean chronology given by Maurice Clavelin, *La Philosophie naturelle de Galilée* (Paris: Armand Colin, 1968), pp. 9-15.

79. Galileo Galilei, *Dialogues concerning Two New Sciences*, trans. Henry Crew and Alfonso de Salvio (Evanston: Northwestern University, 1946), p. 235.

80. Ibid., p. 238.

81. Ibid., p. 265.

82. Ibid.

83. Maurice Clavelin, Introduction to *Discours et démonstrations mathématiques concernant deux sciences nouvelles*, by Galileo trans. Maurice Clavelin, (Paris: Armand Colin, 1970), p. xxiii. (My emphasis.)

84. Ibid., p. xxii.

85. Descartes, *Oeuvres*, ed. Adam and Tannery, vol. 10, p. 362. *Rules for the Direction of the Mind*, trans. Elizabeth Haldane and G. R. T. Ross, (Cambridge U. P., 1911 (1981)), I, p. 3: "Only those objects should engage our attention, to the sure and indubitable knowledge of which our mental powers seem to be adequate."

86. Aristotle, *Metaphysics*, α in *The Basic Works*, ed. McKeon, (The Hague: Nijhoff, 1977), 993a-b.

87. Descartes, *Règles utiles pour la direction de l'esprit*, p. 6. *Rules for the Direction*, p. 5.

88. See A. Ernout and A. Meillet, *Dictionnaire étymologique de la langue latine* (Paris: Klincksieck, 1959), p. 663.

89. Descartes, *Règles utiles*, p. 13. *Rules for the Direction*, p. 11.

90. See Simondon, *Objets techniques, pp. 127-29.*

91. See Joseph Schumpeter, *Capitalism, Socialism and Democracy* (London: Allen and Unwin, 1961), p. 134. Cf. Jon Elster, *Explaining Technical Change* (Cambridge: Cambridge University Press 1983), p. 126.

92. According to him, the two other factors of this qualitative change are chronological acceleration and the considerable extension of the range of social effects of technological progress: See Gabor, *Innovations: Scientific, Technological and Social*, pp. 6-9.

93. Robin Clarke, *La Course à la mort* (Paris: Seuil, 1972), p. 54.

94. Heidegger, *Holzwege*, p. 71. "World Picture," in *Question concerning Technology*, p. 138.

95. See Jean-Jacques Salomon, "Recherche," in *Encyclopaedia Universalis* (Paris: Encyclopaedia Universalis France).

96. Salomon, *Science et politique*, p. 135.

97. René Thom, "La science en crise?" *Le Débat*, no. 18 (1982): p. 38.

98. Nicholas Rescher, *Scientific Progress. A Philosophical Essay on the Economics of Research in Natural Science* (Oxford: Blackwell, 1978), p. 80. I thank Daniel Schulthess for having drawn my attention to this book in his excellent article, "Les limites à l'avancement de la science selon Nicholas Rescher," in *Réflexions en hommage à Philippe Muller* by Michel Schaffter et al. (Neuchatel: Messeiller, 1981), pp 63-84.

99. See Salomon, *Science et politique*, p. 145 ff.

100. Ibid., p. 149.

101. Ibid., p. 147.

102. See "Entretien avec Pierre Aigrain," *Le Courrier du C.N.R.S.*, no. 21 (July 1976): 5.

103. Salomon, *Science et politique*, p. 163. Salomon cites the American specialists, Charles Hitch and Roland McKean (*The Economics of Defense in the Nuclear Age* (Cambridge: Harvard University Press, 1960).

104. Placing mathematics on the side of philosophy, Aigrain sought a symbiosis among fundamental research, pure technology, and a professionalized university: a typical techno-scientific approach (see the interview cited).

105. Yet, this operativity is supposed to be marvelously efficient: witness the unemployment in the east and the labor camps of the former U.S.S.R.

106. See Part III.

107. Martin Heidegger, *Was heisst Denken?* (Tübingen: Niemeyer, 1954), p. 4. *What Is Called Thinking?* trans. J. Glenn Gray (New York: Harper and Row, 1968), p.8

108. Thom, "La Science en crise?" cited 39.

109. Salomon, *Science et politique*, p. 268.

6. Science between Power and Alliance

1. Ilya Prigogine and Isabelle Stengers, *La Nouvelle Alliance: Métamorphose de la science* (Paris: Gallimard, 1979), p. 10.

2. Michel Serres, *Hermès*, p. 104.

3. Edgar Morin, *La Méthode*. Vol. 1, *La Nature de la nature* (Paris: Seuil, 1977), p. 16.

4. Prigogine and Stengers, *La Nouvelle Alliance*, p. 18.

5. Jean Dieudonné, in *Les Grands Courants de la pensée mathématique*, 2d ed., enl. by François Le Lionnais (Paris: Blanchard, 1962), p. 544.

6. Ibid.

7. According to the expression of Jean Ladrière; see his *Les Limitations Internes des Formalismes* his book with this title (Louvain: Nauwelaerts, 1957).

8. Benoit Mandelbrot, *Les Objets fractals* (Paris: Flammarion, 1975), p. 170.

9. Cf. René Thom, *Modèles mathématiques de la morphogenèse* (Paris: U. G. E., 1974).

10. See in particular R. Viallard, in Daumas (ed.), *Histoire de la science*, p. 979.

11. Ibid.

12. Werner Heisenberg, *The Physicist's Conception of Nature*, trans. Arnold J. Pomerans (New York: Harcourt Brace, 1958), p. 39.

13. Ibid., p. 38.

14. The authors of *La Nouvelle Alliance* (p. 232) were commenting on the theory of complementarity proposed by Niels Bohr.

15. Prigogine and Stengers, *La Nouvelle Alliance*, p. 218.

16. Ibid., p. 16.

17. Jean-François Lyotard, *La Condition postmoderne* (Paris: Ed. de Minuit, 1979), p. 7. *The Postmodern Condition*, trans. Geoff Bennington and Brian Massumi (Minneapolis: University of Minnesota Press, 1984), p. xxiii.

18. Ibid.

19. Ibid.

20. Ibid., p. 108/67.

21. Ibid., p. 90/55. (My emphasis.)

22. Prigogine and Stengers, *La Nouvelle Alliance*, p. 15.

23. Ibid., p. 296.

24. Ibid., p. 17.

25. Raymond Ruyer, *La Gnose de Princeton: Des savants à la recherche d'une réligion* (Paris: Fayard, 1974).

26. Morin, *La Méthode*, 1:16.

27. Ibid., p. 42.

28. Jacques Monod, *Le Hasard et la nécessité* (Paris: Seuil, 1970), pp. 194–95.

29. Prigogine and Stengers, *La Nouvelle Alliance*, p. 14.

30. See ibid., p. 16.

31. See Monod, *Le Hasard et la nécessité*, p. 185.

32. See Morin's phrase cited in the exergue (*La Méthode*, p. 16).

33. Thom, *Modèles mathématiques*, p. 20.

34. See Albert Einstein, *Relativity*, trans. Robert Lawson (New York: Crown, 1961), p. 44.

35. See Heisenberg, *The Physicist's Conception*, p. 41.

36. Recall that, since 1874, Shanks calculated π with 707 exact decimals!

37. Immanuel Kant, *Prolegomena to Any Future Metaphysics*, trans. Lewis White Beck (New York: Macmillan, 1950 [1989]), p. 118.

38. Prigogine and Stengers, *La Nouvelle Alliance*, p. 11.

39. See Koyré, "Galilée et le révolution scientifique" and "Galilée et l'experience de Pise: à propos d'une légende," in *Études d'histoire de la pensée scientifique*, pp. 179–80, 192–201.

40. Prigogine and Stengers, *La Nouvelle Alliance*, p. 14.

41. Ibid., p. 64. On this page, Prigogine and Stengers are rather modest: "This is, to be sure, only a possible."

42. Ibid., p. 38.

43. Serres, *Hermès*, p. 78.

44. The first definition given by Lalande, *Vocabulaire technique et critique*, p. 960.

45. Serres, *Hermès*, p. 79.

46. See chapter 5.

47. At least in "The Thanatocracy." It would be another task to determine the role that the idea of the New Science plays in the rest of Serres's work: an inspiring role, no doubt, but one that has the merit of stimulating the poetic complicities discovered by Hermès in the contemporary world, rather than contributing to optimizing the performances of total Administration.

48. Prigogine and Stengers, *La Nouvelle Alliance*, p. 296.

49. See above.

50. Serres, *Hermès*, p. 100.

51. Ibid., p. 84.

52. *"Neu zu sein gehört zur Welt, die zum Bild geworden."* ["To be new is peculiar to the world that has become picture."] (Heidegger, *Holzwege*, p. 85. "World Picture," in *Question concerning Technology*, p. 132.)

53. See Ladrière, *Les Enjeux*, pp. 39-40, 64-65.

7. From Absolute Power to Contiguity

1. See chapter 4.

2. The most accessible documents for me on these complex givens were found in Daumas, *Histoire des sciences et des techniques*, as well as in *Histoire des Techniques*, dir. Bertrand Gille (Paris: Gallimard, Encycl. de la Pléiade, 1978).

3. See chapter 5.

4. Listen for one of the voices in the dialogue, chapter 8.

5. In this regard, see Michel Haar's interesting article on the "Hegelianism" of Heidegger's thinking, "Structure hégélienne dans la pensée heideggérienne de l'Histoire," *Révue de métaphysique et de morale* 85, January-March 1980, pp. 48-59.

6. For example, *Phänomenologie des Geistes* (Hamburg: Meiner, 1952), pp. 29-30; *Die Vernunft in der Geschichte* (Hamburg: Meiner, 1955), p. 113.

7. Hegel, *Werke*, 17, p. 535.

8. See the end of the *Encyclopedia*, and that of the *Wissenschaft der Logik*, ed. Georg Lasson (Hamburg: Meiner, 1934), where the idea of the absolute is *unvergängliches Leben*, p. 484.

9. The idea of philosophy is eternal: see paragraph 577 of the *Encyclopedia*. Already the State, the element of constancy and stability in human relations, is the "rational in and for itself," and its idea is "the eternal and necessary being in and for itself." (*Philosophy of Right*, Section 258; my emphasis.)

10. Hegel, *Phänomenologie des Geistes*, p. 29.

11. Hegel, *Wissenschaft der Logik*, pp. 486-87.

12. Hegel, *Phänomenologie des Geistes*, p. 15.

13. Plato, *Phaedrus*, 2d corr. ed., trans. C. J. Rowe (Warminster, Wiltshire, England: Aris and Phillips, 1988), 247b.

14. Martin Heidegger, *Identität und Differenz* (Pfullingen: Neske, 1957), p. 70; *Identity and Difference*, trans. Joan Stambaugh (New York: Harper and Row, 1969).

15. "But nothing religious is ever destroyed by logic; it is destroyed only by the god's withdrawal." (Heidegger, *Was heisst Denken?* p. 7.; *What Is Called Thinking?* p. 10.)

16. Hegel, *Phänomenologie des Geistes*, p. 563.

17. See Janicaud, *A nouveau la philosophie*, pp. 109-20.

18. "Die Vernunft als die Rose im Kreuze der Gegenwart zu erkennen . . . " Hegel, *Grundlinien der Philosophie des Rechts* (Vorrede: Hoffmeister 1955), p. 16.

19. As, among others, Lyotard's *The Postmodern Condition* clearly shows, but without sufficiently recognizing his debt to Hegel in this matter.

20. Max Horkheimer and Theodor W. Adorno, *Dialektik der Aufklärung* (Amsterdam: Querido, 1947). *Dialectic of Enlightenment*, trans. John Cumming (New York: Seabury Press, 1944 [1969]).

21. I give a (strange) example of the exploitation of this source of inspiration. In *Ulysses and the Sirens* Jon Elster again takes up the famous passage of the *Odyssey*, but fails to make the slightest reference to the *Dialectic of Enlightenment*, for which this episode is the very "allegory" of the dialectic of the rational.

22. Horkheimer and Adorno, *Dialektik*, p. 16, but page 37 specifies it "as every system."

23. Cf. ibid., p. 108, and Jacques Lacan, "Kant avec Sade," *Écrits*, vol. 2 [Paris: Seuil, 1971], pp. 119–48.

24. Horkheimer and Adorno, *Dialektik*, p. 14.

25. Charles Mugler, *Les Origines de la science grecque chez Homère; L'Homme et l'univers physique* (Paris: Klincksieck, 1963).

26. Horkheimer and Adorno, *Dialektik*, p. 48.

27. Ibid., p. 18.

28. Ibid., p. 20.

29. Habermas, "Technology and Science," in *Toward a Rational Society*, p. 118–19.

30. I note that Habermas's thought is now better known in France, thanks to a rather sophisticated body of translation and to some solid contributions, for example: J. Rivelaygue, "Habermas et le maintien de la philosophie," *Archives de philosophie*, April–June. 1982, pp. 257–98; Y. Michaud, "Habermas ou la raison decidée," *Critique*, July 1976, pp. 643–63; Paul-Laurent Assoun and Gerard Raulet, *Marxisme et théorie critique* (Paris: Payot, 1978).

31. See J. Rivelaygue, op. cit. p. 258.

32. See Habermas, "Technology and Science," in *Toward a Rational Society*, p. 104.

33. See Jean-René Ladmiral, Preface to Habermas, *La Technique*, (Paris: Gallimard, 1973), p. xliii.

34. See Jürgen Habermas, *Legitimationsprobleme im Spatkapitalismus* (Frankfurt: Suhrkamp, 1973).

35. See Jürgen Habermas, *Erkenntnis und Interesse* (Frankfurt: Suhrkamp, 1968), p. 244.

36. Cf. my first presentation of "rationalization" according to Weber, chapter 1.

37. Weber, *Le Savant et le politique*, p. 70.

38. Löwith, *Max Weber and Karl Marx*, p. 42.

39. Jürgen Habermas, 2 vols. *Theorie des kommunikativen Handelns* (Frankfurt: Suhrkamp, 1981), 2:449.

40. However, Habermas has himself overevaluated *Zweckrationalität* in understanding technology, in agreeing with Gehlen too quickly on this point (while opposing him elsewhere: see "Technology and Science," in *Toward a Rational Society*, p. 13).

41. Habermas, *Theorie*, 2:23–68.

42. Ibid., 2:453.

43. See ibid., 1:252.

44. See Jürgen Habermas, "Aspects of the Rationality of Action," in Geraets (ed.), *Rationality To-day*, pp. 185–204.

45. Cf. Parsons, *The System of Modern Societies*, p. 6.

46. See Habermas, *Theorie*, 1:226–227.

47. Ibid., 1:228–33.

48. Ibid., 1:234.

49. See ibid., 2:230.

50. Ibid., 2:470.

51. Ibid., 2:471.

52. See ibid., 2:454.

53. See ibid.

54. See ibid., 2:456, 457.

55. See ibid., 2:461–62.

56. Habermas, *Theorie*, vol. 2, *Zur Kritik funktionalistischen Vernunft*. Note that Habermas does not critique *functional* reason, but the ideology which exploits its ends (functionalism).

57. Ibid., vol. 2, p. 549.

58. Ibid., 2:550.

59. See Hans-Georg Gadamer, "Replik," in Karl-Otto Apel, *Hermeneutik und Ideologiekritik* (Frankfurt: Suhrkamp, 1971).

60. See Rivelaygue, "Habermas," p. 298.

61. Rüdiger Bubner, *Modern German Philosophy*, trans. Eric Matthews (Cambridge: Cambridge University Press, 1981), p. 188.

62. Weber, *The Protestant Ethic*, p. 24.

63. Cf. Löwith, *Max Weber and Karl Marx*, p. 41.

64. Martin Heidegger, *Vorträge und Aufsätze*, p.13/3.

65. Ibid., p. 44/35.

66. Ibid., p. 14/4.

67. Ibid., p. 13/3.

68. Please excuse me for being schematic. Apart from the fact that this interpretation is apparently well-known and, thus, it is not a question of taking it up in detail, this slightly abrupt presentation at least obliges me not to delude about the difficulty.

69. Heidegger, *Vorträge und Aufsätze*, p. 21/13.

70. Ibid., p. 22/20.

71. Ibid., p. 28/20.

72. Ibid., p. 41/33.

73. Heidegger, *Die Technik und die Kehre*, p. 45/47.

74. Martin Heidegger, *Vom Wesen der Wahrheit* (Frankfurt: Klostermann, 1954), p. 25; *On the Essence of Truth*, trans. John Sallis, in *Basic Writings*, ed. David Farrell Krell (New York: Harper and Row, 1977), p. 140.

75. Heidegger, *Vorträge und Aufsätze*, p. 33/25.

76. Ibid.

77. Ibid., p. 22/14.

78. Heidegger, *Die Frage nach dem Ding*, p. 78–80; *What Is a Thing?* pp. 101–102.

79. Ibid., pp. 59–62, 66–68/76–78, 85–88.

80. Heidegger, *Vorträge und Aufsätze*, p. 28/20.

81. The theory-practice distinction is no longer fundamental to understanding techno-science. It is from techno-science that one must understand it, not the inverse.

82. Marx, *Capital*, 1:512, 26.

83. In the course of a seminar held by J. P. Larthomas and me at C. U. M. of Nice, in 1967.

84. Heidegger, *Die Technik und die Kehre*, p. 47/49.

85. Heidegger, *Vorträge und Aufsätze*, pp. 35–44/27–35.

86. Ibid., p. 41/33.

87. See P. Bourdieu, "L'Ontologie politique de Heidegger," *Actes de la Recherche en Sciences Sociales*, Nov. 1975, p. 152. *The Political Ontology of Heidegger* (Stanford: Stanford University Press, 1991), p. 52.

88. Heidegger, *Vorträge und Aufsätze*, pp. 35–36/28.

89. Ibid., p. 36/28. (My emphasis.)

90. On the exchanges of these gazes, see ibid., pp. 36, 41/47, 49, among others.

91. See Ysabel de Andia, *Présence et eschatologie dans la pensée de Martin Heidegger* (Lille and Paris: Éditions universitaires, 1975).

92. Heidegger, *Vorträge und Aufsätze*, p. 35/27.

93. Martin Heidegger, *Gelassenheit* (Pfullingen: Neske, 1959), p. 25/54.

94. Friedrich Georg Jünger, *Die Perfektion der Technik*.

95. Martin Heidegger, "Einige Hinweise . . . ," in *Phänomenologie und Theologie*, p. 42. We cite this passage elsewhere in "Métamorphose de l'indécidable," Dominique Janicaud and Jean-François Mattei, *La Métaphysique à la limite* (Paris: P.U.F., 1983), pp. 163–76. Here, we again take up

the last pages of "Face à la domination," in *Martin Heidegger*, dir. Michel Haar (Paris: L'Herne, 1983), pp. 369–78.

96. Heidegger, *Phänomenologie und Theologie*, p. 37.

97. Heidegger, *Holzwege*, p. 32. "The Origin of the Work of Art," in *Poetry, Language, Thought*, trans. and ed. Albert Hofstadter (New York: Harper and Row, 1971), 42.

98. Heidegger, *Vorträge und Aufsätze*, p. 149. "Building, Dwelling, Thinking," in *Poetry, Language, Thought*, p. 148. [English translation modified.]

99. Heidegger, *Zur Sache des Denken*, p. 11. *On Time and Being*, p. 11.

100. Ibid., p. 16/16. [English translation modified.]

101. Friedrich Hölderlin, "In Adorable Blue . . . ".

102. Heidegger, *Vorträge und Aufsätze*, p. 90/102.

103. *Der Spiegel*, p. 209/56.

104. See Janicaud, *A nouveau de la philosophie*, pp. 122–34.

105. Heidegger, *Phänomenologie und Theologie*, p. 41.

106. *Der Spiegel*, op. cit., p. 214/61; *Réponses et questions sur l'histoire et la politique*, trans. Jean Launay (Paris: Mercure de France, 1977), pp. 61–62.

107. On the condition that their specificities not be dissolved into a formless fraternization.

108. Heidegger, *Aus der Erfahrung des Denkens* (Pfullingen: Neske, 1947), p. 15. "The Thinker as Poet," in *Poetry, Language, Thought*, p. 8.

8. Rationality as *Partage*

1. *"We are a sign with no meaning."* Quotation from Hölderlin, "Mnemosyne."

Part III. Of Rationality and the Enigma

1. Gustave Flaubert, letter to Louis Bouilhet, 4 September 1850.

2. Marx, *Manuscripts of 1844*, p. 102.

3. Martin Heidegger, *What Is Philosophy?* trans. William Kluback and Jean T. Wilde (London: Vision Press, 1956), p. 63.

4. Theses 6 and 4 of 24 theses, Gottfried Wilhelm Leibniz, *Die Philosophische Schriften*, ed. C. I. Gerhardt (Hildesheim: Olms, 1961), vol. 7, p. 289.

5. Leon Brillouin, *Science and Information Theory* 2d ed. (New York: Academic Press, 1962), p. 294.

6. See Plato, *The Laws*, 645a.

7. For which X proposed, with the New Meaning, a neo-Hegelian version. (See chapter 8.)

8. Heidegger, *Sein und Zeit*, p. 34/58.

9. See Denis Zaslawsky, *Analyse de l'être (essai de philosophie analytique)* (Paris: Ed. de Minuit, 1982).

10. As is the case with Douglas R. Hofstadter (*Gödel, Escher, Bach* [Penguin, 1979]).

11. Quine, *Methods of Logic*, p. 20. Cf. Willard Quine, *Elementary Logic* (New York: Harper Torchbooks, 1941), p. 9.

12. This is why Fontainier calls attention to the *difference of meaning* that introduces the *chiasmus*: "The reversion causes all words of a proposition, at least the most essential words, to return to themselves with a different and often contrary meaning. One can regard it as the specific space of antithesis." (Pierre Fontainier, *Les Figures du discours* [Paris: Flammarion, 1968], p. 381. Cf. J-Fr. Mattei, "Le chiasme heidéggérien," in Janicaud and Mattei, *La Métaphysique à la limite*, pp. 55–60.)

13. Aristotle, *Metaphysics*, in *The Basic Works*, ed. McKeon, pp. Γ, 4, 1006a 11.

14. Pierre Aubenque, *Le Problème de l'être chez Aristote* (Paris: P.U.F., 1966), p. 124 n. 2.

15. Aristotle, *Rhetoric*, in *The Basic Works*, ed. McKeon, pp. See Jean-Marie Le Blond, *Logique et méthode chez Aristote* (Paris: Vrin, 1939), p. 42.

16. Aristotle, *Metaphysics*, in *The Basic Works*, ed. McKeon, pp. Γ, 1006a, 5.

17. Ibid., pp. 1006a, 16–17.

18. Hegel, *Wissenschaft der Logik*, 2:58.

19. Ibid., p. 59.

20. In the physical world, movement is the most immediate example of the omnipresence of contradiction.

21. Hegel, *Wissenschaft der Logik*, 2:49.

22. Ibid., p. 31.

23. See Taylor, *Scientific Management*, pp. 36–43.

24. Ibid., p. 140.

25. As Lyotard has more recently remarked, "Tombeau de l'intellectuel," *Le Monde*, 8 October 1983.

26. Jacques Ellul, *Changer de révolution: L'Inéluctable prolétariat* (Paris, Seuil, 1982). In a comparable manner, Jean Chesneaux (*De la modernité* [Paris: Maspero, 1983]) is more convincing when he denounces the "perverse effects" of modernity than when he apologizes to the Movement.

27. Undoubtedly, it would be necessary to point out here Jean Baudrillard's *Les Stratégies fatales* (Paris: Grasset, 1983), if their lucidity had not been compromised by the very inflation that they denounce. "The only revolution among things today is no longer in their dialectical overcoming (*Aufhebung*), but in their potentialization, their elevation to the power of 2, to the power *n* . . . " (p. 59). Agreed. But why peremptorily add " . . . that of terrorism, irony or simulation"? Why *in addition*? Baudrillard dogmatically retotalizes this specific saturation that he first knew to capture skillfully in its effects, mediating and otherwise. In not allowing any distance from this, he invalidates his own discourse: "Whatever anomaly must be justified . . . here also, it is terrorism! . . . We are all hostages, we are all terrorists . . . " (pp. 52, 56). If irony still has an edge, would it not be reduced to these questionable stakes? Forget Baudrillard?

28. Is there even divergence? For Heidegger, what matters more than the non-will is the *relation* by which the renouncing of the metaphysical will is made possible. See Heidegger, *Gelassenheit*, p. 33/59.

29. See the conclusion of my essay, "Savoir philosophique et pensée meditante," *Revue de l'enseignement philosophique*, February-March 1977, p. 14.

30. Friedrich Nietzsche, *Human, All Too Human*, trans. Marion Faber, with Stephen Lehmann, (Lincoln: University of Nebraska Press, 1984), pp. 143–54, (Fragment number 251).

31. See the following fragment, number 252, Ibid., p. 154.

32. See Cornelius Castoriadis, *The Imaginary Institution of Society*, trans. Kathleen Blamey, (Cambridge: MIT Press, 1987).

33. Simondon, *Objets techniques*, p. 125.

34. See Detienne and Vernant, *Cunning Intelligence*.

35. Cf. Jorge Luis Borges, "There Are More Things," in *The Book of Sand*, by Jorge Luis Borges, trans. Norman Thomas di Giovanni (New York: Dutton, 1977), pp. 51–59.

36. Immanuel Kant, *Critique of Practical Reason*, trans. Lewis White Beck (Indianapolis: Bobbs-Merrill, 1956), p. 32.

37. See ibid., p. 32.

38. Friedrich Hölderlin's letter to Carl Gock, 1 January 1799. (Friedrich Hölderlin, *Remarques sur Oedipe. Remarques sur Antigone*, trad. François Fedier (Paris: UGE, 1965).

39. Hölderlin, *Remarques sur Oedipe. Remarques sur Antigone*, pp. 86–87.

40. Blaise Pascal, *Pensées*, in *Oeuvres complète* (Paris: Ed. du Seuil, 1963), p. 521.

41. An objection: In the final analysis, is there a great difference between "thought" and the

"rational"? In any case, natural language demands to "substantialize." The very title of this book makes the rational a supreme subject; we come to a point of speaking of the "rational possible" when the possible seems to have finally devolved to "thought as such." This objection could be viewed as negligible only if the long *catharsis* imposed on metaphysical language, and language in general, by Wittgenstein, Heidegger, and Derrida were ignored or rejected. This is not the case here: rationality has been confronted with its phases, and, through this, differentiated, dissociated from its substantial pseudo-unity: thought itself appeared, in the thread of this diacritic, only as the fragile jointure of a perhaps unthinkable Difference. There is not, therefore, in my proposal, a final moment. This is a diacritical, not a metaphysical, phenomenology. The Enigma itself has been "capitalized" only in order to distinguish it from every other ontic curiosity. What remains "in place" after the last moment is not rational. Is this refusal a sign? The sign of the sign, Ariadne's thread in this Labyrinth: Is it not a question of taking "the power of the rational" *at the word* of its antiphrase? Thought: the name of this impertinence.

42. Brillouin, *Science and Information Theory*, pp. vii, 9.

43. von Neumann and Morgenstern, *Theory of Games*, pp. 8–9.

44. L. von Bertallanfy, *General Theory of Systems*.

45. See Elster, *Ulysses and the Sirens*, p. 154.

46. On this point, von Bertallanfy himself has his opinion: *General Theory*, 99–103.

47. Von Neumann (*Theory of Games*, p. 14) notes that phenomena in great numbers are more easily treated than those of "medium" numbers.

48. See Henri Gouhier, Introduction to *Oeuvres*, by Henri Bergson (Paris: P.U.F., 1959), pp. vii–xiv: Bergsonian scientism is treated here in a remarkable manner. Cf. my own treatment on this point: *Une généalogie du spiritualisme français* (The Hague: Nijhoff, 1969), pp. 169–78, 206–207.

49. In order to respond to this type of objection, Bergson assumes that intuition implies the effort and the work of the intellect. See Bergson, *Oeuvres*, p. 1328.

50. Pascal, *Pensées*, p. 540.

51. The mathematical formulation of this distance between the "all" and the "nothing" is the opposition between infinity and zero; the profane values: everything in the order of the body, nothing in the order of the spirit.

52. Pascal, *Pensées*, p. 524.

53. Ibid., p. 622.

54. Dominique Lecourt, *Proletarian Science? The Case of Lysenko*, trans. Ben Brewster (Atlantic Highlands, N.J.: Humanities Press, 1977), p. 77.

55. See chapter 4. In *La societé de maturité*, Gabor thus formulates this rule: "That which *can* be done, *ought* to be done."

56. See chapter 4.

57. Privileges of Phase 1.

58. On the significance of the latin root *sucer*, see Ernout and Meillet, *Dictionnaire étymologique de la langue latine*, p. 586.

59. On the "war of time," see Paul Virilio, *Vitesse et politique* (Paris: Galilée, 1977), pp. 136–37. *Speed and Politics*, trans. Mark Polizzotti (New York: Semiotexte, 1986), pp. 138–39.

60. Pascal, *Pensées*, p. 622.

Index

Dominique Janicaud is Professor of Philosophy at the University of Nice. His publications include *Hegel et le destin de la Grèce*; *La métaphysique à la limite: Cinq études sur Heidegger*; *L'Ombre de cette pensée: Heidegger et la question politique*; and *Le Tournant théologique de la phénoménologie française*.

Peg Birmingham is Associate Professor of Philosophy at DePaul University. She is the author of articles on Hannah Arendt, Martin Heidegger, and Michel Foucault and is completing a book on Heidegger's practical philosophy.

Elizabeth Birmingham studied at the University of Freiburg and at the Centre de Linguistique Appliquée, Besançon. She is a dancer and choreographer in Minneapolis.